twistor theory

PURE AND APPLIED MATHEMATICS

A Program of Monographs, Textbooks, and Lecture Notes

LECTURE NOTES IN PURE AND APPLIED MATHEMATICS

1. *N. Jacobson,* Exceptional Lie Algebras
2. *L.-Å. Lindahl and F. Poulsen,* Thin Sets in Harmonic Analysis
3. *I. Satake,* Classification Theory of Semi-Simple Algebraic Groups
4. *F. Hirzebruch, W. D. Newmann, and S. S. Koh, Differentiable Manifolds and Quadratic Forms*
5. *I. Chavel,* Riemannian Symmetric Spaces of Rank One
6. *R. B. Burckel,* Characterization of C(X) Among Its Subalgebras
7. *B. R. McDonald, A. R. Magid, and K. C. Smith,* Ring Theory: Proceedings of the Oklahoma Conference
8. *Y.-T. Siu,* Techniques of Extension on Analytic Objects
9. *S. R. Caradus, W. E. Pfaffenberger, and B. Yood,* Calkin Algebras and Algebras of Operators on Banach Spaces
10. *E. O. Roxin, P.-T. Liu, and R. L. Sternberg,* Differential Games and Control Theory
11. *M. Orzech and C. Small,* The Brauer Group of Commutative Rings
12. *S. Thomier,* Topology and Its Applications
13. *J. M. Lopez and K. A. Ross,* Sidon Sets
14. *W. W. Comfort and S. Negrepontis,* Continuous Pseudometrics
15. *K. McKennon and J. M. Robertson,* Locally Convex Spaces
16. *M. Carmeli and S. Malin,* Representations of the Rotation and Lorentz Groups: An Introduction
17. *G. B. Seligman,* Rational Methods in Lie Algebras
18. *D. G. de Figueiredo,* Functional Analysis: Proceedings of the Brazilian Mathematical Society Symposium
19. *L. Cesari, R. Kannan, and J. D. Schuur,* Nonlinear Functional Analysis and Differential Equations: Proceedings of the Michigan State University Conference
20. *J. J. Schäffer,* Geometry of Spheres in Normed Spaces
21. *K. Yano and M. Kon,* Anti-Invariant Submanifolds
22. *W. V. Vasconcelos,* The Rings of Dimension Two
23. *R. E. Chandler,* Hausdorff Compactifications
24. *S. P. Franklin and B. V. S. Thomas,* Topology: Proceedings of the Memphis State University Conference
25. *S. K. Jain,* Ring Theory: Proceedings of the Ohio University Conference
26. *B. R. McDonald and R. A. Morris,* Ring Theory II: Proceedings of the Second Oklahoma Conference
27. *R. B. Mura and A. Rhemtulla,* Orderable Groups
28. *J. R. Graef,* Stability of Dynamical Systems: Theory and Applications
29. *H.-C. Wang,* Homogeneous Branch Algebras
30. *E. O. Roxin, P.-T. Liu, and R. L. Sternberg,* Differential Games and Control Theory II
31. *R. D. Porter,* Introduction to Fibre Bundles
32. *M. Altman,* Contractors and Contractor Directions Theory and Applications
33. *J. S. Golan,* Decomposition and Dimension in Module Categories
34. *G. Fairweather,* Finite Element Galerkin Methods for Differential Equations
35. *J. D. Sally,* Numbers of Generators of Ideals in Local Rings
36. *S. S. Miller,* Complex Analysis: Proceedings of the S.U.N.Y. Brockport Conference
37. *R. Gordon,* Representation Theory of Algebras: Proceedings of the Philadelphia Conference
38. *M. Goto and F. D. Grosshans,* Semisimple Lie Algebras
39. *A. I. Arruda, N. C. A. da Costa, and R. Chuaqui,* Mathematical Logic: Proceedings of the First Brazilian Conference
40. *F. Van Oystaeyen,* Ring Theory: Proceedings of the 1977 Antwerp Conference
41. *F. Van Oystaeyen and A. Verschoren,* Reflectors and Localization: Application to Sheaf Theory
42. *M. Satyanarayana,* Positively Ordered Semigroups
43. *D. L Russell,* Mathematics of Finite-Dimensional Control Systems
44. *P.-T. Liu and E. Roxin,* Differential Games and Control Theory III: Proceedings of the Third Kingston Conference, Part A
45. *A. Geramita and J. Seberry,* Orthogonal Designs: Quadratic Forms and Hadamard Matrices
46. *J. Cigler, V. Losert, and P. Michor,* Banach Modules and Functors on Categories of Banach Spaces

47. *P.-T. Liu and J. G. Sutinen,* Control Theory in Mathematical Economics: Proceedings of the Third Kingston Conference, Part B
48. *C. Byrnes,* Partial Differential Equations and Geometry
49. *G. Klambauer,* Problems and Propositions in Analysis
50. *J. Knopfmacher,* Analytic Arithmetic of Algebraic Function Fields
51. *F. Van Oystaeyen,* Ring Theory: Proceedings of the 1978 Antwerp Conference
52. *B. Kadem,* Binary Time Series
53. *J. Barros-Neto and R. A. Artino,* Hypoelliptic Boundary-Value Problems
54. *R. L. Sternberg, A. J. Kalinowski, and J. S. Papadakis,* Nonlinear Partial Differential Equations in Engineering and Applied Science
55. *B. R. McDonald,* Ring Theory and Algebra III: Proceedngs of the Third Oklahoma Conference
56. *J. S. Golan,* Structure Sheaves Over a Noncommutative Ring
57. *T. V. Narayana, J. G. Williams, and R. M. Mathsen,* Combinatorics, Representation Theory and Statistical Methods in Groups: YOUNG DAY Proceedings
58. *T. A. Burton,* Modeling and Differential Equations in Biology
59. *K. H. Kim and F. W. Roush,* Introduction to Mathematical Consensus Theory
60. *J. Banas and K. Goebel,* Measures of Noncompactness in Banach Spaces
61. *O. A. Nielson,* Direct Integral Theory
62. *J. E. Smith, G. O. Kenny, and R. N. Ball,* Ordered Groups: Proceedings of the Boise State Conference
63. *J. Cronin,* Mathematics of Cell Electrophysiology
64. *J. W. Brewer,* Power Series Over Commutative Rings
65. *P. K. Kamthan and M. Gupta,* Sequence Spaces and Series
66. *T. G. McLaughlin,* Regressive Sets and the Theory of Isols
67. *T. L. Herdman, S. M. Rankin III, and H. W. Stech,* Integral and Functional Differential Equations
68. *R. Draper,* Commutative Algebra: Analytic Methods
69. *W. G. McKay and J. Patera,* Tables of Dimensions, Indices, and Branching Rules for Representations of Simple Lie Algebras
70. *R. L. Devaney and Z. H. Nitecki,* Classical Mechanics and Dynamical Systems
71. *J. Van Geel,* Places and Valuations in Noncommutative Ring Theory
72. *C. Faith,* Injective Modules and Injective Quotient Rings
73. *A. Fiacco,* Mathematical Programming with Data Perturbations I
74. *P. Schultz, C. Praeger, and R. Sullivan,* Algebraic Structures and Applications: Proceedings of the First Western Australian Conference on Algebra
75. *L Bican, T. Kepka, and P. Nemec,* Rings, Modules, and Preradicals
76. *D. C. Kay and M. Breen,* Convexity and Related Combinatorial Geometry: Proceedings of the Second University of Oklahoma Conference
77. *P. Fletcher and W. F. Lindgren,* Quasi-Uniform Spaces
78. *C.-C. Yang,* Factorization Theory of Meromorphic Functions
79. *O. Taussky,* Ternary Quadratic Forms and Norms
80. *S. P. Singh and J. H. Burry,* Nonlinear Analysis and Applications
81. *K. B. Hannsgen, T. L. Herdman, H. W. Stech, and R. L. Wheeler,* Volterra and Functional Differential Equations
82. *N. L. Johnson, M. J. Kallaher, and C. T. Long,* Finite Geometries: Proceedings of a Conference in Honor of T. G. Ostrom
83. *G. I. Zapata,* Functional Analysis, Holomorphy, and Approximation Theory
84. *S. Greco and G. Valla,* Commutative Algebra: Proceedings of the Trento Conference
85. *A. V. Fiacco,* Mathematical Programming with Data Perturbations II
86. *J.-B. Hiriart-Urruty, W. Oettli, and J. Stoer,* Optimization: Theory and Algorithms
87. *A. Figa Talamanca and M. A. Picardello,* Harmonic Analysis on Free Groups
88. *M. Harada,* Factor Categories with Applications to Direct Decomposition of Modules
89. *V. I. Istrăţescu,* Strict Convexity and Complex Strict Convexity
90. *V. Lakshmikantham,* Trends in Theory and Practice of Nonlinear Differential Equations
91. *H. L. Manocha and J. B. Srivastava,* Algebra and Its Applications
92. *D. V. Chudnovsky and G. V. Chudnovsky,* Classical and Quantum Models and Arithmetic Problems
93. *J. W. Longley,* Least Squares Computations Using Orthogonalization Methods
94. *L. P. de Alcantara,* Mathematical Logic and Formal Systems
95. *C. E. Aull,* Rings of Continuous Functions
96. *R. Chuaqui,* Analysis, Geometry, and Probability
97. *L. Fuchs and L. Salce,* Modules Over Valuation Domains

98. *P. Fischer and W. R. Smith,* Chaos, Fractals, and Dynamics
99. *W. B. Powell and C. Tsinakis,* Ordered Algebraic Structures
100. *G. M. Rassias and T. M. Rassias,* Differential Geometry, Calculus of Variations, and Their Applications
101. *R.-E. Hoffmann and K. H. Hofmann,* Continuous Lattices and Their Applications
102. *J. H. Lightbourne III and S. M. Rankin III,* Physical Mathematics and Nonlinear Partial Differential Equations
103. *C. A. Baker and L, M. Batten,* Finite Geometrics
104. *J. W. Brewer, J. W. Bunce, and F. S. Van Vleck,* Linear Systems Over Commutative Rings
105. *C. McCrory and T. Shifrin,* Geometry and Topology: Manifolds, Varieties, and Knots
106. *D. W. Kueker, E. G. K. Lopez-Escobar, and C. H. Smith,* Mathematical Logic and Theoretical Computer Science
107. *B.-L. Lin and S. Simons,* Nonlinear and Convex Analysis: Proceedings in Honor of Ky Fan
108. *S. J. Lee,* Operator Methods for Optimal Control Problems
109. *V. Lakshmikantham,* Nonlinear Analysis and Applications
110. *S. F. McCormick,* Multigrid Methods: Theory, Applications, and Supercomputing
111. *M. C. Tangora,* Computers in Algebra
112. *D. V. Chudnovsky and G. V. Chudnovsky,* Search Theory: Some Recent Developments
113. *D. V. Chudnovsky and R. D. Jenks,* Computer Algebra
114. *M. C. Tangora,* Computers in Geometry and Topology
115. *P. Nelson, V. Faber, T. A. Manteuffel, D. L. Seth, and A. B. White, Jr.,* Transport Theory, Invariant Imbedding, and Integral Equations: Proceedings in Honor of G. M. Wing's 65th Birthday
116. *P. Clément, S. Invernizzi, E. Mitidieri, and I. I. Vrabie,* Semigroup Theory and Applications
117. *J. Vinuesa,* Orthogonal Polynomials and Their Applications: Proceedings of the International Congress
118. *C. M. Dafermos, G. Ladas, and G. Papanicolaou,* Differential Equations: Proceedings of the EQUADIFF Conference
119. *E. O. Roxin,* Modern Optimal Control: A Conference in Honor of Solomon Lefschetz and Joseph P. Lasalle
120. *J. C. Díaz,* Mathematics for Large Scale Computing
121. *P. S. Milojević,* Nonlinear Functional Analysis
122. *C. Sadosky,* Analysis and Partial Differential Equations: A Collection of Papers Dedicated to Mischa Cotlar
123. *R. M. Shortt,* General Topology and Applications: Proceedings of the 1988 Northeast Conference
124. *R. Wong,* Asymptotic and Computational Analysis: Conference in Honor of Frank W. J. Olver's 65th Birthday
125. *D. V. Chudnovsky and R. D. Jenks,* Computers in Mathematics
126. *W. D. Wallis, H. Shen, W. Wei, and L. Zhu,* Combinatorial Designs and Applications
127. *S. Elaydi,* Differential Equations: Stability and Control
128. *G. Chen, E. B. Lee, W. Littman, and L. Markus,* Distributed Parameter Control Systems: New Trends and Applications
129. *W. N. Everitt,* Inequalities: Fifty Years On from Hardy, Littlewood and Pólya
130. *H. G. Kaper and M. Garbey,* Asymptotic Analysis and the Numerical Solution of Partial Differential Equations
131. *O. Arino, D. E. Axelrod, and M. Kimmel,* Mathematical Population Dynamics: Proceedings of the Second International Conference
132. *S. Coen,* Geometry and Complex Variables
133. *J. A. Goldstein, F. Kappel, and W. Schappacher,* Differential Equations with Applications in Biology, Physics, and Engineering
134. *S. J. Andima, R. Kopperman, P. R. Misra, J. Z. Reichman, and A. R. Todd,* General Topology and Applications
135. *P Clément, E. Mitidieri, B. de Pagter,* Semigroup Theory and Evolution Equations: The Second International Conference
136. *K. Jarosz,* Function Spaces
137. *J. M. Bayod, N. De Grande-De Kimpe, and J. Martínez-Maurica,* *p*-adic Functional Analysis
138. *G. A. Anastassiou,* Approximation Theory: Proceedings of the Sixth Southeastern Approximation Theorists Annual Conference
139. *R. S. Rees,* Graphs, Matrices, and Designs: Festschrift in Honor of Norman J. Pullman
140. *G. Abrams, J. Haefner, and K. M. Rangaswamy,* Methods in Module Theory

141. *G. L. Mullen and P. J.-S. Shiue*, Finite Fields, Coding Theory, and Advances in Communications and Computing
142. *M. C. Joshi and A. V. Balakrishnan*, Mathematical Theory of Control: Proceedings of the International Conference
143. *G. Komatsu and Y. Sakane*, Complex Geometry: Proceedings of the Osaka International Conference
144. *I. J. Bakelman*, Geometric Analysis and Nonlinear Partial Differential Equations
145. *T. Mabuchi and S. Mukai*, Einstein Metrics and Yang–Mills Connections: Proceedings of the 27th Taniguchi International Symposium
146. *L. Fuchs and R. Göbel*, Abelian Groups: Proceedings of the 1991 Curaçao Conference
147. *A. D. Pollington and W. Moran*, Number Theory with an Emphasis on the Markoff Spectrum
148. *G. Dore, A. Favini, E. Obrecht, and A. Venni*, Differential Equations in Banach Spaces
149. *T. West*, Continuum Theory and Dynamical Systems
150. *K. D. Bierstedt, A. Pietsch, W. Ruess, and D. Vogt*, Functional Analysis
151. *K. G. Fischer, P. Loustaunau, J. Shapiro, E. L. Green, and D. Farkas*, Computational Algebra
152. *K. D. Elworthy, W. N. Everitt, and E. B. Lee*, Differential Equations, Dynamical Systems, and Control Science
153. *P.-J. Cahen, D. L. Costa, M. Fontana, and S.-E. Kabbaj*, Commutative Ring Theory
154. *S. C. Cooper and W. J. Thron*, Continued Fractions and Orthogonal Functions: Theory and Applications
155. *P. Clément and G. Lumer*, Evolution Equations, Control Theory, and Biomathematics
156. *M. Gyllenberg and L. Persson*, Analysis, Algebra, and Computers in Mathematical Research: Proceedings of the Twenty-First Nordic Congress of Mathematicians
157. *W. O. Bray, P. S. Milojević, and Č. V. Stanojević*, Fourier Analysis: Analytic and Geometric Aspects
158. *J. Bergen and S. Montgomery*, Advances in Hopf Algebras
159. *A. R. Magid*, Rings, Extensions, and Cohomology
160. *N. H. Pavel*, Optimal Control of Differential Equations
161. *M. Ikawa*, Spectral and Scattering Theory: Proceedings of the Taniguchi International Workshop
162. *X. Liu and D. Siegel*, Comparison Methods and Stability Theory
163. *J.-P. Zolésio*, Boundary Control and Variation
164. *M. Křížek, P. Neittaanmäki, and R. Stenberg*, Finite Element Methods: Fifty Years of the Courant Element
165. *G. Da Prato and L. Tubaro*, Control of Partial Differential Equations
166. *E. Ballico*, Projective Geometry with Applications
167. *M. Costabel, M. Dauge, and S. Nicaise*, Boundary Value Problems and Integral Equations in Nonsmooth Domains
168. *G. Ferreyra, G. R. Goldstein, and F. Neubrander*, Evolution Equations
169. *S. Huggett*, Twistor Theory

Additional Volumes in Preparation

twistor theory

edited by
Stephen Huggett
School of Mathematics and Statistics
University of Plymouth
Plymouth, England

CRC Press
Taylor & Francis Group
Boca Raton London New York

CRC Press is an imprint of the
Taylor & Francis Group, an **informa** business

CRC Press
Taylor & Francis Group
6000 Broken Sound Parkway NW, Suite 300
Boca Raton, FL 33487-2742

First issued in hardback 2017

ISBN-13: 978-0-8247-9321-0 (pbk)
ISBN-13: 978-1-138-44214-6 (hbk)

Library of Congress Cataloging-in-Publication Data

Twistor theory / edited by Stephen Huggett.
p. cm. — (Lecture notes in pure and applied mathematics ; v. 169)
Includes bibliographical references and index.
ISBN 0-8247-9321-8
1. Twistor theory. I. Huggett, S. A. II. Series.
QC173.75.T85T85 1995
516.3'62—dc20 94-37775
CIP

Visit the Taylor & Francis Web site at
http://www.taylorandfrancis.com

and the CRC Press Web site at
http://www.crcpress.com

Preface

The original motivation behind twistor theory was to exploit complex holomorphic geometry in a detailed study of the relationship between general relativity and quantum physics. Early successes included a twistor description of zero-rest-mass fields in Minkowski space (which yields a nice geometrical description of the distinction between positive and negative frequency) and a twistor solution of the (half-flat and vacuum) Einstein equations.

How is the complex holomorphic geometry exploited, and how were these successes obtained? The basic structure is in two parts. There is a geometrical correspondence between compactified complexified Minkowski space and complex projective three-space (twistor space), which can be set up by writing the former as the Grassmannian $Gr_2(C^4)$ and the latter as $Gr_1(C^4)$. Then there is an integral transform, the Penrose transform, between solutions of field equations in space-time and classes of holomorphic functions on twistor space.

This transform (which itself can be regarded as a generalization of the Radon transform) has been much studied and generalized, as has the underlying geometrical correspondence. The Penrose transform actually relates analytic cohomology on the twistor space to the cohomology of certain differential operators on space-time, and it can be generalized to other geometrical correspondences between homogenous spaces (and not just these Grassmannians). Conversely, work in representation theory influences developments in the Penrose transform. For example, the subject of Eastwood's chapter is the description of the Barchini-Knapp-Zierau transform in terms of twistor geometry.

Most of the other chapters focus on the study of the geometrical correspondence. The twistor construction of manifolds with self-dual Weyl tensor was achieved when Penrose showed that a deformation of twistor space could still have an appropriate four-dimensional family of holomorphic curves. The chapters here illustrate the richness of this idea and the diversity of its ramifications. LeBrun considers the result (of

Taubes) that, given a smooth oriented compact manifold M, the space $M\#mCP^2$ has an anti-self-dual metric if m is sufficiently big, and shows that for $M = CP^2$, m has to be 14. Pedersen gives a sufficient condition for a self-dual four manifold to be diffeomorphic to nCP^2, and then studies self-dual structures with symmetries on nCP^2. Merkulov observes that Penrose's original "non-linear graviton" is an example of a class of "relative deformation" problems, in which he studies the moduli space M of complex compact submanifolds X (with given normal bundle N) in a complex manifold Y, and finds conditions on the cohomology of N that ensure that M always has a family of torsion-free connections. (Extensive details of this construction can be found in the last two chapters in this volume.)

Penrose's chapter starts from the two observations that the consistency condition for massless helicity 3/2 fields is that the space-time should be Ricci-flat, and that twistors are the space of conserved charges for massless helicity 3/2 fields in Minkowski space. So a possible curved twistor space for a general Ricci-flat space-time is suggested: a non-linear version of the space of charges for massless helicity 3/2 fields. Many other fascinating topics were discussed at the conference, such as the relationship between integrable systems and symmetry reductions of the self-dual Yang-Mills equations, and I am very grateful to all the contributors. Thanks to their efforts these proceedings present the very latest work in this rapidly moving field. (The last collection of articles achieving much the same aim was the very good volume *Twistors in Mathematics and Physics* edited by T. N. Bailey and R. J. Baston, Cambridge University Press, 1990.) As such, this volume will be of unique value to graduate students and research workers in mathematical physics and complex differential geometry.

I am also grateful to Domonic Green and Robin Horan, who helped me to organize the conference itself, and to my new colleagues Richard Jozsa, David McMullan, and Sergey Merkulov, who helped me edit these proceedings. Finally, I would like to say thank you to the London Mathematical Society, which provided the crucial financial support.

Stephen Huggett

Contents

Contributors

T. N. Bailey Department of Mathematics, University of Edinburgh, Edinburgh, Scotland

Andrew S. Dancer Max-Planck-Institut für Mathematik, Bonn, Germany

Michael Eastwood Department of Pure Mathematics, University of Adelaide, Adelaide, South Australia, Australia

Giampiero Esposito Istituto Nazionale di Fisica Nucleare and Dipartimento di Scienze Fisiche, Naples, Italy

Robin Horan School of Mathematics and Statistics, University of Plymouth, Plymouth, Devon, United Kingdom

L. P. Hughston Merrill Lynch International Limited, London, United Kingdom

Claude LeBrun Mathematics Department, State University of New York at Stony Brook, Stony Brook, New York

L. J. Mason Mathematical Institute, University of Oxford, Oxford, United Kingdom

Sergey A. Merkulov School of Mathematics and Statistics, University of Plymouth, Plymouth, Devon, United Kingdom

Henrik Pedersen Department of Mathematics and Computer Science, Odense University, Odense, Denmark

Roger Penrose Mathematical Institute, University of Oxford, Oxford, United Kingdom

Giuseppe Pollifrone Istituto Nazionale di Fisica Nucleare, Naples and Dipartimento di Fisica, Università di Roma "La Sapienza", Rome, Italy

Michael Singer Department of Mathematics, University of Edinburgh, Edinburgh, Scotland

Ian A. B. Strachan Department of Mathematics and Statistics, University of Newcastle, Newcastle-upon-Tyne, England

K. P. Tod Mathematical Institute and St. John's College, Oxford, United Kingdom

N. M. J. Woodhouse Wadham College, Oxford, United Kingdom

twistor theory

1
Thomas's *D*-Calculus, Parabolic Invariant Theory, and Conformal Invariants

T. N. Bailey Department of Mathematics, University of Edinburgh, Edinburgh, Scotland

1 Introduction

In this talk, I want to draw attention to the connections between, and consequences of, several pieces of recent work concerning conformally invariant differential operators.

A *conformal manifold* is a smooth n-dimensional manifold M equipped with an equivalence class of metrics $[g]$, the elements of which are related by conformal rescalings $g \mapsto \Omega^2 g$, where Ω is a nowhere vanishing smooth function. For definiteness, we shall assume that our metrics are Riemannian, although everything we say is true also in the pseudo-Riemannian case. On a conformal manifold, one has line bundles $\mathcal{E}[w]$ where $w \in \mathbb{R}$, sections of which are conformally weighted functions—i.e., functions f which transform according to $f \mapsto \hat{f} := \Omega^w f$ under the conformal rescaling $g \mapsto \hat{g} := \Omega^2 g$. I will use the abstract index notation [PR] for tensor fields, so that for example, $\mathcal{E}^a$ denotes the tangent bundle of M, and U^a a section thereof. I will write $\mathcal{E}^a[w]$ for the tensor product of $\mathcal{E}^a$ with $\mathcal{E}[w]$, etc.

The problem which I wish to address is that of finding conformally invariant differential operators between the line bundles of conformally weighted functions—i.e. operators $L : \mathcal{E}[w] \to \mathcal{E}[q]$ satisfying $\widehat{Lf} = L\hat{f}$. I will sometimes refer to such operators as *invariants.* One can restrict one's attention to linear differential operators, but even then there is no complete theory. A simpler problem is to consider the special case of the "flat model" for conformal geometry, and there the case of linear operators is well understood, and there has been much progress recently on the general problem.

2 The flat model

Consider $\mathbb{R}^{n+2}$, where $n > 1$, together with a signature $(n+1, 1)$ bilinear form $\tilde{g}$. Write X^I for coordinates on $\mathbb{R}^{n+2}$. Let e_0 be a non-zero element of $\mathbb{R}^{n+2}$, null with respect to $\tilde{g}$. Let G denote the identity-connected component of the pseudo-orthogonal group which preserves g. Let Q denote the connected component of the set of non-zero null vectors which contains e_0.

The group G acts linearly on $\mathbb{R}^{n+2}$ preserving Q, and hence it acts on the space of generators of Q. This space of generators is an n-dimensional sphere, which comes equipped with a conformal structure, with G acting by conformal automorphisms ("Möbius transformations").

Sections of the line bundles $\mathcal{E}[w]$ over S associated with the conformal structure can be identified with functions on Q homogeneous of degree w, in the sense that

$$f(\lambda X) = \lambda^w f(X), \text{ for } X \in Q, \quad 0 < \lambda \in \mathbb{R}.$$

The condition that $L : \mathcal{E}[w] \to \mathcal{E}[q]$ be conformally invariant is equivalent to its being equivariant with respect to the G-action. The problem of finding all conformally invariant *linear* operators is completely solved: it is a fairly standard exercise in representation theory (see [BE] for a review).

Let us now turn to the general problem of invariant differential operators. One way of constructing these is to use the manifestly G invariant operations of differentiating with respect to coordinates on $\mathbb{R}^{n+2}$, and forming contractions of the resulting tensors with $\tilde{g}$, the volume form $\tilde{\epsilon}$, and the coordinate vector X^I. To do this however one needs a way of taking the homogeneous function on Q representing a section of $\mathcal{E}[w]$ and extending it (at least formally, to as many orders as one is intending to differentiate) to a function on a neighbourhood of Q in $\mathbb{R}^{n+2}$. One does this by making the extension "harmonic" with respect to the indefinite-signature Laplacian

$$\Delta := \tilde{g}^{IJ} \frac{\partial}{\partial X^I} \frac{\partial}{\partial X^J}.$$

It is shown in [EG] that such extensions exist and are unique to all orders, except when $w + n/2 \in \mathbb{Z}_{\geq 0}$, in which cases one can extend only to order $w + n/2 - 1$.

As an example, suppose w is such that the second order harmonic extension $\tilde{f}$ of a function f on Q exists. An invariant operator $\mathcal{E}[w] \to \mathcal{E}[w-4]$ is given by

$$f \mapsto \left(\tilde{g}^{IK} \tilde{g}^{JL} \partial^2_{IJ} \tilde{f}\, \partial^2_{KL} \tilde{f}\right)\Big|_Q.$$

(The image takes values in $\mathcal{E}[w-4]$ since differentiation with respect to coordinates lowers homogeneity by one unit.) Invariants which arise in this way are called *Weyl Invariants*, and the others are termed *exceptional.*

3 Parabolic Invariant Theory

One would like to know to what extent are all invariants Weyl invariants. The sphere S^n is the homogeneous space G/P, where P is the stabiliser of the generator of Q containing e_0. A G-equivariant operator is determined by its action at the identity coset. The space of infinite jets of sections of $\mathcal{E}[w]$ is naturally a P-module, and the problem of finding invariant scalar-valued operators is equivalent to finding polynomials on this space of jets which are invariant, in the sense that they simply rescale according to a character under the P-action. The problem is therefore reduced to the algebraic problem of finding invariants of this P-module. Let us henceforth restrict attention to the cases where $w + n/2 \notin \mathbb{N}$, so that one has unique harmonic extension to all orders, in which case, this module is isomorphic to the module $\mathcal{H}(w)$ of jets at e_0 of homogeneity w harmonic functions on $\mathbb{R}^{n+2}$.

If $w \notin \mathbb{Z}_{\geq 0}$ (where $\mathbb{Z}_{\geq 0}$ denotes the non-negative integers), then relatively straightforward arguments show that all invariants are Weyl invariants (see e.g. [EG]). If $w \in \mathbb{Z}_{\geq 0}$, then $\mathcal{H}(w)$ decomposes as a direct sum $\mathcal{H}(w) = \mathcal{H}^w \oplus \mathcal{H}_w$, where $\mathcal{H}^w$ denotes the (finite-dimensional) space of jets of homogeneous harmonic polynomials, and $\mathcal{H}_w$ denotes the space of jets that vanish to order $w + 1$ at e_0. For $w \in \mathbb{Z}_{\geq 0}$, the problem of finding invariants of $\mathcal{H}(w)$ has proved intractable, but progress can been made by restricting attention to $\mathcal{H}_w$.

Every invariant is a sum of invariants, each of which is a homogeneous polynomial of some *degree d*. Also, each invariant is uniquely a sum of an odd invariant and an even invariant, where "odd" and "even" refer to behaviour under reflections. Thus it suffices to consider invariants homogeneous of some degree and of definite parity, and we will do so henceforth.

The results of Fefferman [F], strengthened by [BEGr], are that for $w \in \mathbb{Z}_{\geq 0}$, an invariant of $\mathcal{H}_w$ is exceptional iff it is odd and of degree n. (For trivial reasons, there are no odd invariants of degree $< n$.)

Converting this back into the original problem, we have the following results for $w + n/2 \notin \mathbb{N}$: If $w \notin \mathbb{Z}_{\geq 0}$ then every invariant is a Weyl invariant; if $w \in \mathbb{Z}_{\geq 0}$, then every invariant which is independent of a certain finite-dimensional subspace of low order jets is a Weyl invariant, except those which are odd and of degree n.

4 General conformal manifolds

Let us now return to the general problem of conformally invariant differential operators on a conformal manifold. The question on which most attention has been focused is that of the existence of curved analogues of the invariant operators that exist in the flat case. A *curved analogue* of a given linear invariant operator in the flat model is a conformally invariant operator defined on a general conformal manifold which reduces to the given operator in the flat case, and whose principal symbol agrees with that of the original operator. Such an operator will look like the original operator, but with "curvature correction terms". The best known example is that of the Laplace operator on S^n, which is an invariant operator $\mathcal{E}[1-n/2] \to \mathcal{E}[-1-n/2]$ whose curved analogue is the "conformally invariant Laplacian" or "Yamabe operator"

$$\Delta + \tfrac{1}{2(n-1)}R,$$

where R is the scalar curvature. For non-linear operators, a curved analogue is an invariant operator on a general conformal manifold, of the same order as the original, and which reduces to the original operator in the flat case.

Even for the linear operators, the question of when curved analogues exist is not fully answered. Such a linear operator between line bundles in the flat case is necessarily some power of the Laplacian. Indeed,

$$\Delta^k : \mathcal{E}[k-n/2] \to \mathcal{E}[-k-n/2]$$

is invariant. It is known that in odd dimensions, all these operators have curved analogues, and in even dimensions curved analogues exist for $k \leq n/2$. It is conjectured that the other operators do not have curved analogues, but this is proved only in the case $k = 3, n = 4$ [Gr]. It is however the case that the Weyl invariants have curved analogues, as we shall see below.

4.1 Thomas's D-operator

One can construct curved analogues of the Weyl invariants in §2 by constructing an analogue of differentiation with respect to coordinates in $\mathbb{R}^{n+2}$ and $\tilde{g}, \tilde{\epsilon}$. There are two essentially equivalent ways of achieving this. One of these is to construct an analogue of the space $\mathbb{R}^{n+2}$ itself. This was done by Fefferman and Graham [FG], in the sense of formal power series, and the resulting "ambient space" exists and is unique up to infinite order in the odd-dimensional case, but only to a given finite order in the even dimensional case. This method gives a nice proof of the existence of the curved analogues of Δ^k for n odd and for n even and $k \leq n/2$ [GJMS].

The other, closely related, possibility is to construct a vector bundle over M which generalises the product vector bundle $S^n \times \mathbb{R}^{n+2} \to S^n$ in the flat case. This construction has been "discovered" independently by several people in the last few years, as a generalisation of the theory of "local twistors". It turns out however that T.Y. Thomas [T] first produced a version of it at about the same time as Cartan was developing his "conformal connection", to which it is intimately related.

An account in modern language of Thomas's work and some related matters can be found in [BEGo]. Very briefly, given a choice of metric g_{ij} in the conformal class, define a vector bundle $\mathcal{E}^I = \mathcal{E}[1] \oplus \mathcal{E}^i[-1] \oplus \mathcal{E}[-1]$. Under conformal rescaling, one identifies the two direct sum decompositions according to

$$\begin{pmatrix} \hat{\sigma} \\ \hat{\mu}^i \\ \hat{\rho} \end{pmatrix} = \begin{pmatrix} \sigma \\ \mu^i + \Upsilon^i\sigma \\ \rho - \Upsilon_j\mu^j - \frac{1}{2}\Upsilon_j\Upsilon^j\sigma \end{pmatrix},$$

where $\Upsilon = \nabla \log \Omega$. This defines a rank $n+2$ vector bundle over M in a conformally invariant way, but note that it is not invariantly a direct sum. There is a conformally invariant metric $\tilde{g}$ on $\mathcal{E}^I$, with associated quadratic form given by

$$\tilde{g}_{IJ}U^IU^J = 2\sigma\rho + \mu_i\mu^i, \quad \text{where} \quad U^I = \begin{pmatrix} \sigma \\ \mu^i \\ \rho \end{pmatrix},$$

and also a volume form $\tilde{\epsilon}$. There is also a preferred section of $\mathcal{E}^I[1]$, denoted X^I, with components $(0,0,1)$.

The vector bundle $\mathcal{E}^I$ has a conformally invariant connection ∇, defined by

$$\nabla_j \begin{pmatrix} \sigma \\ \mu^i \\ \rho \end{pmatrix} = \begin{pmatrix} \nabla_j\sigma - \mu_j \\ \nabla_j\mu^i + \delta^i_j\rho + P^i_j\sigma \\ \nabla_j\rho - P_{ij}\mu^i \end{pmatrix}, \tag{1}$$

where P_{ij} is a trace-adjusted version of the Ricci curvature tensor, which appears often in conformal differential geometry.

Finally, there is a "differentiation operator" $D : \mathcal{E}[w] \to \mathcal{E}_I[w-1]$ defined by

$$D_I f = \begin{pmatrix} w(n+2w-2)f \\ (n+2w-2)\nabla_i f \\ -(\Delta + wP_j{}^j)f \end{pmatrix}.$$

This differentiation operator can also be defined on conformally weighted fields taking values in tensor powers of $\mathcal{E}^I$, using the same formula as above, but with ∇ and Δ being induced from the connection defined in (1).

In the flat model, the map $f \to D_I f$ is given, when $n+2w-2 \neq 0$, by taking $n+2w-2$ times the first order harmonic extension of f, differentiating with respect to coordinates in $\mathbb{R}^{n+2}$, and restricting back to Q. In the case $w + n/2 \notin \mathbb{N}$, it is not hard to see that if one takes a Weyl invariant, and writes down for a general conformal manifold the identical formula with differentiation with respect to coordinates replaced by the D-operator, one obtains a curved analogue.

5 The Exceptional Invariants

Recently, a way of listing the exceptional invariants of $\mathcal{H}_w$ has been found [BG]. Let $T_{AB\ldots K}$ denote the $w+1$-fold coordinate derivative of f. (This is in fact the lowest order derivative which is not forced to be zero in $\mathcal{H}_w$.) Choose a fixed element $B_I \in \mathbb{R}^{n+2*}$ such that $B_I e_0^I \neq 0$, and set $\eta^I = (B_J X^J)^{-1} B^I$. Finally, set $\eta_{C \ldots F} = \tilde{\epsilon}_{ABC \ldots F} X^A \eta^B$.

Now let $L_{A \ldots C}$ be formed as follows. Take the tensor product of n copies of T and 1 copy of η. Now use $\tilde{g}^{IJ}$ to contract each index of η with an index of a T, in such a way that each T has exactly one index contracted with η. Now use $\tilde{g}^{IJ}$ to contract off some pairs of indices on the T's. ("Some" here means anything from performing no contractions to contracting all the remaining free indices.) Then it can be shown that $D_A \ldots D_C L^{A \ldots C}$ is (if it is non-zero), an odd invariant of degree n. Furthermore, every exceptional invariant is a linear combination of these *basic exceptional invariants.*

These invariants also have curved analogues. One chooses η^I to be a section of $\mathcal{E}^I[-1]$ such that $\eta_I X^I = 1$. Then it is not hard to see that writing down essentially the same formula in the curved case yields a curved analogue.

References

[BG] T.N. Bailey and A.R. Gover, *Exceptional invariants in the parabolic invariant theory of conformal geometry*, Proc. A.M.S., to appear.

[BEGo] T.N. Bailey, M.G. Eastwood & A.R. Gover, *The Thomas structure bundle for conformal, projective and related structures*, Rocky Mountain Journal of Math, to appear.

[BEGr] T.N. Bailey, M.G. Eastwood & C.R. Graham, *Invariant theory for conformal and CR geometry*, Annals of Math, To appear.

[BE] R.J. Baston and M.G. Eastwood, *Invariant Operators*, in "Twistors in Mathematics and Physics", ed. T.N. Bailey & R.J. Baston, Cambridge University Press, 1990.

[EG] M.G. Eastwood and C.R. Graham, *Invariants of conformal densities*, Duke Math. Jour. **63** (1991), 633–671.

[F] C. Fefferman *Parabolic invariant theory in complex analysis*, Adv. in Math. **31** (1979), 131–262.

[FG] C. Fefferman and C.R. Graham, *Conformal invariants*, in: *Élie Cartan et les Mathématiques d'Aujourdui*, Astérisque (1985), 95–116.

[GJMS] C.R. Graham, et al, *Conformally invariant powers of the Laplacian, I: Existence,* J. London Math. Soc. **46** (1992), 557–565.

[Gr] C.R. Graham, *Conformally invariant powers of the Laplacian, II: Nonexistence,* J. London Math. Soc. **46** (1992), 566–576.

[PR] R. Penrose and W. Rindler, *Spinors and space-time, Vol. 1*, Cambridge University Press, 1984.

[T] T.Y. Thomas, *On conformal geometry*, Proc. N.A.S. **12** (1926) 352–359.

2
Cohomogeneity-One Kähler Metrics

ANDREW S. DANCER, Max-Planck-Institut für Mathematik, Gottfried-Claren-Strasse 26, 53225 Bonn, Germany.
(Address from Oct.'94: I.H.E.S., 35 route de Chartres, 91440 Bures-sur-Yvette, France.)

IAN A.B. STRACHAN, Department of Mathematics and Statistics, University of Newcastle, Newcastle-upon-Tyne, NE1 7RU, United Kingdom.

1. Introduction

The subject of this article is the study of real 4-dimensional Kähler-Einstein and scalar-flat Kähler metrics with a 3-dimensional symmetry group which is transitive on 3-surfaces. Such symmetries have been classified, and the resulting metrics are often referred to as Bianchi metrics. Explicitly, the metrics that will be considered are of the form

$$\begin{aligned} g &= (abc)^2dt^2 + a^2\sigma_1^2 + b^2\sigma_2^2 + c^2\sigma_3^2\,, \\ &\equiv w_1w_2w_3dt^2 + \frac{w_2w_3}{w_1}\sigma_1^2 + \frac{w_3w_1}{w_2}\sigma_2^2 + \frac{w_1w_2}{w_3}\sigma_3^2\,, \end{aligned} \tag{1}$$

where a, b, c (or equivalently, $w_1\,, w_2\,, w_3$) are functions of t, the coordinate transverse to the group orbits. The σ_i are the invariant one-forms satisfying

$$\begin{aligned} d\sigma_1 &= n_1\sigma_2\wedge\sigma_3\,, \\ d\sigma_2 &= n_2\sigma_3\wedge\sigma_1 - \tilde{a}\sigma_1\wedge\sigma_2\,, \\ d\sigma_3 &= n_3\sigma_1\wedge\sigma_2 + \tilde{a}\sigma_3\wedge\sigma_1\,. \end{aligned}$$

Dual to these one-forms are the vector fields $\frac{\partial}{\partial t}, X_1, X_2, X_3$, where

$$\begin{aligned} [X_2, X_3] &= -n_1 X_1 , \\ [X_3, X_1] &= -n_2 X_2 - \tilde{a} X_3 , \\ [X_1, X_2] &= -n_3 X_3 + \tilde{a} X_2 , \\ \left[\frac{\partial}{\partial t}, X_i \right] &= 0 . \end{aligned}$$

The constants $\tilde{a}$ and n_i are given in Table 1. Note that in each case $\tilde{a} n_1 = 0$, and so the classification divides into two: those metrics with $\tilde{a} = 0$ are said to belong to Bianchi class A, and those with $\tilde{a} \neq 0$ are said to belong to Bianchi class B. For more details of this classification see [1].

The field equations for Kähler-Einstein or scalar-flat Kähler metrics reduce to systems of first order ordinary differential equations for the metric coefficients for the Bianchi class A metrics, and to algebraic equations for the Bianchi class B metrics. Some of these systems may be solved explicitly in terms of known functions. For the special case of Bianchi IX metrics (for which the symmetry is $SU(2)$) the equations will be studied in more detail to determine which trajectories give rise to complete metrics. Before this, however, it is necessary to determine the conditions needed for the metric (1) to be Kähler (but not hyperKähler, in which case the metric is automatically vacuum).

One main assumption throughout this article is that the metric is diagonal. One may always diagonalize the metric at a particular time $t = t_0$, and for the metric to remain diagonal for $t > t_0$ one requires that the evolution equations imply the diagonal components of the metric remain zero. For Bianchi types VIII and IX Einstein metrics (irrespective of any further condition on the metric, such as being Kähler) this is the case, and (1) is the general form of the metric (see [1]). For other Einstein metric types, and more pronouncedly, for all scalar flat Kähler metrics (including types VIII and IX), taking the metric to be diagonal is an assumption; in general one should study non-diagonal metrics (see [2] for some recent progress in this direction).

Another assumption is that the metric is non-hyperKähler and non-reducible. This allows a result of Lichnerowicz [3] to be used which implies, under these conditions, that the Kähler form must be group invariant. In the Bianchi IX case, however, a representation theory argument gives the invariance of the Kähler form from the assumption that the metric is non-hyperKähler alone (see [4]). In order to ensure that a metric is not hyperKähler it is sufficient to assume that the Ricci curvature is not identically zero.

2. Kähler Structure

In this section the conditions needed for (1) to be Kähler will be derived. This derivation follows that in [4], where more details may be found. With the orthonormal coframe for g given by $e_0 = abc\, dt, e_1 = a\sigma_1, e_2 = b\sigma_2, e_3 = c\sigma_3$, the orientation may be chosen so that

Class	Bianchi Type	$\tilde{a}$	n_1	n_2	n_3
A	I	0	0	0	0
	II	0	1	0	0
	VI_0	0	1	-1	0
	VII_0	0	1	1	0
	VIII	0	1	1	-1
	IX	0	1	1	1
B	V	1	0	0	0
	IV	1	0	0	1
	III	1	0	1	-1
	VI_h	$\sqrt{-h}$	0	1	-1
	VII_h	$\sqrt{+h}$	0	1	1

Table 1: Structure constants for Bianchi classification

$$\begin{aligned} \Omega_1^+ &= e_0 \wedge e_1 + e_2 \wedge e_3 \\ \Omega_2^+ &= e_0 \wedge e_2 + e_3 \wedge e_1 \\ \Omega_3^+ &= e_0 \wedge e_3 + e_1 \wedge e_2. \end{aligned}$$

are self-dual two-forms.

Now the Kähler form on a complex surface is always self-dual, and (assuming that the metric is non-reducible and non-hyperKähler) must be invariant under the group action [**3**] (in the Bianchi IX case the invariance of the Kähler form follows automatically from the representation theory of $SU(2)$, given only the assumption that the metric is non-hyperKähler [**4**]). Thus the Kähler form Ω is given by

$$\Omega = S_1(t)\Omega_1^+ + S_2(t)\Omega_2^+ + S_3(t)\Omega_3^+ \tag{2}$$

where S_i are functions of t only. Now the Kähler condition $d\Omega = 0$ is equivalent to the equations (where the prime denotes differentiation with respect to t)

$$\begin{aligned} \tilde{a}S_1 &= 0\,, & (3) \\ S_1' + \alpha_1 S_1 &= 0\,, & (4) \\ S_2' + \alpha_2 S_2 &= +\tilde{a}w_1 S_3\,, & (5) \\ S_3' + \alpha_3 S_3 &= -\tilde{a}w_1 S_2\,, & (6) \end{aligned}$$

where the functions α_i are defined by

$$\begin{aligned} w_1' &= n_1 w_2 w_3 + \alpha_1 w_1\,, & (7) \\ w_2' &= n_2 w_3 w_1 + \alpha_2 w_2\,, & (8) \\ w_3' &= n_3 w_1 w_2 + \alpha_3 w_3\,. & (9) \end{aligned}$$

Thus for each metric g we have a number of closed, self-dual, group-invariant two-forms which are candidates for Kähler forms. One now needs to check which of these two-forms are in fact Kähler forms for some choice of complex structure.

Given such a form Ω one can use the metric to define an endomorphism I of the tangent bundle by

$$g(IX,Y) = \Omega(X,Y).$$

With Ω given by (2) the endomorphism is defined by

$$I\frac{\partial}{\partial t} = S_1w_1X_1 + S_2w_2X_2 + S_3w_3X_3\,, \tag{10}$$

$$IX_1 = -\frac{S_1}{w_1}\frac{\partial}{\partial t} + \frac{S_3w_2}{w_1}X_2 - \frac{S_2w_3}{w_1}X_3\,, \tag{11}$$

$$IX_2 = -\frac{S_2}{w_2}\frac{\partial}{\partial t} - \frac{S_3w_1}{w_2}X_1 + \frac{S_1w_3}{w_2}X_3\,, \tag{12}$$

$$IX_3 = -\frac{S_3}{w_3}\frac{\partial}{\partial t} + \frac{S_2w_1}{w_3}X_1 - \frac{S_1w_2}{w_3}X_2\,. \tag{13}$$

Moreover I is an almost complex structure if and only if $S_1^2 + S_2^2 + S_3^2 = 1$. If S_1, S_2, S_3 satisfy this constraint, as well as equations (3-6), then Ω is a Kähler form precisely when I is integrable.

Assume first that S_2 is not identically ± 1. Then

$$\chi_1 = \frac{\partial}{\partial t} - iS_1w_1X_1 - iS_2w_2X_2 - iS_3w_3X_3$$

and

$$\chi_2 = \frac{iS_2}{w_2}\frac{\partial}{\partial t} + \frac{iS_3w_1}{w_2}X_1 + X_2 - \frac{iS_1w_3}{w_2}X_3$$

are independent (1,0) vector fields (that is, vector fields such that $I\chi = i\chi$). Their Lie bracket is given by

$$[\chi_1, \chi_2] = W_0\frac{\partial}{\partial t} + W_1X_1 + W_2X_2 + W_3X_3$$

where

$$\begin{aligned}
W_0 &= i(\frac{S_2}{w_2})'\,,\\
W_1 &= i(\frac{S_3w_1}{w_2})' - \frac{S_2}{w_2}(S_1w_1)' + (S_1S_2 - iS_3)n_1w_3\,,\\
W_2 &= -n_2(S_1^2 + S_3^2)\frac{w_1w_3}{w_2} - \frac{S_2}{w_2}(S_2w_2)' - \tilde{a}w_1(S_2S_3 + iS_1)\,,\\
W_3 &= -i(\frac{S_1w_3}{w_2})' + (S_2S_3 + iS_1)n_3w_1 - \frac{S_2}{w_2}(S_3w_3)' - \tilde{a}(S_1^2 + S_3^2)\frac{w_1w_3}{w_2}\,.
\end{aligned}$$

Using equations (3-9) these expressions simplify to

$$
\begin{aligned}
W_0 &= i\tilde{a}\frac{w_1 S_3}{w_2} - \frac{iS_2}{w_2}(2\alpha_2 + n_2\frac{w_3 w_1}{w_2}), \\
W_1 &= i\frac{w_1 S_3}{w_2}(\alpha_1 - \alpha_2 - \alpha_3 - n_2\frac{w_1 w_3}{w_2}) - i\tilde{a}\frac{w_1^2 S_2}{w_2}, \\
W_2 &= -2\tilde{a}w_1 S_2 S_3 - n_2\frac{w_3 w_1}{w_2}, \\
W_3 &= -i\frac{S_1 w_3}{w_2}(-\alpha_1 - \alpha_2 + \alpha_3 - n_2\frac{w_1 w_3}{w_2}) + \tilde{a}\frac{w_3 w_1}{w_2}(S_2^2 - S_3^2).
\end{aligned}
$$

Now I is integrable if and only if $[\chi_1, \chi_2]$ is a (1,0) vector field, and this condition is equivalent to the following equations.

$$
\begin{aligned}
-iW_0 - \frac{S_1}{w_1}W_1 - \frac{S_2}{w_2}W_2 - \frac{S_3}{w_3}W_3 &= 0, \\
S_1 w_1 P - iW_1 - \frac{S_3 w_1}{w_2}W_2 + \frac{S_2 w_1}{w_3}W_3 &= 0, \\
S_2 w_2 W_0 + \frac{S_3 w_2}{w_1}W_1 - iW_2 - \frac{S_1 w_2}{w_3}W_3 &= 0, \\
S_3 w_3 W_0 - \frac{S_2 w_3}{w_1}W_1 + \frac{S_1 w_3}{w_2}W_2 - iW_3 &= 0.
\end{aligned}
$$

On substituting the simplified expressions for W_i these equations reduce to

$$\alpha_2 S_2 - i(\alpha_3 - \alpha_1)S_3 S_1 = \tilde{a}w_1 S_3, \tag{14}$$
$$(S_3 - iS_1 S_2)(\alpha_1 - \alpha_2 - \alpha_3) = 0, \tag{15}$$
$$\alpha_2(1 + S_2^2) + (S_3^2 - S_1^2)(\alpha_3 - \alpha_1) = 2\tilde{a}w_1 S_2 S_3, \tag{16}$$
$$(S_2 S_3 + iS_1)(\alpha_1 + \alpha_2 - \alpha_3) = 2\tilde{a}w_1 S_3^2. \tag{17}$$

Similar equations may be derived in the cases $S_2 = \pm 1$ (when the vector fields χ_i are no longer independent) by using a different pair of independent $(1,0)$ vector fields. To analyse the equations further one has to consider the cases $\tilde{a} = 0$ and $\tilde{a} \neq 0$ separately.

Bianchi class A Kähler metrics

Recall that these metrics have $\tilde{a} = 0$. From equations (14-17) (with $\tilde{a} = 0$) it is easy to show that two of the α_i must be equal, and the third must be zero. Also, if $\alpha_1 = 0, \alpha_2 = \alpha_3$, one must have $S_2 \equiv S_3 \equiv 0$, or else the metric is hyperKähler. Similar results hold, by cyclic permutation, in the other two cases. Thus one has the following Theorem:

THEOREM 1

If the metric (1) belongs to Class A, and (with our choice of orientation) is Kähler and not hyperKähler or reducible, then one of the following three conditions holds:

(i) $\alpha_1 = 0$, and $\alpha_2 = \alpha_3$

(ii) $\alpha_2 = 0$, and $\alpha_3 = \alpha_1$

(iii) $\alpha_3 = 0$, and $\alpha_1 = \alpha_2$.

Conversely if (i),(ii) or (iii) is true then the metric is Kähler. The Kähler forms are:

(i) $\Omega = w_2 w_3 dt \wedge \sigma_1 + w_1 \sigma_2 \wedge \sigma_3$

(ii) $\Omega = w_3 w_1 dt \wedge \sigma_2 + w_2 \sigma_3 \wedge \sigma_1$

(iii) $\Omega = w_1 w_2 dt \wedge \sigma_3 + w_3 \sigma_1 \wedge \sigma_2$

respectively. □

Bianchi class B Kähler metrics

Class B metrics have $\tilde{a} \neq 0$, and so from (3) one has $S_1 = 0$ (so $S_2^2 + S_3^2 = 1$), and from (5) and (14) $S_2' = 0$. Thus S_2 and S_3 must be constants. Moreover, if $S_2 = 0$ or $S_3 = 0$ one obtains, by similar manipulations, the condition $\tilde{a} = 0$. Thus genuine Kähler Bianchi B metrics have S_2 and S_3 equal to non-zero constants. The functions α_2 and α_3 may be obtained from equations (5) and (6), and (15-17) may then be used to obtain α_1.

THEOREM 2

If metric (1) belongs to Class B, and (with our choice of orientation) is Kähler and not hyperKähler or reducible, then:

$$w_1' = +\tilde{a}\left(\frac{S_3}{S_2} - \frac{S_2}{S_3}\right) w_1^2 , \tag{18}$$

$$w_2' = n_2 w_3 w_1 + \tilde{a}\frac{S_3}{S_2} w_2 w_1 , \tag{19}$$

$$w_3' = n_3 w_1 w_2 - \tilde{a}\frac{S_2}{S_3} w_3 w_1 , \tag{20}$$

where S_2 and S_3 are non-zero constants satisfying $S_2^2 + S_3^2 = 1$. Conversely, if (18-20) hold, with $S_2^2 + S_3^2 = 1$, then the metric is Kähler. The Kähler form is

$$\Omega = S_2(w_3 w_1 dt \wedge \sigma_2 + w_2 \sigma_3 \wedge \sigma_1) + S_3(w_1 w_2 dt \wedge \sigma_3 + w_3 \sigma_1 \wedge \sigma_2) .$$

□

This system of equations may be solved by quadrature without any further conditions imposed on the metric.

Note the difference between Bianchi class A and class B Kähler metrics. In the former the Kähler condition involves a single arbitrary function, while in the latter it involves a single arbitrary constant.

3. Kähler-Einstein metrics

Before deriving the Kähler-Einstein equations with non-zero cosmological constant ($\Lambda \neq 0$) the $\Lambda = 0$ case will be considered. For such metrics, one has the following equivalence:

$$\begin{pmatrix}\text{Vacuum}\\ \text{Kähler}\end{pmatrix} \equiv \begin{pmatrix}\text{Vacuum}\\ \text{anti-self-dual Weyl tensor}\end{pmatrix}$$

and hence the equations are integrable via, say, Penrose's nonlinear graviton construction [5]. The solutions to the field equations have been studied for the Bianchi metrics by Lorenz [6] and, for the Bianchi IX metrics, by [7],[8], and [9].

For $\Lambda \neq 0$ this equivalence breaks down, there being two natural generalizations:

$$\begin{pmatrix}\text{Einstein}\\ \text{Kähler}\end{pmatrix} \text{ and } \begin{pmatrix}\text{Einstein}\\ \text{anti-self-dual Weyl tensor}\end{pmatrix}.$$

In the second case the equations are again, a priori, integrable, via Ward's modification of the nonlinear graviton construction [10]. The equations for the Bianchi class A metrics have recently been solved by Tod [1] and, for Bianchi IX metrics, by N.J. Hitchin (unpublished).

Here the first generalization will be considered, i.e. Kähler-Einstein metrics. From Theorems 1 and 2 one knows which metrics of the form (1) are Kähler (and not hyperKähler or reducible); imposing the Einstein equations fixes the otherwise free constants and functions which appear there.

Class A Kähler-Einstein metrics

The Einstein equations for Bianchi class A metrics are (in terms of a, b and c):

$$\begin{aligned}
2(\frac{a'}{a})' + 2\Lambda(abc)^2 &= a^4 n_1^2 - (b^2 n_2 - c^2 n_3)^2 ,\\
2(\frac{b'}{b})' + 2\Lambda(abc)^2 &= b^4 n_2^2 - (c^2 n_3 - a^2 n_1)^2 ,\\
2(\frac{c'}{c})' + 2\Lambda(abc)^2 &= c^4 n_3^2 - (a^2 n_1 - b^2 n_2)^2 ,\\
4(\frac{a'b'}{ab} + \frac{b'c'}{bc} + \frac{c'a'}{ca}) + 4\Lambda(abc)^2 &= -a^4 n_1^2 - b^4 n_2^2 - c^4 n_3^2 + 2a^2b^2 n_1 n_2 + 2b^2c^2 n_2 n_3 + 2c^2a^2 n_3 n_1 ,
\end{aligned}$$

where Λ is the Einstein constant.

Corresponding to the three possible Kähler structures in Theorem 1 the Einstein equations reduce to one of the relations:

(i) $n_1\alpha_2 = -\Lambda w_1^2$,
(ii) $n_2\alpha_3 = -\Lambda w_2^2$,
(iii) $n_3\alpha_1 = -\Lambda w_3^2$.

Note, for example, that for Bianchi I metrics (where all the $n_i = 0$) there are no solutions to these equations for non-zero Λ.

Bianchi Type	Field Equations
II	$w_1' = w_2w_3$ $w_2' = -\Lambda w_1^2 w_2$ $w_3' = -\Lambda w_1^2 w_3$
VI$_0$	$w_1' = w_2w_3$ $w_2' = -w_3w_1 - \Lambda w_1^2 w_2$ $w_3' = -\Lambda w_1^2 w_3$
	$w_1' = w_2w_3 + \Lambda w_2^2 w_1$ $w_2' = -w_3w_1$ $w_3' = +\Lambda w_2^2 w_3$
VII$_0$	$w_1' = w_2w_3$ $w_2' = w_3w_1 - \Lambda w_1^2 w_2$ $w_3' = -\Lambda w_1^2 w_3$
	$w_1' = w_2w_3 - \Lambda w_2^2 w_1$ $w_2' = w_3w_1$ $w_3' = -\Lambda w_2^2 w_3$

Bianchi Type	Field Equations
VIII	$w_1' = w_2w_3$ $w_2' = w_3w_1 - \Lambda w_1^2 w_2$ $w_3' = -w_1w_2 - \Lambda w_1^2 w_3$
	$w_1' = w_2w_3 + \Lambda w_3^2 w_1$ $w_2' = w_3w_1 + \Lambda w_3^2 w_2$ $w_3' = -w_1w_2$
IX	$w_1' = w_2w_3 - \Lambda w_3^2 w_1$ $w_2' = w_3w_1 - \Lambda w_3^2 w_2$ $w_3' = w_1w_2$

Table 2: Field Equations for Bianchi Class A Kähler-Einstein metrics

The field equations for the various choices of Bianchi type and Kähler structure are shown[1] in Table 2. In some cases the equations in Table 2 may be solved explicitly:

- **Bianchi II**

This is the simplest case, and the equations may be integrated directly, giving the metric

$$g = \Delta_{\rm II}^{-1} dr^2 + \frac{1}{4} r^2 (\Delta_{\rm II}\sigma_1^2 + \sigma_2^2 + \sigma_3^2)$$

where

$$\Delta_{\rm II} = -\frac{\Lambda}{6} r^2 + \frac{16\delta}{r^4}$$

and δ is a constant. Here r is a new radial coordinate defined by $r = 2\sqrt{w_1}$.

- **Bianchi VIII**

It is only possible to impose the extra symmetry consisting of two of the w_i being equal for two of the possible Kähler structures. In these cases the metric (in terms of a new radial coordinate ρ defined by $\rho = 2\sqrt{w_3}$) is

$$g = \Delta_{\rm VIII}^{-1} d\rho^2 + \frac{1}{4}\rho^2 (\sigma_1^2 + \sigma_2^2 + \Delta_{\rm VIII}\sigma_3^2)$$

where

[1] In some cases there are other equations obtained by cyclic permutations of the indices. Only inequivalent forms are given here

$$\Delta_{\rm VIII} = -1 - \frac{\Lambda}{6}\rho^2 - \frac{16\delta}{\rho^4}$$

and δ is a constant. This metric was obtained in [**11**].

- **Bianchi IX**

In the next section this system will be considered in more detail, so for convenience the field equations will be written out again:

$$\begin{aligned} w_1' &= w_2 w_3 - \Lambda w_3^2 w_1 \\ w_2' &= w_3 w_1 - \Lambda w_3^2 w_2 \\ w_3' &= w_1 w_2 \end{aligned}$$

or, in terms of the variables a, b, c :

$$a' = \frac{1}{2}a(b^2 + c^2 - a^2) \tag{21}$$

$$b' = \frac{1}{2}b(c^2 + a^2 - b^2) \tag{22}$$

$$c' = \frac{1}{2}c(a^2 + b^2 - c^2 - 2\Lambda a^2 b^2). \tag{23}$$

It is not known to us if these equations can be solved explicitly. The fact that the metrics will not in general be self-dual or anti-self-dual suggests that this may not be possible. A further indication of this is that the equations do not satisfy either the strong Painlevé property or the weak Painlevé property[2]. However neither of these points is conclusive, and the solvability or otherwise of the equations remains an open question. However, one can easily recover the known Kähler-Einstein Bianchi IX metrics from these equations.

(i) If two of the w_i are equal (which is possible for all Kähler structures) one obtains the metric:

$$g = \Delta_{\rm IX}^{-1} d\rho^2 + \frac{1}{4}\rho^2(\sigma_1^2 + \sigma_2^2 + \Delta_{\rm IX}\sigma_3^2) \tag{24}$$

where

$$\Delta_{\rm IX} = 1 - \frac{\Lambda}{6}\rho^2 + \frac{16\delta}{\rho^4}$$

and again δ is a constant. This metric was obtained in [**9**] and [**12**]. The extra symmetry means the metric has isometry group $U(2)$ rather than $SU(2)$.

[2]Andrew Pickering; private communication.

(ii) One can also obtain the Bianchi IX form of the standard product metric on $CP^1 \times CP^1$ by making the ansatz

$$a = \sqrt{\frac{2}{\Lambda}},\ b = \sqrt{\frac{2}{\Lambda}} \sin F,\ c = \sqrt{\frac{2}{\Lambda}} \cos F$$

where F is a function of t. After taking F as the new radial coordinate we can write the metric as

$$\frac{2}{\Lambda}(dF^2 + \sigma_1^2 + \sin^2 F\, \sigma_2^2 + \cos^2 F\, \sigma_3^2).$$

(iii) Finally, the ansatz

$$a^2 = \frac{3}{\Lambda}(1 + \cos G),\ b^2 = \frac{3}{\Lambda}(1 - \cos G),\ c^2 = \frac{6}{\Lambda} \cos^2 G$$

leads to the metric

$$\frac{6}{\Lambda}\,(d(\frac{G}{2}))^2 + \frac{1}{2}(1 + \cos G)\sigma_1^2 + \frac{1}{2}(1 - \cos G)\sigma_2^2 + \cos^2 G\sigma_3^2)$$

which is the form of the Fubini-Study metric on CP^2 obtained by writing it in an $SO(3)$-invariant form with respect to the maximal embedding of $SO(3)$ in the isometry group $SU(3)$ **[13]**.

Class B Kähler-Einstein metrics

Imposing the Einstein condition on metrics given by Theorem 2 gives a number of algebraic conditions on the w_i, and these are consistent if and only if $n_2 = n_3 = 0$ and $S_2 = \pm S_3$. From Table 1 the metric must be of Bianchi type V (and hence have $\tilde{a} = 1$). Solving the remaining equations yields

$$\begin{aligned} w_1 &= A\,, \\ w_2 &= B\exp(\pm At)\,, \\ w_3 &= C\exp(\mp At)\,, \end{aligned}$$

where A, B and C are non-zero constants. Representing the invariant one-forms in a coordinate basis ($\sigma_1 = dx\,, \sigma_2 = \exp(-x)dy\,, \sigma_3 = \exp(-x)dz$) and rescaling the variables gives the following diagonal Bianchi V Kähler-Einstein metric:

$$g = dT^2 + dX^2 + \exp(\mu(X - T))dY^2 + \exp(\mu(X + T))dZ^2\,,$$

where $\Lambda = -2\mu^2$.

4. Complete Bianchi IX Kähler Einstein Metrics

In this section we consider which solutions of equations (21-23) give rise to complete metrics. As remarked above, the equations are completely understood when $\Lambda = 0$ [7],[8],[9].

If one takes the Einstein constant Λ to be positive then one can argue as follows, without needing to study the equations directly.

Any complete Kähler-Einstein manifold M of real dimension four with $\Lambda > 0$ is compact, by Myers's Theorem **[14]**. Moreover, considered as a complex manifold M has $c_1 > 0$, so is isomorphic to either $P^1 \times P^1$ or the blowup of P^2 at n points, where $0 \leq n \leq 8$. If the metric is $SU(2)$-invariant then, as explained earlier, the Kähler form is also $SU(2)$-invariant so the $SU(2)$ action is holomorphic. Hence the complex automorphism group of M has a subgroup isomorphic to $SL(2, C)$. Using the descriptions of the automorphism groups of blowups of P^2 in Chapter VII of **[15]** we see that n must be less than 3. In fact the blowup of P^2 at one or two points cannot admit any Kähler-Einstein metric **[15]** as the Lie algebra of the automorphism group is not reductive in these two cases. So M must be isomorphic to P^2 or $P^1 \times P^1$.

Now it is a theorem of Matsushima **[16]**,**[17]** that the Lie algebra of the isometry group of a compact Kähler-Einstein space with positive Einstein constant is a compact real form of the Lie algebra of the automorphism group. It follows that any such metric on P^2 has isometry group $SU(3)$ (or a finite quotient) so is the Fubini-Study metric (using the classification of compact Einstein four-manifolds with isometry group of dimension at least four given in **[18]**). Similarly any such metric on $P^1 \times P^1$ is the standard product metric.

From now on, therefore, the Einstein constant will be taken to be negative. It is clear that by rescaling t, a, b, c one can set Λ equal to -1.

Consider a solution of (21-23), analytic on a maximal interval (ξ, η). It is immediate from the equations that if a is zero at any point of (ξ, η) then a is identically zero throughout this interval. The corresponding statements also hold for b, c. As we are assuming that $SU(2)$ acts with generically three-dimensional orbits we can exclude this case, so we can assume that a, b, c are nowhere zero in the interval. The metric will therefore be defined on (ξ, η); to check whether it is complete one needs to study its behaviour as t tends to ξ from above and as t tends to η from below. Now the expressions for the metric, as well as the equations, are invariant under changes of sign of a, b or c, so we shall take a, b, c to be positive.

Although we have not been able to obtain the general solution to (21-23), one may obtain qualitative properties of the solution by analysing the differential equations directly. How this is done will be apparent from the following Lemma (the proof of which is by direct computation).

LEMMA 3

$$\begin{aligned}
(ab)' &= abc^2\,, \\
(bc)' &= bca^2(1+b^2)\,, \\
(ac)' &= acb^2(1+a^2)\,, \\
(\frac{a}{b})' &= \frac{a}{b}(b^2-a^2)\,, \\
(a^2-b^2)' &= (a^2-b^2)(c^2-a^2-b^2)\,, \\
(a^2-c^2)' &= (a^2-c^2)(b^2-a^2-c^2)-2(abc)^2\,.
\end{aligned}$$

□

It follows that ab, bc, ac are increasing on (ξ,η). Moreover either a is identically equal to b in (ξ,η) or else a is never equal to b in this open interval. In the latter case the symmetry between a and b means that we can take $a > b$. Therefore we can assume from now on that $a \geq b$ and hence that b is increasing and $\frac{a}{b}$ is decreasing.

By studying various expressions as t tends to ξ (for full details of this, and for the rest of this section, see **[4]**) one obtains:

PROPOSITION 4

If the maximal interval is (ξ,η) where ξ is finite, then the metric is incomplete. □

This result shows that to study complete metrics one needs only to consider solutions defined on $(-\infty,\eta)$. Such solutions certainly exist, because the unstable curves of critical points of the equations provide examples. The critical points of the equations (21-23) are the points where the functions a, b and c satisfy $a' = b' = c' = 0$, namely the points with

$a = b, c = 0$ or

$b = c, a = 0$ or

$a = c, b = 0.$

The linearisation about a critical point which is not the origin has one positive, one negative and one zero eigenvalue. Hence there is at least one unstable curve for such a critical point. Thus there is at least one family of trajectories defined on $(-\infty,\eta)$, the ones emerging from unstable critical points.

In fact these are the only such trajectories:

PROPOSITION 5

Trajectories of solutions to (21-23) (with $a, b, c \geq 0$ and $a \geq b$) defined on $(-\infty,\eta)$ are precisely the unstable curves of critical points $(q,0,q)$ or $(q,q,0)$. If the critical point is $(q,0,q)$ where q is positive then we have $a > b$. Otherwise we have $a \equiv b$. □

Thus the analysis splits into two subcases:

Case 1: Critical points $(q,q,0)\,, q>0$.

The metric defined by this trajectory will have $a \equiv b$ and hence will be one of the $U(2)$-invariant metrics of **[9]**,**[12]**.

For t large and negative

$$\begin{aligned} a &\simeq q \\ b &\simeq q \\ c &\simeq ke^{(q^2+q^4)t}. \end{aligned}$$

for some constant k, so the metric is asymptotically

$$q^4k^2e^{2(q^2+q^4)t}dt^2 + q^2(\sigma_1^2+\sigma_2^2) + k^2e^{2(q^2+q^4)t}\sigma_3^2.$$

With

$$y = \frac{k}{1+q^2}e^{(q^2+q^4)t}$$

the metric is asymptotically

$$dy^2 + q^2(\sigma_1^2+\sigma_2^2) + (1+q^2)^2y^2\sigma_3^2$$

as y tends to zero. A complete metric is obtained only if $1+q^2$ is a half-integer; if this condition holds one can complete the metric by adding in a *bolt* (two-dimensional orbit of $SU(2)$) at $y=0$.

It is easy to check that the metric corresponding to a trajectory converging to the origin can be completed by adding in a *nut* (point orbit of $SU(2)$) at $t=-\infty$. The resulting metric is the pseudo-Fubini-Study metric on the open ball in C^2. (See **[9]** for a detailed discussion of nuts and bolts). It can be checked that in the above examples the Kähler structure extends to the manifold obtained by adding in the nut or bolt. Moreover, using the form (24) of the metric, it follows that there are no completeness problems for any of the above metrics as t approaches η. Thus we have the following result:

THEOREM 6

The trajectories of solutions to (21-23) which give complete metrics with a identically equal to b are the unstable curves of critical points $(q,q,0)$ where $1+q^2$ is a half-integer. □

On putting these metrics in the form (24) one finds that the complete examples are those with

$$\delta = -\frac{(n-2)^2(n+1)}{12}$$

where $n=2(1+q^2)\geq 2$ is an integer. If $n>2$ the underlying complex manifold is the total space of the complex line bundle over CP^1 of degree $-n$. The case $n=2$ gives the pseudo-Fubini-Study metric on the open ball in C^2. This analysis agrees with the discussion of completeness in [**12**].

Case 2: Critical points $(q, 0, q), q > 0$.

Fot t large and negative

$$\begin{aligned} a &\simeq q \\ b &\simeq ke^{q^2 t} \\ c &\simeq q \end{aligned}$$

for some constant k, so the metric is asymptotically

$$q^4 k^2 e^{2q^2 t} dt^2 + q^2(\sigma_1^2 + \sigma_3^2) + k^2 e^{2q^2 t} \sigma_2^2.$$

Putting $v = ke^{q^2 t}$ brings the metric asymptotically into the form

$$dv^2 + q^2(\sigma_1^2 + \sigma_3^2) + v^2 \sigma_2^2$$

as $v \longrightarrow 0$, so we obtain a complete metric by adding a bolt.

In fact as we approach the bolt the metric is asymptotic to the Eguchi-Hanson metric. Moreover our Kähler form Ω is asymptotic to one of the two-sphere of Kähler forms on Eguchi-Hanson; this is *not* the form corresponding to the complex structure on Eguchi-Hanson with respect to which the bolt is a complex submanifold. Thus the Kähler structure extends over the bolt, but the bolt is not a complex manifold. Hence, as t tends to $-\infty$, the metric and Kähler structure are well-behaved. Consider next the behaviour as t tends to η.

Recall from Lemma 3 that $(a^2 - c^2)' = (a^2 - c^2)(b^2 - a^2 - c^2) - 2(abc)^2$, so if $a^2 - c^2$ is positive at u, then $(a^2 - c^2)'$ is negative, so $a^2 - c^2$ is bounded away from zero on $(-\infty, u)$. On the unstable curve of $(q, 0, q)$, however, $a^2 - c^2$ tends to zero as t tends to $-\infty$, so for this trajectory $a^2 \leq c^2$ for all t. It is clear that one cannot have $a \equiv c$ so the metric is *triaxial*, that is, no two of a, b, c are identically equal.

Recalling that $a > b$ it follows from the equations that

$$b' \geq \frac{1}{2} b^3$$

so for each $s \in (-\infty, \eta)$ there exists K_s such that

$$b \geq (K_s - t)^{-\frac{1}{2}} \text{ for } t \geq s.$$

Hence our solution becomes infinite at a finite value of t and the upper bound η of our interval is finite.

By adapting an argument of Pedersen and Poon [**19**] the asymptotics of the metric may be found as $t \longrightarrow \eta$. In terms of $r = 2\sqrt{ab}$ the metric becomes

$$g = W^{-1} dr^2 + \frac{1}{4} r^2 (V\sigma_1^2 + V^{-1}\sigma_2^2 + W\sigma_3^2)$$

where $W = \frac{c^2}{ab}$ and $V = \frac{a}{b}$. As ab is strictly increasing this is an allowable change of variables. Moreover, it can be shown that r tends to ∞ as t tends to η.

Now

$$\begin{aligned}\frac{dW}{dr} &= \frac{dW}{dt}\frac{dt}{dr} \\ &= (\frac{c^2}{ab})'(\frac{dr}{dt})^{-1} \\ &= (\frac{c^2}{ab})'(ab)^{-\frac{1}{2}}c^{-2}\end{aligned}$$

and after simplifying we arrive at

$$\frac{dW}{dr} = \frac{2}{r}(\frac{a}{b} + \frac{b}{a}) + r - \frac{4}{r}W.$$

Next, recall that $\frac{a}{b}$ is decreasing , so tends to a finite limit $L \geq 1$ as t tends to η. So asymptotically

$$\frac{dW}{dr} = \frac{2}{r}(L + L^{-1}) + r - \frac{4}{r}W,$$

therefore the metric is asymptotically

$$g = W^{-1}dr^2 + \frac{1}{4}r^2(L\sigma_1^2 + L^{-1}\sigma_2^2 + W\sigma_3^2)$$

where

$$W = \frac{1}{2}(L + L^{-1}) + \frac{r^2}{6} + \frac{\kappa}{r^4}$$

for some constant of integration κ. The geodesic distance to $t = \eta$ is therefore infinite and so the metric is complete. Hence:

THEOREM 7

The trajectories of solutions to (21-23) which give complete metrics with $a > b$ are precisely the unstable curves of the critical points $(q, 0, q)$ where q is positive. □

The form of the metric on the bolt means that the underlying manifold is topologically the total space of the degree two R^2 bundle over the two-sphere.

Thus there are precisely two families of complete $SU(2)$-invariant Kähler-Einstein metrics in four real dimensions with negative Einstein constant, where the generic $SU(2)$ orbit is three-dimensional. One of these families, given by Theorem 6, consists of the $U(2)$-invariant metrics of **[9]**,**[12]**. The other family, given by Theorem 7, consists of triaxial metrics.

Bianchi Type	Field Equations
I	$w_1' = \quad + \kappa w_1$ $w_2' = \quad + \kappa w_2$ $w_3' = 0$
II	$w_1' = w_2 w_3$ $w_2' = \quad + \kappa w_2$ $w_3' = \quad + \kappa w_3$
	$w_1' = w_2 w_3 + \kappa w_1$ $w_2' = \quad + \kappa w_2$ $w_3' = 0$
VI_0	$w_1' = w_2 w_3$ $w_2' = -w_3 w_1 + \kappa w_2$ $w_3' = \quad + \kappa w_3$
	$w_1' = w_2 w_3 + \kappa w_1$ $w_2' = -w_3 w_1$ $w_3' = \quad + \kappa w_3$
	$w_1' = w_2 w_3 + \kappa w_1$ $w_2' = -w_3 w_1 + \kappa w_2$ $w_3' = 0$

Bianchi Type	Field Equations
VII_0	$w_1' = w_2 w_3 + \kappa w_1$ $w_2' = w_3 w_1 + \kappa w_2$ $w_3' = 0$
	$w_1' = w_2 w_3 + \kappa w_1$ $w_2' = w_3 w_1$ $w_3' = \quad + \kappa w_3$
VIII	$w_1' = w_2 w_3$ $w_2' = w_3 w_1 + \kappa w_2$ $w_3' = -w_1 w_2 + \kappa w_3$
	$w_1' = w_2 w_3 + \kappa w_1$ $w_2' = w_3 w_1 + \kappa w_2$ $w_3' = -w_1 w_2$
IX	$w_1' = w_2 w_3 + \kappa w_1$ $w_2' = w_3 w_1 + \kappa w_2$ $w_3' = w_1 w_2$

Table 3: Field Equations for Bianchi Class A scalar flat Kähler metrics

Jensen **[20]** has classified the homogeneous four-dimensional Einstein spaces, and the only Kähler examples with negative Einstein constant on his list are the pseudo-Fubini-Study metric on the homogeneous space $SU(1,2)/U(2)$ and the product metric on $H^2 \times H^2$, where H^2 denotes hyperbolic 2-space. In the case of pseudo-Fubini-Study the only way of embedding $SU(2)$ in the isometry group is via a $U(2)$ embedding, so this cannot lead to triaxial metrics. Moreover, the isometry group of $H^2 \times H^2$ admits no $SU(2)$ embedding. Thus the metrics given in Theorem 7 are not homogeneous.

5. Scalar-Flat Kähler metrics

Rather than imposing the Einstein condition on the general Kähler metrics given by Theorems 1 and 2, one may impose other conditions, such as the scalar curvature being zero. Any such scalar-flat Kähler metric will automatically have an anti-self-dual Weyl tensor and hence an associated twistor space (see **[1]** for the field equations for metrics with anti-self-dual Weyl tensor and vanishing scalar curvature); via the integrability 'meta-theorem' one would expect the resulting equations to be integrable.

Class A scalar-flat Kähler metrics

The expression for the scalar curvature of a diagonal Bianchi class A metric is

$$R_{scalar} = \frac{-1}{w_1 w_2 w_3}(2(\alpha_1' + \alpha_2' + \alpha_3') + \alpha_1^2 + \alpha_2^2 + \alpha_3^2 - 2(\alpha_1\alpha_2 + \alpha_2\alpha_3 + \alpha_3\alpha_1))$$

where the α_i are the functions defined by equations (7-9). Thus, using Theorem 1, the scalar-flat Kähler metrics in this class are given by $\alpha_1 = \text{constant}\,, \alpha_2 = \alpha_3 = 0$ (and cyclically). The field equations for these metrics are given[3] in Table 3, and, as expected, these are all integrable and may be solved in terms of elementary functions (including Painlevé transcendents) **[1]**, **[19]**.

As for the Kähler-Einstein systems, one may ask which trajectories give rise to complete metrics, this not being clear from the form of the general solution. However, one may repeat the analysis of the differential equations along the same lines as in section 4, and for the Bianchi IX metrics this analysis has been carried out by Dancer **[21]** (this also includes details of the corresponding twistor space). The same generic features persist – the trajectories which correspond to complete metrics emerge from certain unstable critical points of the defining system of equations. In this case the only complete metrics are

$$g = \Delta^{-1}dr^2 + \frac{1}{4}r^2(\sigma_1^2 + \sigma_2^2 + \Delta\sigma_3^2)$$

where

$$\Delta = 1 + \frac{4\mu^2(n-2)}{r^2} - \frac{16\mu^4(n-1)}{r^4}$$

and μ is a constant. These are the $U(2)$ invariant metrics of **[22]**.

Class B scalar-flat Kähler metrics

The expression for the scalar curvature of the class B Kähler metrics is

$$R_{\text{scalar}} = \frac{-2\tilde{a}^2 w_1^2}{w_1 w_2 w_3}\left(\frac{S_2}{S_3} + \frac{S_3}{S_2}\right)^2 .$$

So, for this to vanish one requires $S_2^2 + S_3^2 = 0$. However, this contradicts the requirement that I is a complex structure (i.e. $S_2^2 + S_3^2 = 1$). Thus there are no genuine real scalar-flat Kähler metrics in class B. However, with $S_2/S_3 = \pm i$ one may still integrate equations (21-23) and the solution gives complex metrics with anti-self-dual Weyl tensor and vanishing scalar curvature **[1]**.

Finally, it should be remarked that taking the metric (1) to be diagonal is a definite assumption for scalar flat Kähler metrics of all Bianchi types; in general one should study non-diagonal metrics. For such metrics direct computation (along the lines presented here) is problematical, and perhaps indirect approaches (based on twistor space and isomonodromic constructions) will be more useful **[2]**,**[21]**.

Acknowledgements

This work was carried out while A.S.D. was a Stone Research Fellow at Peterhouse, Cambridge, and I.A.B.S. was an SERC Postdoctoral Fellow at the Mathematical Institute, Oxford. We would like to thank Gary Gibbons, Paul Tod and Henrik Pedersen for various useful discussions.

[3]The same remarks apply here as for Table 2.

References

[1] K.P. Tod. Cohomogeneity-one metrics with self-dual Weyl tensor. (This volume).

[2] R. Maszczyk, L.J. Mason and N.M.J. Woodhouse. Self-dual Bianchi metrics and Painlevé transcendents. (Oxford Preprint).

[3] A. Lichnerowicz. Sur les groupes d'automorphismes de certaines variétés Kähleriennes. C. R. Acad. Sci. Paris 239 (1954) 1344-1346.

[4] A.S. Dancer and I.A.B. Strachan. Kähler-Einstein metrics with $SU(2)$ action. To appear: Math. Proc. Camb. Phil. Soc., March 1994.

[5] R. Penrose. Nonlinear gravitons and curved twistor theory. Gen. Rel. Grav. 7 (1976) 31-52.

[6] D. Lorenz. Gravitional instanton solutions for Bianchi types I-IX. Acta Phys. Polon. B14 (1983) 791-805.

Positive-definite self-dual solutions of Einstein's field equations. J. Math. Phys. 24 (1983) 2632-2634.

[7] M.F. Atiyah and N.J. Hitchin. The geometry and dynamics of magnetic monopoles. (Princeton University Press, 1988).

[8] V.A. Belinskii, G.W. Gibbons, D.N. Page and C.N. Pope. Asymptotically Euclidean Bianchi IX metrics in quantum gravity. Phys. Lett. B76 (1978) 433-435.

[9] G.W. Gibbons and C.N. Pope. The positive action conjecture and asymptotically Euclidean metrics in quantum gravity Commun. Math. Phys. 66 (1979) 267-290.

[10] R.S. Ward. Self-dual space-times with cosmological constant. Commun. Math. Phys. 78 (1980) 1-17.

[11] H. Pedersen and Y.Sun Poon. Hamiltonian construction of Kähler-Einstein metrics and Kähler metrics of constant scalar curvature. Commun. Math. Phys. 136 (1991) 309-326.

[12] H. Pedersen. Eguchi-Hanson metrics with cosmological constant. Classical and Quantum Gravity 2 (1985) 579-587.

[13] C. Bouchiat and G.W. Gibbons. Non-integrable quantum phase in the evolution of a spin-1 system: a physical consequence of the non-trivial topology of the quantum state-space. J. Phys. France 49 (1988) 187-199.

[14] S.B. Myers. Riemannian geometry in the large. Duke Math. J. 1 (1935) 39-49.

[15] A. Futaki. Kähler-Einstein metrics and integral invariants. Springer Lecture Notes in Mathematics 1314. (Springer-Verlag Berlin-Heidelberg-New York-London- Paris-Tokyo, 1988).

[16] A. Besse. Einstein Manifolds. Ergebnisse der Math. und ihrer Grenzgebiete. (Springer-Verlag Berlin-Heidelberg-New York-London-Paris-Tokyo, 1987).

[17] Y. Matsushima. Remarks on Kähler-Einstein manifolds. Nagoya Math. J. 46 (1972) 161-173.

[18] L. Bérard Bergery. Sur des nouvelles variétés riemanniennes d'Einstein. Publications de l'Institut E. Cartan (Nancy) No. 4 (1982) 1-60.

[19] H. Pedersen and Y.S. Poon. Kähler surfaces with zero scalar curvature. Classical and Quantum Gravity 7 (1990) 1707-1719.

[20] G.R. Jensen. Homogeneous Einstein spaces of dimension four. J. Differential Geometry 3 (1969) 309-349.

[21] A.S. Dancer. Scalar-flat Kähler metrics with $SU(2)$ symmetry. Cambridge preprint.

[22] C. LeBrun. Counter-examples to the generalized positive action conjecture. Commun. Math. Phys. 118 (1988) 591-596.

3
Another Integral Transform in Twistor Theory

Michael Eastwood Department of Pure Mathematics, University of Adelaide, South Australia 5005, Australia

1 INTRODUCTION

This article will describe a case of the Barchini-Knapp-Zierau transform (denoted by $\mathcal{S}$ in Barchini et al., 1992). As explained by these authors, the transform composes well with the Penrose transform. Whereas Barchini, Knapp, and Zierau define their transform in terms of representation theory and integrals over groups, this article will present a geometric version. Only the simplest case, namely that for 'hyperbolic minitwistors' (see Atiyah 1987), will be described in detail. In this simplest case, it is just a matter of translating the representation-theoretic language. This is not so for the general case which has been worked out jointly with Leticia Barchini and Rod Gover. In general, the geometric viewpoint is an aid to proving some of the properties of the transform. The details will appear elsewhere.

I am grateful to Leticia Barchini, Toby Bailey, Rod Gover, Tony Knapp, and Michael Singer for many useful conversations.

2 GENERALITIES AND THE PENROSE TRANSFORM

The sort of geometry involved in the BKZ-transform should be already familiar to twistor theorists. Certainly, the starting point is extremely familiar:

$$\begin{array}{ccc} & Y & \\ \swarrow & & \searrow \\ X & & Z \end{array}$$

a *double fibration* or *correspondence.*

In classical (complex) twistor theory the basic example is

$$\begin{array}{ccc} & \mathbb{F}_{1,2}(\mathbb{C}^4) & \\ {}^{\eta}\swarrow & & \searrow^{\tau} \\ \mathbb{P} = \mathbb{F}_1(\mathbb{C}^4) & & \mathbb{F}_2(\mathbb{C}^4) = \mathbb{M} \end{array} \qquad (*)$$

but there are also non-holomorphic examples from 'Euclidean' twistor theory

$$\begin{array}{ccc} & \mathbb{CP}_3 & \\ \swarrow\!\!/\!/ & & \searrow^{\tau} \\ \mathbb{CP}_3 & & S^4 \end{array} \qquad \begin{array}{ccc} & Y & \\ {}^{\eta}\swarrow & & \searrow^{\tau} \\ \mathbb{CP}_1 \times \overline{\mathbb{CP}_1} \setminus \text{diagonal} & & \mathcal{H} = \text{hyperbolic 3-space} \end{array}$$

where the latter diagram is the one for hyperbolic minitwistors. If X, Y, and Z are just smooth manifolds with no extra structure, then one is in the realm of the *Radon transform* (e.g. X-ray transform for the real version of $(*)$) and the double fibration formulation is due to Helgason. Of course, if X is a complex manifold, then one has the *Penrose transform* relating analytic cohomology on X to solutions of linear PDE on Z (assuming η has contractible fibres and τ has compact complex fibres). There is a spectral sequence (due to Schmid 1967, Salamon 1983, Singer 1987, Wong 1992) which is computable fairly explicitly in the homogeneous case. For example, the hyperbolic minitwistor correspondence is homogeneous under $\mathrm{SL}(2,\mathbb{C})$ with

$$\mathbb{CP}_1 \times \overline{\mathbb{CP}_1} \setminus \text{diag} = G/L \quad \text{for} \quad L = \left\{ \begin{pmatrix} \lambda & 0 \\ 0 & \lambda^{-1} \end{pmatrix} \text{ for } \lambda \in \mathbb{C}^* \right\}$$
$$\mathcal{H} = G/K \quad \text{for} \quad K = \mathrm{SU}(2) \text{ (this is maximal compact).}$$

Here, G is $\mathrm{SL}(2,\mathbb{C})$. Of course, $Y = G/(K \cap L)$. Every smooth homogeneous line bundle on G/L admits an invariant holomorphic structure (see Tirao et al., 1970 for a general discussion). Let us write $\mathcal{E}(m,n)$ for the smooth line bundle corresponding to the representation

$$\begin{pmatrix} \lambda & 0 \\ 0 & \lambda^{-1} \end{pmatrix} \mapsto \lambda^m \overline{\lambda}^{-n} = \lambda^{m+n}(\lambda\overline{\lambda})^{-n}$$

of L and $\mathcal{O}(m,n)$ for the same line bundle equipped with its invariant holomorphic structure. (<u>Warning</u>: Only $m+n$ need be assumed integral. The parameter $m-n \in \mathbb{C}$ may be arbitrary. When m and n are integral, these line bundles are exactly what one would expect on $\mathbb{CP}_1 \times \overline{\mathbb{CP}_1}$.) The Dolbeault resolution on G/L may be written:

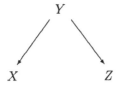

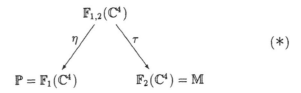

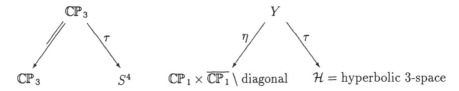

$$\begin{array}{ccccccc} & & & \mathcal{E}(m,n+2) & & & \\ 0 \longrightarrow & \mathcal{E}(m,n) & \nearrow\!\!\searrow & \oplus & \searrow\!\!\nearrow & \mathcal{E}(m+2,n+2) & \longrightarrow 0 \\ & & & \mathcal{E}(m+2,n) & & & \\ & \| & & \| & & \| & \\ & \mathcal{E}(m,n) & \xrightarrow{\bar\partial} & \mathcal{E}^{0,1}(m,n) & \xrightarrow{\bar\partial} & \mathcal{E}^{0,2}(m,n). & \end{array}$$

As an example of the Penrose transform, Toby Bailey has computed (using the alternative general method of complexification and proceeding as in Chapter 7 of Baston et al., 1989):

$$\mathcal{P} : H^1(G/L, \mathcal{O}(-l-1, l-1)) \xrightarrow{\cong} \left\{\text{smooth } \phi \text{ on } G/K \text{ s.t. } (\Delta + \tfrac{1}{2}(l-1)(l+1))\phi = 0\right\}$$

where Δ is the hyperbolic Laplacian. (In fact, he has identified the Dolbeault cohomology on G/L of all the homogeneous line bundles.)

3 THE BKZ TRANSFORM

There are also some special constructions which can be carried out when the manifold Z in the original correspondence is complex (and some other compatibility). In the homogeneous case Barchini, Knapp and Zierau introduced a transform which has an interesting geometric interpretation. In the simplest case, this is as follows. Take $G = \mathrm{SL}(2, \mathbb{C})$ as above and

$$P = \left\{ \begin{pmatrix} * & * \\ 0 & * \end{pmatrix} \right\}, \quad \text{the upper triangular matrices.}$$

Then $G/P = \mathbb{CP}_1$ and $\rho : G/L \to G/P$ is given by projection onto the first factor:

$$\begin{array}{ccc} & \mathbb{CP}_1 \times \overline{\mathbb{CP}_1} \setminus \text{diag} & \\ {}^{\rho}\swarrow & & \searrow\!\!\searrow{}^{\pi} \\ \mathbb{CP}_1 & & \mathbb{CP}_1 \times \overline{\mathbb{CP}_1} \setminus \text{diag}\,. \end{array}$$

On $\mathbb{CP}_1$ with homogeneous coördinates $[u_1, u_2]$, let $\mathcal{E}(m,n)$ for $m+n \in \mathbb{Z}$ be the smooth line bundle whose local sections $f(u)$ satisfy $f(\lambda u) = \lambda^m \overline{\lambda}^{-n} f(u)$. These are the homogeneous line-bundles. The notation has been arranged so that $\rho^* \mathcal{E}(m,n) = \mathcal{E}(m,n)$ but notice that only for $n = 0$ does $\mathcal{E}(m,n)$ on G/P admit an invariant holomorphic structure. A principal series representation of $\mathrm{SL}(2,\mathbb{C})$ is simply a space of sections $\Gamma(\mathbb{CP}_1, \mathcal{E}(m,n))$ with $\mathrm{SL}(2,\mathbb{C})$ acting in the obvious way. To obtain $\mathcal{S}$ notice that

- $\mathcal{E}(m,n) \hookrightarrow \mathcal{E}^{0,1}(m, n-2)$ on G/L;
- the $\mathcal{E}(m,n) \to \mathcal{E}(m+2,n) = \mathcal{E}^{0,2}(m,n-2)$ part of the $\bar\partial$-operator is precisely d_ρ, the relative exterior derivative.

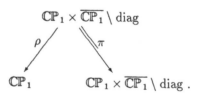

Hence, if $f \in \Gamma(\mathbb{CP}_1, \mathcal{E}(m,n))$, then $\rho^* f$ may be regarded as a $\overline{\partial}$-closed $(0,1)$-form with coefficients in $\mathcal{E}(m, n-2)$. Thus, we obtain

$$\mathcal{S} : \Gamma(G/P, \mathcal{E}(m,n)) \longrightarrow H^1(G/L, \mathcal{O}(m, n-2)).$$

To study $\mathcal{S}$ further it is useful to compose it with $\mathcal{P}$ (since $\mathcal{P}$ is an isomorphism, if we know $\mathcal{P} \circ \mathcal{S}$, then we know $\mathcal{S}$). There are three typical cases:

CASE 1 $\Gamma(G/P, \mathcal{E}(-2,2)) \xrightarrow{\mathcal{S}} H^1(G/L, \mathcal{O}(-2,0)) \xrightarrow{\mathcal{P}} \{\phi \text{ on } G/K \text{ s.t. } \Delta\phi = 0\}$.

In this case, $\mathcal{E}(-2,2) = \Lambda^2$, the bundle of smooth 2-forms on G/P and $\mathcal{P} \circ \mathcal{S}$ is simply given by integration over the sphere G/P to give a constant function ϕ. Notice that the range of $\mathcal{P} \circ \mathcal{S}$ is 1-dimensional in this case.

CASE 2 $\Gamma(G/P, \mathcal{E}(-1,1)) \xrightarrow{\mathcal{S}} H^1(G/L, \mathcal{O}(-1,-1)) \xrightarrow{\mathcal{P}} \left\{\phi \text{ on } G/K \text{ s.t. } \Delta\phi = \tfrac{1}{2}\phi\right\}$.

By conformal rescaling, solutions of $\Delta\phi = \frac{1}{2}\phi$ on $G/K = \mathcal{H}$ may be identified with ordinary harmonic functions on the ball. With this interpretation, $\mathcal{P} \circ \mathcal{S}$ is just the usual Poisson integral (and, hence, a monomorphism).

CASE 3 $\Gamma(G/P, \mathcal{E}(0,0)) \xrightarrow{\mathcal{S}} H^1(G/L, \mathcal{O}(0,-2)) \xrightarrow{\mathcal{P}} \{\phi \text{ on } G/K \text{ s.t. } \Delta\phi = 0\}$.

In this case, $\mathcal{P} \circ \mathcal{S}$ is a Szegö integral operator, again a monomorphism.

Notice that in cases 1 and 3, the constants are naturally sitting inside the cohomology as $H^1(\mathbb{CP}_1 \times \overline{\mathbb{CP}_1}, \mathcal{O}(-2,0))$ or $H^1(\mathbb{CP}_1 \times \overline{\mathbb{CP}_1}, \mathcal{O}(0,-2))$, respectively. An explicit formula for

$$\mathcal{P} \circ \mathcal{S} : \Gamma(G/P, \mathcal{E}(-l-1, l+1)) \to \left\{\phi \text{ on } G/K \text{ s.t. } (\Delta + \tfrac{1}{2}(l-1)(l+1))\phi = 0\right\}$$

is

$$f \mapsto \int_{\zeta \in \mathbb{CP}_1} |\zeta|^{2(l-1)} f\,[a\zeta_1 + b\zeta_2, c\zeta_1 + d\zeta_2]\, \zeta \wedge d\zeta\; \overline{\zeta} \wedge d\overline{\zeta}\,.$$

A more interesting case is to take $G = \mathrm{SO}_e(4,1)$ with subgroups

$$P = \text{stabiliser of} \begin{pmatrix} 0 \\ 0 \\ 0 \\ 1 \\ 1 \end{pmatrix}, \qquad L = \left\{ \begin{pmatrix} * & * & 0 & 0 & 0 \\ * & * & 0 & 0 & 0 \\ 0 & 0 & * & * & * \\ 0 & 0 & * & * & * \\ 0 & 0 & * & * & * \end{pmatrix} \right\},$$

$$\text{and} \quad G^{\mathbb{C}} > Q = \text{stabilizer of} \begin{pmatrix} 1 \\ i \\ 0 \\ 0 \\ 0 \end{pmatrix}.$$

In contrast with the previous case, $P \not> L$ so we are obliged to consider

$$\begin{array}{ccc} & G/(P \cap L) & \\ {}^{\rho}\swarrow & & \searrow^{\pi} \\ G/P & & G/L \end{array} \quad \text{with real dimensions} \quad \begin{array}{ccc} & 7 & \\ \swarrow & & \searrow \\ 3 & & 6 \end{array}$$

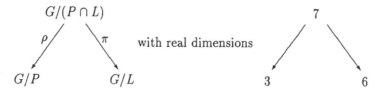

but again G/L has a complex structure endowed by $G/L \hookrightarrow G^{\mathbb{C}}/Q$. The space G/P is a 3-sphere and G/L is the complement of a 3-sphere in a complex quadric and the fibres of π are circles. The correspondence space $G/(P \cap L)$ may be regarded as pairs of points in a non-singular complex 3-quadric one lying on the embedded S^3 and one not (these are the two $\mathrm{SO_e}(4,1)$ orbits) and such that the complex line joining them lies entirely in the quadric (there is only one family of such lines in 3 dimensions).

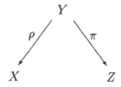

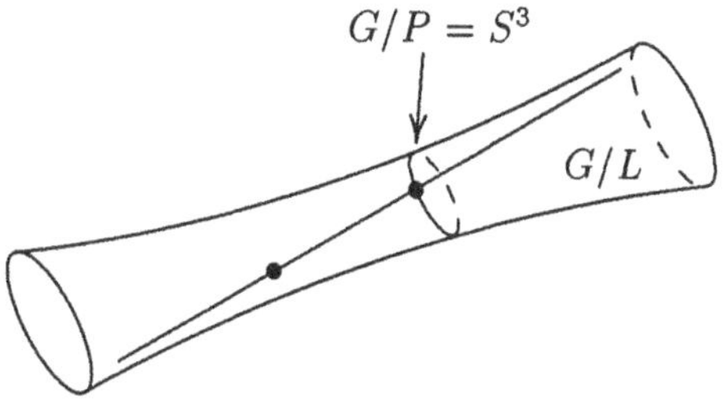

schematic diagram only

Roughly speaking, the BKZ transform may be described geometrically as follows. In the double fibration

$$\begin{array}{ccc} & Y & \\ {\scriptstyle \rho}\swarrow & & \searrow{\scriptstyle \pi} \\ X & & Z \end{array}$$

it is assumed that the fibres of π are compact, say of dimension n. (For $\mathrm{SO_e}(4,1)$ as above, $n = 1$.) A smooth vector bundle V is given on X and a holomorphic line bundle $\mathcal{L}$ on Z, a complex manifold with $\Lambda^{p,q}$ denoting the (p,q)-forms. The other piece of information is a homomorphism $\Phi : \rho^*V \to \Lambda^n_\pi \otimes \pi^*(\Lambda^{0,s} \otimes \mathcal{L})$ where Λ^n_π is the relative n-forms along π. We obtain

$$\begin{array}{ccc} \mathcal{S} : \Gamma(X, V) & \longrightarrow & \Gamma(Z, \Lambda^{0,s} \otimes \mathcal{L}) \\ \Cup & & \Cup \\ f & \longmapsto & \int_\pi \Phi(\rho^* f)\,. \end{array}$$

In the homogeneous case, there are natural choices for V, $\mathcal{L}$, and Φ. For example, if $G = \mathrm{SO_e}(4,1)$ as above, for $m \le -2$ there is a homogeneous bundle V on G/P of rank $-2m-3$ with

$$V \to \overset{m+4\ \ -4}{\times\!\!\Leftarrow\!\!\Rightarrow\!\!\times} \hookrightarrow \overset{1\ \ -2}{\times\!\!\Leftarrow\!\!\Rightarrow\!\!\times} \otimes \left(\overset{m+1\ \ 2}{\times\!\!\Leftarrow\!\!\Rightarrow\!\!\times} + \overset{m+2\ \ 0}{\times\!\!\Leftarrow\!\!\Rightarrow\!\!\times} + \overset{m+3\ \ -2}{\times\!\!\Leftarrow\!\!\Rightarrow\!\!\times} \right) = \Lambda^1_\pi \otimes \left(\Lambda^{0,2}\left(\overset{m\ \ \ 0}{\times\!\!\Leftarrow\!\!\Rightarrow\!\!\bullet}\right)\right);$$

in this case notice that $s = 2$ (just right to get interesting cohomology); the notation here is from Baston et al., 1989. In order that $\mathcal{S}$ map into Dolbeault cohomology, it is necessary that the $(0,s)$-form it produces be $\bar\partial$-closed. The further geometrical ingredients needed in order to see this in general are discussed in Barchini et al., 1993.

It is, again, constancy along the fibres of ρ which is responsible but an Ehresmann connection for π is needed in order to relate d_ρ on Y to $\overline{\partial}$ on Z.

REFERENCES

Atiyah M. F. (1987). Magnetic monopoles in hyperbolic space, Proceedings of Bombay Colloquium 1984 on Vector Bundles on Algebraic Varieties, Oxford University Press, pp. 1–34.

Barchini, L., Knapp, A. W., Zierau, R. (1992). Intertwining operators into Dolbeault cohomology, *Jour. Funct. Anal.* **107**: 302–341.

Barchini, L., Eastwood, M. G., Gover, A. R. (1993). An integral operator into Dolbeault cohomology, in preparation.

Baston, R. J. and Eastwood, M. G. (1989). The Penrose transform: its interaction with representation theory, Oxford University Press.

Salamon, S. M. (1983). Topics in four-dimensional Riemannian geometry, Geometry Seminar 'Luigi Bianchi', Lecture Notes in Math., vol. 1022, Springer, pp. 33–124.

Schmid W. (1967). Homogeneous complex manifolds and representations of semisimple Lie groups, Ph.D. dissertation, University of California, Berkeley, reprinted in Representation Theory and Harmonic Analysis on Semisimple Lie Groups, Math. Surveys and Monographs, vol. 31, Amer. Math. Soc., 1989, pp. 223–286.

Singer, M. A. (1987). A general theory of global twistor descriptions, D.Phil. thesis, Oxford University.

Tirao, J. A. and Wolf, J. A. (1970). Homogeneous holomorphic vector bundles, *Indiana Univ. Math. Jour.* **20**: 15–31.

Wong, H.-W. (1992) Dolbeault cohomologies and Zuckerman modules associated with finite rank representations, Ph.D. dissertation, Harvard University.

4
Twistors and Spin-3/2 Potentials in Quantum Gravity

Giampiero Esposito[1,2] **and Giuseppe Pollifrone**[1,3]

[1]*Istituto Nazionale di Fisica Nucleare*

Mostra d'Oltremare Padiglione 20, 80125 Napoli, Italy;

[2]*Dipartimento di Scienze Fisiche*

Mostra d'Oltremare Padiglione 19, 80125 Napoli, Italy;

[3]*Dipartimento di Fisica, Università di Roma "La Sapienza"*

Piazzale Aldo Moro 2, 00185 Roma, Italy.

Abstract. Local boundary conditions involving field strengths and the normal to the boundary, originally studied in anti-de Sitter space-time, have been recently considered in one-loop quantum cosmology. This paper derives the conditions under which spin-lowering and spin-raising operators preserve these local boundary conditions on a 3-sphere for fields of spin $0, \frac{1}{2}, 1, \frac{3}{2}$ and 2. Moreover, the two-component spinor analysis of the four potentials of the totally symmetric and independent field strengths for spin $\frac{3}{2}$ is applied to the case of a 3-sphere boundary. It is shown that such boundary conditions can only be imposed in a flat Euclidean background, for which the gauge freedom in the choice of the potentials remains. Alternative boundary conditions for supergravity involving the spinor-valued 1-forms for gravitinos and the normal to the boundary are also studied.

1. Introduction

Recent work in the literature has studied the quantization of gauge theories and supersymmetric field theories in the presence of boundaries, with application to one-loop quantum cosmology [1-9]. In particular, in the work described in [9], two possible sets of local boundary conditions were studied. One of these, first proposed in anti-de Sitter space-time [10-11], involves the normal to the boundary and Dirichlet or Neumann conditions for spin 0, the normal and the field for massless spin-$\frac{1}{2}$ fermions, and the normal and totally symmetric field strengths for spins $1, \frac{3}{2}$ and 2. Although more attention has been paid to alternative local boundary conditions motivated by supersymmetry, as in [2-3,8-9], the analysis of the former boundary conditions remains of mathematical and physical interest by virtue of its links with twistor theory [9]. The aim of this paper is to derive the mathematical properties of the corresponding boundary-value problems in both cases, since these are relevant for quantum cosmology and twistor theory.

For this purpose, sections 2-3 derive the conditions under which spin-lowering and spin-raising operators preserve local boundary conditions involving field strengths and normals. Section 4 applies the 2-spinor form of spin-$\frac{3}{2}$ potentials to Riemannian 4-geometries with a 3-sphere boundary. Boundary conditions on spinor-valued 1-forms describing gravitino fields are studied in section 5. Concluding remarks and open problems are presented in section 6.

2. Spin-lowering operators in cosmology

In section 5.7 of [9], a flat Euclidean background bounded by a 3-sphere was studied. On the bounding S^3, the following boundary conditions for a spin-s field were required:

$$2^s \, {}_e n^{AA'} \ldots {}_e n^{LL'} \, \phi_{A\ldots L} = \epsilon \, \widetilde{\phi}^{A'\ldots L'} \quad . \tag{2.1}$$

With our notation, ${}_e n^{AA'}$ is the Euclidean normal to S^3 [3,9], $\phi_{A\ldots L} = \phi_{(A\ldots L)}$ and $\widetilde{\phi}_{A'\ldots L'} = \widetilde{\phi}_{(A'\ldots L')}$ are totally symmetric and independent (i.e. not related by any conjugation) field strengths, which reduce to the massless spin-$\frac{1}{2}$ field for $s = \frac{1}{2}$. Moreover, the complex scalar field ϕ is such that its real part obeys Dirichlet conditions on S^3 and its imaginary part obeys Neumann conditions on S^3, or the other way around, according to the value of the parameter $\epsilon \equiv \pm 1$ occurring in (2.1), as described in [9].

In flat Euclidean 4-space, we write the solutions of the twistor equations [9,12]

$$D_{A'}^{\ (A} \, \omega^{B)} = 0 \quad , \tag{2.2}$$

$$D_{A}^{\ (A'} \, \widetilde{\omega}^{B')} = 0 \quad , \tag{2.3}$$

as [9]

$$\omega^A = (\omega^o)^A - i\left({}_e x^{AA'}\right)\pi^o_{A'} \quad , \tag{2.4}$$

$$\widetilde{\omega}^{A'} = (\widetilde{\omega}^o)^{A'} - i\left({}_e x^{AA'}\right)\widetilde{\pi}^o_{A} \quad . \tag{2.5}$$

Note that, since unprimed and primed spin-spaces are no longer isomorphic in the case of Riemannian 4-metrics, Eq. (2.3) is not obtained by complex conjugation of Eq. (2.2).

Hence the spinor field $\widetilde{\omega}^{B'}$ is independent of ω^B. This leads to distinct solutions (2.4)-(2.5), where the spinor fields $\omega^o_A, \widetilde{\omega}^o_{A'}, \widetilde{\pi}^o_A, \pi^o_{A'}$ are covariantly constant with respect to the flat connection D, whose corresponding spinor covariant derivative is here denoted by $D_{AB'}$. The following theorem can be now proved:

Theorem 2.1 Let ω^D be a solution of the twistor equation (2.2) in flat Euclidean space with a 3-sphere boundary, and let $\widetilde{\omega}^{D'}$ be the solution of the independent equation (2.3) in the same 4-geometry with boundary. Then a form exists of the spin-lowering operator which preserves the local boundary conditions on S^3:

$$4 \; {}_en^{AA'} \; {}_en^{BB'} \; {}_en^{CC'} \; {}_en^{DD'} \; \phi_{ABCD} = \epsilon \, \widetilde{\phi}^{A'B'C'D'} \quad , \tag{2.6}$$

$$2^{\frac{3}{2}} \; {}_en^{AA'} \; {}_en^{BB'} \; {}_en^{CC'} \; \phi_{ABC} = \epsilon \, \widetilde{\phi}^{A'B'C'} \quad . \tag{2.7}$$

Of course, the independent field strengths appearing in (2.6)-(2.7) are assumed to satisfy the corresponding massless free-field equations.

Proof. Multiplying both sides of (2.6) by ${}_en_{FD'}$ one gets

$$-2 \; {}_en^{AA'} \; {}_en^{BB'} \; {}_en^{CC'} \; \phi_{ABCF} = \epsilon \, \widetilde{\phi}^{A'B'C'D'} \; {}_en_{FD'} \quad . \tag{2.8}$$

Taking into account the total symmetry of the field strengths, putting $F = D$ and multiplying both sides of (2.8) by $\sqrt{2}\,\omega^D$ one finally gets

$$-2^{\frac{3}{2}} \; {}_en^{AA'} \; {}_en^{BB'} \; {}_en^{CC'} \; \phi_{ABCD} \, \omega^D = \epsilon \, \sqrt{2} \, \widetilde{\phi}^{A'B'C'D'} \; {}_en_{DD'} \, \omega^D \quad , \tag{2.9}$$

$$2^{\frac{3}{2}} \; {}_en^{AA'} \; {}_en^{BB'} \; {}_en^{CC'} \; \phi_{ABCD} \, \omega^D = \epsilon \, \widetilde{\phi}^{A'B'C'D'} \, \widetilde{\omega}_{D'} \quad , \tag{2.10}$$

where (2.10) is obtained by inserting into (2.7) the definition of the spin-lowering operator. The comparison of (2.9) and (2.10) yields the preservation condition

$$\sqrt{2}\; {}_e n_{DA'}\, \omega^D = -\widetilde{\omega}_{A'} \quad . \tag{2.11}$$

In the light of (2.4)-(2.5), equation (2.11) is found to imply

$$\sqrt{2}\; {}_e n_{DA'}\, (\omega^o)^D - i\sqrt{2}\; {}_e n_{DA'}\; {}_e x^{DD'} = -\widetilde{\omega}^o_{A'} - i\; {}_e x_{DA'}\, (\widetilde{\pi}^o)^D \quad . \tag{2.12}$$

Requiring that (2.12) should be identically satisfied, and using the identity ${}_e n^{AA'} = \frac{1}{r}\; {}_e x^{AA'}$ on a 3-sphere of radius r, one finds

$$\widetilde{\omega}^o_{A'} = i\sqrt{2}\, r\; {}_e n_{DA'}\; {}_e n^{DD'}\, \pi^o_{D'} = -\frac{ir}{\sqrt{2}}\, \pi^o_{A'} \quad , \tag{2.13}$$

$$-\sqrt{2}\; {}_e n_{DA'}\, (\omega^o)^D = ir\; {}_e n_{DA'}\, (\widetilde{\pi}^o)^D \quad . \tag{2.14}$$

Multiplying both sides of (2.14) by ${}_e n^{BA'}$, and then acting with ϵ_{BA} on both sides of the resulting relation, one gets

$$\omega^o_A = -\frac{ir}{\sqrt{2}}\, \widetilde{\pi}^o_A \quad . \tag{2.15}$$

The equations (2.11), (2.13) and (2.15) completely solve the problem of finding a spin-lowering operator which preserves the boundary conditions (2.6)-(2.7) on S^3. Q.E.D.

If one requires local boundary conditions on S^3 involving field strengths and normals also for lower spins (i.e. spin $\frac{3}{2}$ vs spin 1, spin 1 vs spin $\frac{1}{2}$, spin $\frac{1}{2}$ vs spin 0), then by using the same technique of the theorem just proved, one finds that the preservation condition obeyed by the spin-lowering operator is still expressed by (2.13) and (2.15).

3. Spin-raising operators in cosmology

To derive the corresponding preservation condition for spin-raising operators [12], we begin by studying the relation between spin-$\frac{1}{2}$ and spin-1 fields. In this case, the independent spin-1 field strengths take the form [9,11-12]

$$\psi_{AB} = i\, \widetilde{\omega}^{L'} \Bigl(D_{BL'}\, \chi_A\Bigr) - 2\chi_{(A}\, \widetilde{\pi}^{o}_{B)} \quad , \tag{3.1}$$

$$\widetilde{\psi}_{A'B'} = -i\, \omega^{L} \Bigl(D_{LB'}\, \widetilde{\chi}_{A'}\Bigr) - 2\widetilde{\chi}_{(A'}\, \pi^{o}_{B')} \quad , \tag{3.2}$$

where the independent spinor fields $\left(\chi_A, \widetilde{\chi}_{A'}\right)$ represent a massless spin-$\frac{1}{2}$ field obeying the Weyl equations on flat Euclidean 4-space and subject to the boundary conditions

$$\sqrt{2}\; {}_{e}n^{AA'}\; \chi_A = \epsilon\, \widetilde{\chi}^{A'} \tag{3.3}$$

on a 3-sphere of radius r. Thus, by requiring that (3.1) and (3.2) should obey (2.1) on S^3 with $s = 1$, and bearing in mind (3.3), one finds

$$2\epsilon\Bigl[\sqrt{2}\, \widetilde{\pi}^{o}_{A}\, \widetilde{\chi}^{(A'}\; {}_{e}n^{AB')} - \widetilde{\chi}^{(A'}\; \pi^{o\; B')}\Bigr] = i\Bigl[2\; {}_{e}n^{AA'}\; {}_{e}n^{BB'}\; \widetilde{\omega}^{L'}\; D_{L'(B}\, \chi_{A)}$$
$$+\, \epsilon\, \omega^{L}\; D_{L}^{\;(B'}\; \widetilde{\chi}^{A')}\Bigr] \tag{3.4}$$

on the bounding S^3. It is now clear how to carry out the calculation for higher spins. Denoting by s the spin obtained by spin-raising, and defining $n \equiv 2s$, one finds

$$n\epsilon\Bigl[\sqrt{2}\, \widetilde{\pi}^{o}_{A}\; {}_{e}n^{A(A'}\; \widetilde{\chi}^{B'...K')} - \widetilde{\chi}^{(A'...D'}\; \pi^{o\; K')}\Bigr] = i\Bigl[2^{\frac{n}{2}}\; {}_{e}n^{AA'} ... {}_{e}n^{KK'}\; \widetilde{\omega}^{L'}\; D_{L'(K}\, \chi_{A...D)}$$
$$+\, \epsilon\, \omega^{L}\; D_{L}^{\;(K'}\; \widetilde{\chi}^{A'...D')}\Bigr] \tag{3.5}$$

on the 3-sphere boundary. In the comparison spin-0 vs spin-$\frac{1}{2}$, the preservation condition is not obviously obtained from (3.5). The desired result is here found by applying the spin-raising operators [12] to the independent scalar fields ϕ and $\widetilde{\phi}$ (see below) and bearing in mind (2.4)-(2.5) and the boundary conditions

$$\phi = \epsilon \, \widetilde{\phi} \quad \text{on} \quad S^3 \quad , \tag{3.6}$$

$$ {}_e n^{AA'} D_{AA'} \phi = -\epsilon \, {}_e n^{BB'} D_{BB'} \widetilde{\phi} \quad \text{on} \quad S^3 \quad . \tag{3.7}$$

This leads to the following condition on S^3 (cf. equation (5.7.23) of [9]):

$$0 = i\phi \left[\frac{\widetilde{\pi}^o_A}{\sqrt{2}} - \pi^o_{A'} \, {}_e n_A{}^{A'} \right] - \left[\frac{\widetilde{\omega}^{K'}}{\sqrt{2}} \left(D_{AK'} \phi \right) - \frac{\omega_A}{2} \, {}_e n_C{}^{K'} \left(D^C{}_{K'} \phi \right) \right]$$

$$+ \epsilon \, {}_e n_{(A}{}^{A'} \, \omega^B \, D_{B)A'} \, \widetilde{\phi} \quad . \tag{3.8}$$

Note that, while the preservation conditions (2.13) and (2.15) for spin-lowering operators are purely algebraic, the preservation conditions (3.5) and (3.8) for spin-raising operators are more complicated, since they also involve the value at the boundary of four-dimensional covariant derivatives of spinor fields or scalar fields. Two independent scalar fields have been introduced, since the spinor fields obtained by applying the spin-raising operators to ϕ and $\widetilde{\phi}$ respectively are independent as well in our case.

4. Spin-$\frac{3}{2}$ potentials in cosmology

In this section we focus on the totally symmetric field strengths ϕ_{ABC} and $\widetilde{\phi}_{A'B'C'}$ for spin-$\frac{3}{2}$ fields, and we express them in terms of their potentials, rather than using spin-raising (or spin-lowering) operators. The corresponding theory in Minkowski space-time (and curved space-time) is described in [13-16], and adapted here to the case of flat Euclidean 4-space with flat connection D. It turns out that $\widetilde{\phi}_{A'B'C'}$ can then be obtained from two potentials defined as follows. The first potential satisfies the properties [13-16]

$$\gamma_{A'B'}^{C} = \gamma_{(A'B')}^{C} \quad , \tag{4.1}$$

$$D^{AA'} \, \gamma_{A'B'}^{C} = 0 \quad , \tag{4.2}$$

$$\widetilde{\phi}_{A'B'C'} = D_{CC'} \, \gamma_{A'B'}^{C} \quad , \tag{4.3}$$

with the gauge freedom of replacing it by

$$\widehat{\gamma}_{A'B'}^{C} \equiv \gamma_{A'B'}^{C} + D^{C}{}_{B'} \, \widetilde{\nu}_{A'} \quad , \tag{4.4}$$

where $\widetilde{\nu}_{A'}$ satisfies the positive-helicity Weyl equation

$$D^{AA'} \, \widetilde{\nu}_{A'} = 0 \quad . \tag{4.5}$$

The second potential is defined by the conditions [13-16]

$$\rho_{A'}^{BC} = \rho_{A'}^{(BC)} \quad , \tag{4.6}$$

$$D^{AA'} \, \rho_{A'}^{BC} = 0 \quad , \tag{4.7}$$

$$\gamma_{A'B'}^{C} = D_{BB'} \, \rho_{A'}^{BC} \quad , \tag{4.8}$$

with the gauge freedom of being replaced by

$$\widehat{\rho}_{A'}^{BC} \equiv \rho_{A'}^{BC} + D^{C}{}_{A'} \, \chi^{B} \quad , \tag{4.9}$$

where χ^B satisfies the negative-helicity Weyl equation

$$D_{BB'} \, \chi^B = 0 \quad . \tag{4.10}$$

Moreover, in flat Euclidean 4-space the field strength ϕ_{ABC} is expressed in terms of the potential $\Gamma^{C'}_{AB} = \Gamma^{C'}_{(AB)}$, independent of $\gamma^{C}_{A'B'}$, as

$$\phi_{ABC} = D_{CC'} \, \Gamma^{C'}_{AB} \quad , \tag{4.11}$$

with gauge freedom

$$\widehat{\Gamma}^{C'}_{AB} \equiv \Gamma^{C'}_{AB} + D^{C'}_{\;\;B} \, \nu_A \quad . \tag{4.12}$$

Thus, if we insert (4.3) and (4.11) into the boundary conditions (2.1) with $s = \frac{3}{2}$, and require that also the gauge-equivalent potentials (4.4) and (4.12) should obey such boundary conditions on S^3, we find that

$$2^{\frac{3}{2}} \, {}_e n^A_{\;\;A'} \, {}_e n^B_{\;\;B'} \, {}_e n^C_{\;\;C'} \, D_{CL'} \, D^{L'}_{\;\;B} \, \nu_A = \epsilon \, D_{LC'} \, D^L_{\;\;B'} \, \widetilde{\nu}_{A'} \tag{4.13}$$

on the 3-sphere. Note that, from now on (as already done in (3.5) and (3.8)), covariant derivatives appearing in boundary conditions are first taken on the background and then evaluated on S^3. In the case of our flat background, (4.13) is identically satisfied since $D_{CL'} \, D^{L'}_{\;\;B} \, \nu_A$ and $D_{LC'} \, D^L_{\;\;B'} \, \widetilde{\nu}_{A'}$ vanish by virtue of spinor Ricci identities [17-18]. In a curved background, however, denoting by ∇ the corresponding curved connection, and defining $\square_{AB} \equiv \nabla_{M'(A} \nabla^{M'}_{\;\;B)}$, $\square_{A'B'} \equiv \nabla_{X(A'} \, \nabla^X_{\;\;B')}$, since the spinor Ricci identities we need are [17]

$$\square_{AB} \, \nu_C = \psi_{ABDC} \, \nu^D - 2\Lambda \, \nu_{(A} \, \epsilon_{B)C} \quad , \tag{4.14}$$

$$\Box_{A'B'}\, \widetilde{\nu}_{C'} = \widetilde{\psi}_{A'B'D'C'}\, \widetilde{\nu}^{D'} - 2\widetilde{\Lambda}\, \widetilde{\nu}_{(A'}\, \epsilon_{B')C'} \quad , \tag{4.15}$$

one finds that the corresponding boundary conditions

$$2^{\frac{3}{2}}\, {}_e n^A{}_{A'}\, {}_e n^B{}_{B'}\, {}_e n^C{}_{C'}\, \nabla_{CL'}\, \nabla^{L'}{}_B\, \nu_A = \epsilon\, \nabla_{LC'}\, \nabla^L{}_{B'}\, \widetilde{\nu}_{A'} \tag{4.16}$$

are identically satisfied if and only if one of the following conditions holds: (i) $\nu_A = \widetilde{\nu}_{A'} = 0$; (ii) the Weyl spinors $\psi_{ABCD}, \widetilde{\psi}_{A'B'C'D'}$ and the scalars $\Lambda, \widetilde{\Lambda}$ vanish everywhere. However, since in a curved space-time with vanishing $\Lambda, \widetilde{\Lambda}$, the potentials with the gauge freedoms (4.4) and (4.12) only exist provided D is replaced by ∇ and the trace-free part Φ_{ab} of the Ricci tensor vanishes as well [19], the background 4-geometry is actually flat Euclidean 4-space. Note that we require that (4.16) should be identically satisfied to avoid, after a gauge transformation, obtaining more boundary conditions than the ones originally imposed. The curvature of the background should not, itself, be subject to a boundary condition.

The same result can be derived by using the potential $\rho_{A'}^{BC}$ and its independent counterpart $\Lambda_A^{B'C'}$. This spinor field yields the $\Gamma_{AB}^{C'}$ potential by means of

$$\Gamma_{AB}^{C'} = D_{BB'}\, \Lambda_A^{B'C'} \quad , \tag{4.17}$$

and has the gauge freedom

$$\widehat{\Lambda}_A^{B'C'} \equiv \Lambda_A^{B'C'} + D^{C'}{}_A\, \widetilde{\chi}^{B'} \quad , \tag{4.18}$$

where $\widetilde{\chi}^{B'}$ satisfies the positive-helicity Weyl equation

$$D_{BF'}\, \widetilde{\chi}^{F'} = 0 \quad . \tag{4.19}$$

Thus, if also the gauge-equivalent potentials (4.9) and (4.18) have to satisfy the boundary conditions (2.1) on S^3, one finds

$$2^{\frac{3}{2}} \; {}_e n^A{}_{A'} \; {}_e n^B{}_{B'} \; {}_e n^C{}_{C'} \; D_{CL'} \; D_{BF'} \; D^{L'}{}_A \; \widetilde{\chi}^{F'} = \epsilon \; D_{LC'} \; D_{MB'} \; D^L{}_{A'} \; \chi^M \tag{4.20}$$

on the 3-sphere. In our flat background, covariant derivatives commute, hence (4.20) is identically satisfied by virtue of (4.10) and (4.19). However, in the curved case the boundary conditions (4.20) are replaced by

$$2^{\frac{3}{2}} \; {}_e n^A{}_{A'} \; {}_e n^B{}_{B'} \; {}_e n^C{}_{C'} \; \nabla_{CL'} \; \nabla_{BF'} \; \nabla^{L'}{}_A \; \widetilde{\chi}^{F'} = \epsilon \; \nabla_{LC'} \; \nabla_{MB'} \; \nabla^L{}_{A'} \; \chi^M \tag{4.21}$$

on S^3, if the *local* expressions of ϕ_{ABC} and $\widetilde{\phi}_{A'B'C'}$ in terms of potentials still hold [13-16]. By virtue of (4.14)-(4.15), where ν_C is replaced by χ_C and $\widetilde{\nu}_{C'}$ is replaced by $\widetilde{\chi}_{C'}$, this means that the Weyl spinors $\psi_{ABCD}, \widetilde{\psi}_{A'B'C'D'}$ and the scalars $\Lambda, \widetilde{\Lambda}$ should vanish, since one should find

$$\nabla^{AA'} \; \widehat{\rho}^{BC}_{A'} = 0 \quad , \quad \nabla^{AA'} \; \widehat{\Lambda}^{B'C'}_{A} = 0 \quad . \tag{4.22}$$

If we assume that $\nabla_{BF'} \; \widetilde{\chi}^{F'} = 0$ and $\nabla_{MB'} \; \chi^M = 0$, we have to show that (4.21) differs from (4.20) by terms involving a part of the curvature that is vanishing everywhere. This is proved by using the basic rules of 2-spinor calculus and spinor Ricci identities [17-18]. Thus, bearing in mind that [17]

$$\Box^{AB} \; \widetilde{\chi}_{B'} = \Phi^{AB}{}_{L'B'} \; \widetilde{\chi}^{L'} \quad , \tag{4.23}$$

$$\Box^{A'B'} \; \chi_B = \widetilde{\Phi}^{A'B'}{}_{LB} \; \chi^L \quad , \tag{4.24}$$

one finds

$$\nabla^{BB'} \nabla^{CA'} \chi_B = \nabla^{(BB'} \nabla^{C)A'} \chi_B + \nabla^{[BB'} \nabla^{C]A'} \chi_B$$

$$= -\frac{1}{2} \nabla_B^{\ B'} \nabla^{CA'} \chi^B + \frac{1}{2} \widetilde{\Phi}^{A'B'LC} \chi_L \quad . \tag{4.25}$$

Thus, if $\widetilde{\Phi}^{A'B'LC}$ vanishes, also the left-hand side of (4.25) has to vanish since this leads to the equation $\nabla^{BB'} \nabla^{CA'} \chi_B = \frac{1}{2} \nabla^{BB'} \nabla^{CA'} \chi_B$. Hence (4.25) is identically satisfied. Similarly, the left-hand side of (4.21) can be made to vanish identically provided the additional condition $\Phi^{CDF'M'} = 0$ holds. The conditions

$$\Phi^{CDF'M'} = 0 \quad , \quad \widetilde{\Phi}^{A'B'CL} = 0 \quad , \tag{4.26}$$

when combined with the conditions

$$\psi_{ABCD} = \widetilde{\psi}_{A'B'C'D'} = 0 \quad , \quad \Lambda = \widetilde{\Lambda} = 0 \quad , \tag{4.27}$$

arising from (4.22) for the local existence of $\rho_{A'}^{BC}$ and $\Lambda_A^{B'C'}$ potentials, imply that the whole Riemann curvature should vanish. Hence, in the boundary-value problems we are interested in, the only admissible background 4-geometry (of the Einstein type [20]) is flat Euclidean 4-space.

5. Boundary conditions in supergravity

The boundary conditions studied in the previous sections are not appropriate if one studies supergravity multiplets and supersymmetry transformations at the boundary [9]. By contrast, it turns out one has to impose another set of locally supersymmetric boundary conditions, first proposed in [21]. These are in general mixed, and involve in particular Dirichlet conditions for the transverse modes of the vector potential of electromagnetism, a mixture of Dirichlet and Neumann conditions for scalar fields, and local boundary conditions for the spin-$\frac{1}{2}$ field and the spin-$\frac{3}{2}$ potential. Using two-component spinor notation for supergravity [9,22], the spin-$\frac{3}{2}$ boundary conditions take the form

$$\sqrt{2}\ {}_{e}n_{A}^{\ \ A'}\ \psi^{A}_{\ \ i} = \epsilon\ \widetilde{\psi}^{A'}_{\ \ i} \quad \text{on} \quad S^3 \quad . \tag{5.1}$$

With our notation, $\epsilon \equiv \pm 1$, ${}_{e}n_{A}^{\ \ A'}$ is the Euclidean normal to S^3, and $\left(\psi^{A}_{\ \ i}, \widetilde{\psi}^{A'}_{\ \ i}\right)$ are the *independent* (i.e. not related by any conjugation) spatial components (hence $i = 1, 2, 3$) of the spinor-valued 1-forms appearing in the action functional of Euclidean supergravity [9,22].

It appears necessary to understand whether the analysis in the previous section and in [23] can be used to derive restrictions on the classical boundary-value problem corresponding to (5.1). For this purpose, we study a Riemannian background 4-geometry, and we use the decompositions of the spinor-valued 1-forms in such a background, i.e. [9]

$$\psi^{A}_{\ \ i} = h^{-\frac{1}{4}} \left[\chi^{(AB)B'} + \epsilon^{AB}\ \widetilde{\phi}^{B'} \right] e_{BB'i} \quad , \tag{5.2}$$

$$\widetilde{\psi}^{A'}_{\ \ i} = h^{-\frac{1}{4}} \left[\widetilde{\chi}^{(A'B')B} + \epsilon^{A'B'}\ \phi^{B} \right] e_{BB'i} \quad , \tag{5.3}$$

where h is the determinant of the 3-metric on S^3, and $e_{BB'i}$ is the spatial component of the tetrad, written in 2-spinor language. If we now reduce the classical theory of simple supergravity to its physical degrees of freedom by imposing the gauge conditions [9]

$$e_{AA'}{}^{i}\,\psi^{A}{}_{i}=0 \quad , \tag{5.4}$$

$$e_{AA'}{}^{i}\,\widetilde{\psi}^{A'}{}_{i}=0 \quad , \tag{5.5}$$

we find that the expansions of (5.2)-(5.3) on a family of 3-spheres centred on the origin take the forms [9]

$$\psi^{A}{}_{i}=\frac{h^{-\frac{1}{4}}}{2\pi}\sum_{n=0}^{\infty}\sum_{p,q=1}^{(n+1)(n+4)}\alpha_{n}^{pq}\Big[m_{np}^{(\beta)}(\tau)\,\beta^{nqABB'}+\widetilde{r}_{np}^{(\mu)}(\tau)\,\overline{\mu}^{nqABB'}\Big]e_{BB'i} \quad , \tag{5.6}$$

$$\widetilde{\psi}^{A'}{}_{i}=\frac{h^{-\frac{1}{4}}}{2\pi}\sum_{n=0}^{\infty}\sum_{p,q=1}^{(n+1)(n+4)}\alpha_{n}^{pq}\Big[\widetilde{m}_{np}^{(\beta)}(\tau)\,\overline{\beta}^{nqA'B'B}+r_{np}^{(\mu)}(\tau)\,\mu^{nqA'B'B}\Big]e_{BB'i} \quad . \tag{5.7}$$

With our notation, α_{n}^{pq} are block-diagonal matrices with blocks $\begin{pmatrix}1 & 1\\ 1 & -1\end{pmatrix}$, and the β- and μ-harmonics on S^3 are given by [9]

$$\beta^{nq}{}_{ACC'}=\rho^{nq}{}_{(ACD)}\,n^{D}{}_{C'} \quad , \tag{5.8}$$

$$\mu^{nq}{}_{A'B'B}=\sigma^{nq}{}_{(A'B'C')}\,n_{B}{}^{C'} \quad . \tag{5.9}$$

In the light of (5.6)-(5.9), one gets the following physical-degrees-of-freedom form of the spinor-valued 1-forms of supergravity (cf. [9,22,24]):

$$\psi^{A}{}_{i}=h^{-\frac{1}{4}}\,\phi^{(ABC)}\;{}_{e}n_{C}{}^{B'}\,e_{BB'i} \quad , \tag{5.10}$$

$$\widetilde{\psi}^{A'}{}_{i}=h^{-\frac{1}{4}}\,\widetilde{\phi}^{(A'B'C')}\;{}_{e}n^{B}{}_{C'}\,e_{BB'i} \quad , \tag{5.11}$$

where $\phi^{(ABC)}$ and $\widetilde{\phi}^{(A'B'C')}$ are totally symmetric and independent spinor fields.

Within this framework, a *sufficient* condition for the validity of the boundary conditions (5.1) on S^3 is

$$\sqrt{2}\ {}_e n_A^{\ A'}\ {}_e n_C^{\ B'}\ \phi^{(ABC)} = \epsilon\ {}_e n^B_{\ C'}\ \widetilde{\phi}^{(A'B'C')} \quad . \tag{5.12}$$

From now on, one can again try to express *locally* $\phi^{(ABC)}$ and $\widetilde{\phi}^{(A'B'C')}$ in terms of four potentials as in section 4 and in [23], providing they are solutions of massless free-field equations. The alternative possibility is to consider the Rarita-Schwinger form of the field strength, written in 2-spinor language. The corresponding potential is no longer symmetric as in (4.1), and is instead subject to the equations (cf. [13-16,25])

$$\epsilon^{B'C'}\ \nabla_{A(A'}\ \gamma^A_{\ \ B')C'} = 0 \quad , \tag{5.13}$$

$$\nabla^{B'(B}\ \gamma^{A)}_{\ \ B'C'} = 0 \quad . \tag{5.14}$$

Moreover, the spinor field $\widetilde{\nu}_{A'}$ appearing in the gauge transformation (4.4) is no longer taken to be a solution of the positive-helicity Weyl equation (4.5). Hence the classical boundary-value problem might have new features with respect to the analysis of section 4 and [23].

Indeed, the investigation appearing in this section is incomplete, and it relies in part on the unfinished work in [26]. Moreover, it should be emphasized that our analysis, although motivated by quantum cosmology, is entirely classical. Hence we have not discussed ghost modes. The theory has been reduced to its physical degrees of freedom to make a comparison with the results in [23], but totally symmetric field strengths do not enable one to recover the full physical content of simple supergravity. Hence the 4-sphere background studied in [2] is not ruled out by our work [26].

6. Results and open problems

Following [9] and [23], we have derived the conditions (2.13), (2.15), (3.5), and (3.8) under which spin-lowering and spin-raising operators preserve the local boundary conditions studied in [9-11]. Note that, for spin 0, we have introduced a pair of independent scalar fields on the real Riemannian section of a complex space-time, following [27], rather than a single scalar field, as done in [9]. Setting $\phi \equiv \phi_1 + i\phi_2, \widetilde{\phi} \equiv \phi_3 + i\phi_4$, this choice leads to the boundary conditions

$$\phi_1 = \epsilon \, \phi_3 \quad \phi_2 = \epsilon \, \phi_4 \quad \text{on} \quad S^3 \quad , \tag{6.1}$$

$${}_{e}n^{AA'} \, D_{AA'} \, \phi_1 = -\epsilon \, {}_{e}n^{AA'} \, D_{AA'} \, \phi_3 \quad \text{on} \quad S^3 \quad , \tag{6.2}$$

$${}_{e}n^{AA'} \, D_{AA'} \, \phi_2 = -\epsilon \, {}_{e}n^{AA'} \, D_{AA'} \, \phi_4 \quad \text{on} \quad S^3 \quad , \tag{6.3}$$

and it deserves further study.

We have then focused on the potentials for spin-$\frac{3}{2}$ field strengths in flat or curved Riemannian 4-space bounded by a 3-sphere. Remarkably, it turns out that local boundary conditions involving field strengths and normals can only be imposed in a flat Euclidean background, for which the gauge freedom in the choice of the potentials remains. In [16] it was found that ρ potentials exist *locally* only in the self-dual Ricci-flat case, whereas γ potentials may be introduced in the anti-self-dual case. Our result may be interpreted as a further restriction provided by (quantum) cosmology. What happens is that the boundary conditions (2.1) fix at the boundary a spinor field involving *both* the field strength ϕ_{ABC} and the field strength $\widetilde{\phi}_{A'B'C'}$. The local existence of potentials for the field strength ϕ_{ABC}, jointly with the occurrence of a boundary, forces half of the Riemann curvature of the background to vanish. Similarly, the remaining half of such Riemann curvature has to vanish on considering the field strength $\widetilde{\phi}_{A'B'C'}$. Hence the background 4-geometry can only be flat Euclidean space. This is different from the analysis in [13-16], since in that case one is not dealing with boundary conditions forcing us to consider both ϕ_{ABC} and $\widetilde{\phi}_{A'B'C'}$.

A naturally occurring question is whether the potentials studied in this paper can be used to perform one-loop calculations for spin-$\frac{3}{2}$ field strengths subject to (2.1) on S^3. This problem may provide another example (cf. [9]) of the fertile interplay between twistor theory and quantum cosmology [26], and its solution might shed new light on one-loop quantum cosmology and on the quantization program for gauge theories in the presence of boundaries [1-9]. For this purpose, as shown in recent papers by ourselves and other co-authors [28-30], it is necessary to study Riemannian background 4-geometries bounded by two concentric 3-spheres (cf. sections 2-5). Moreover, the consideration of non-physical degrees of freedom of gauge fields, set to zero in our classical analysis, is necessary to achieve a covariant quantization scheme.

Acknowledgments

We are indebted to Stephen Huggett for suggesting we should prepare our contribution to this volume, and to Roger Penrose for bringing Refs. [13-16] to our attention. Many conversations with Alexander Yu. Kamenshchik on one-loop quantum cosmology and related problems have stimulated our research.

References

[1] Moss I. G. and Poletti S. (1990) *Nucl. Phys.* B **341**, 155.

[2] Poletti S. (1990) *Phys. Lett.* **249B**, 249.

[3] D'Eath P. D. and Esposito G. (1991) *Phys. Rev.* D **43**, 3234.

[4] D'Eath P. D. and Esposito G (1991) *Phys. Rev.* D **44**, 1713.

[5] Barvinsky A. O., Kamenshchik A. Yu., Karmazin I. P. and Mishakov I. V. (1992) *Class. Quantum Grav.* **9**, L27.

[6] Kamenshchik A. Yu. and Mishakov I. V. (1992) *Int. J. Mod. Phys.* A **7**, 3713.

[7] Barvinsky A. O., Kamenshchik A. Yu. and Karmazin I. P. (1992) *Ann. Phys., N.Y.* **219**, 201.

[8] Kamenshchik A. Yu. and Mishakov I. V. (1993) *Phys. Rev.* D **47**, 1380.

[9] Esposito G. (1994) *Quantum Gravity, Quantum Cosmology and Lorentzian Geometries* Lecture Notes in Physics, New Series m: Monographs vol m12 second corrected and enlarged edition (Berlin: Springer).

[10] Breitenlohner P. and Freedman D. Z. (1982) *Ann. Phys., N.Y.* **144**, 249.

[11] Hawking S. W. (1983) *Phys. Lett.* **126B**, 175.

[12] Penrose R. and Rindler W. (1986) *Spinors and Space-Time, Vol. 2: Spinor and Twistor Methods in Space-Time Geometry* (Cambridge: Cambridge University Press).

[13] Penrose R. (1990) *Twistor Newsletter* **n 31**, 6.

[14] Penrose R. (1991) *Twistor Newsletter* **n 32**, 1.

[15] Penrose R. (1991) *Twistor Newsletter* **n 33**, 1.

[16] Penrose R. (1991) Twistors as Spin-$\frac{3}{2}$ Charges *Gravitation and Modern Cosmology* eds A. Zichichi, V. de Sabbata and N. Sánchez (New York: Plenum Press).

[17] Ward R. S. and Wells R. O. (1990) *Twistor Geometry and Field Theory* (Cambridge: Cambridge University Press).

[18] Esposito G. (1993) *Nuovo Cimento* B **108**, 123.

[19] Buchdahl H. A. (1958) *Nuovo Cim.* **10**, 96.

[20] Besse A. L. (1987) *Einstein Manifolds* (Berlin: Springer).

[21] Luckock H. C. and Moss I. G. (1989) *Class. Quantum Grav.* **6**, 1993.

[22] D'Eath P. D. (1984) *Phys. Rev.* D **29**, 2199.

[23] Esposito G. and Pollifrone G. (1994) *Class. Quantum Grav.* **11**, 897.

[24] Sen A. (1981) *J. Math. Phys.* **22**, 1781.

[25] Esposito G. (1994) *Complex General Relativity* (book in preparation).

[26] Esposito G., Kamenshchik A. Yu., Mishakov I. V. and Pollifrone G. (1994) *Supersymmetric Boundary Conditions in Quantum Cosmology*, work in progress.

[27] Hawking S. W. (1979) The path integral approach to quantum gravity *General Relativity, an Einstein Centenary Survey* eds S. W. Hawking and W. Israel (Cambridge: Cambridge University Press).

[28] Esposito G. (1994) *Class. Quantum Grav.* **11**, 905.

[29] Esposito G., Kamenshchik A. Yu., Mishakov I. V. and Pollifrone G. (1994) *Euclidean Maxwell Theory in the Presence of Boundaries, Part II*, DSF preprint 94/4.

[30] Esposito G., Kamenshchik A. Yu., Mishakov I. V. and Pollifrone G. (1994) *Gravitons in One-Loop Quantum Cosmology: Correspondence Between Covariant and Non-Covariant Formalisms*, DSF preprint 94/14.

5
Analytic Cohomology of Blown-Up Twistor Spaces

Robin Horan
School of Mathematics and Statistics
University of Plymouth
Plymouth PL4 8AA
Devon, U.K.

In [6], Michael Singer proposed a definition for a four dimensional conformal field theory. In this theory the role of compact Riemann surfaces, which occur in standard conformal field theory, is played instead by compact flat twistor spaces. It is therefore tempting to ask questions about these twistor spaces which are, in some way, natural extensions from Riemann surfaces. The properties of compact Riemann surfaces are well known and have been extensively documented and so there is an immensely rich source of possible questions that can be asked about flat twistor spaces.

One such property of compact Riemann surfaces concerns meromorphic functions having poles of prescribed maximum order at given points: Let X be a compact Riemann surface with distinct points $P_1,\ldots,P_k$ on X. If $n_1,\ldots,n_k$ are arbitrary positive integers, how many linearly independent meromorphic functions are there on X, which have poles of order at most n_i at P_i, and no others?

To answer this question one can use the following strategy:

(a) Convert the question to one involving global data. This is achieved through the introduction of line bundles and divisors. The problem then becomes one of determining the dimension of the cohomology group $H^0(X,\mathcal{O}[D])$ where [D] is the line bundle of the divisor D, which for the above problem is $\Sigma_{i=1}^{k} n_i P_i$.

(b) Use the Riemann–Roch theorem. This enables holomorphic data to be calculated in terms of topological data:

$$\dim H^0(X,\mathcal{O}[D]) - \dim H^1(X,\mathcal{O}[D]) = \deg D + 1 - g,$$

where g is the genus of X.

(c) Use vanishing theorems for $H^1(X,\mathcal{O}[D])$ in order to eliminate the unwanted term. One such is the Kodaira theorem: if $\deg(D) > 2g-2$ then $H^1(X,\mathcal{O}[D]) = 0$.

To extend this problem to flat twistor spaces we need a good analogue of meromorphic functions with prescribed singularities. Fortunately a candidate for these exists, the elementary states based on a line, but one first has to decide what is meant by a prescribed order of singularity, and then how to extend this notion to a general flat twistor space.

The first part of this question was answered by Eastwood and Hughston in [1]. If $f(Z)$ is a homogeneous polynomial of degree m in Z, and if $f(Z)$ is coprime to Z_0 and Z_1, then $\frac{f(Z)}{Z_0 Z_1}$ is a representative of an elementary state of homogeneity $m-2$, with a pole of order 1 on the line $Z_0 = Z_1 = 0$. Similarly, $\frac{f(Z)}{Z_0 Z_1^2} = \frac{Z_0 f(Z)}{(Z_0 Z_1)^2}$ has homogeneity $(m+1)-4 = m-3$, and a pole of order 2 on $Z_0 = Z_1 = 0$.

Thus, at least in the case of $\mathbf{P}^3$, one can formulate the question: given a fixed line L in $\mathbf{P}^3$, a given homogeneity n, and a positive integer k, how many elementary states based on L are there, with homogeneity n, and with a singularity on L of order at most k?

One way of converting this question, at least in $\mathbf{P}^3$, to one involving global data, was given in [1]. Blow-up $\mathbf{P}^3$ along L, to obtain the complex manifold $\tilde{\mathbf{P}}^3$, which in this case is actually a submanifold of $\mathbf{P}^3 \times \mathbf{P}^1$. Now let $a \geq 0, b \leq -2$ be integers, and form the bundle $\mathcal{O}(a) \otimes \mathcal{O}(b)$ on $\mathbf{P}^3 \times \mathbf{P}^1$, from $\mathcal{O}(a)$ on $\mathbf{P}^3$ and $\mathcal{O}(b)$ on $\mathbf{P}^1$. Let $\mathcal{O}(a,b)$ be the restriction of this bundle to $\tilde{\mathbf{P}}^3$. Then it is shown in [1], that elements of $H^1(\tilde{\mathbf{P}}^3, \mathcal{O}(a,b))$, when restricted away from the blown-up line, are representatives for elementary states based on L, of homogeneity $a+b$, and order of singularity at most $-b-1$. The question now concerns the dimension of $H^1(\tilde{\mathbf{P}}^3, \mathcal{O}(\mathrm{a}, \mathrm{b}))$.

This gives a way of extending the definition of elementary states based on a line, to flat twistor-spaces, since such twistor spaces have the property that each projective line (fibre) has a neighbourhood which is biholomorphic to $\mathbf{P}^-$.

Given a flat twistor space Z, and distinguished line L, one can then define the blow-up of Z along L, say $\tilde{Z}$, and the bundle $\mathcal{O}(a,b)$ can be defined on $\tilde{Z}$ in such a way as to preserve the essential properties of $\mathcal{O}(a,b)$ in $\tilde{\mathbf{P}}^3$. Elements of $H^1(\tilde{Z}, \mathcal{O}(a,b))$ then have homogeneity $a+b$ and a 'pole' of order at most $-b-1$ on L, when restricted away from L, but in a neighbourhood of L.

We then define the elements of this group to be our elementary states based on L, with the prescribed conditions, and the problem we wish to solve is to find the dimension of $H^1(\tilde{Z}, \mathcal{O}(a,b))$. This will give a partial answer to the equivalent problem in Riemann surfaces which formed the motivation for this work, and indeed the strategy for the solution follows closely that for Riemann surfaces:

(a) The question has already been converted to one involving global data on a compact manifold, though this time the manifold is $\tilde{Z}$, not Z.

(b) The Hirzebruch–Riemann–Roch theorem enables the calculation of the holomorphic Euler characteristic of $\mathcal{O}(a,b)$ on $\tilde{Z}$, using topological data [8]. It states that the holomorphic Euler characteristic $\chi(\tilde{Z}, \mathcal{O}(a,b))$, which is defined by

$$\chi(\tilde{Z}, \mathcal{O}(a,b)) = \Sigma_{i=0}^{3}(-1)^i \dim H^i(\tilde{Z}, \mathcal{O}(a,b)),$$

is given in terms of Chern classes of $\tilde{Z}$, i.e.

$$\dim H^0(\tilde{Z},\mathcal{O}(a,b)) - \dim H^1(\tilde{Z},\mathcal{O}(a,b)) + \dim H^2(\tilde{Z},\mathcal{O}(a,b)) - \dim H^3(\tilde{Z},\mathcal{O}(a,b)) = [\mathrm{Ch}(\mathcal{O}(a,b))\mathrm{Td}(\tilde{Z})][\tilde{Z}]$$

where $\mathrm{Ch}(\mathcal{O}(a,b))$ is the Chern character of the bundle $\mathcal{O}(a,b)$, and $\mathrm{Td}(\tilde{Z})$ is the Todd class of (the holomorphic tangent bundle of) $\tilde{Z}$. I was able to calculate this when the Riemannian four-manifold M was compact, oriented, and conformally flat, and Z was its twistor space [3]. The method involved using Poincaré duality and intersection of homology classes. I obtained the following results:

$$\chi(\tilde{Z},\mathcal{O}(a,b)) = \frac{1}{12}(a+b+1)(a+b+2)(a+b+3)\chi - \frac{1}{6}b(b+1)(b+3a+5),$$

where χ is the Euler characteristic. It is an easy exercise to extend this result to the case where the single line L is replaced by a set $L_1,\ldots,L_r$ of non-intersecting lines. In this case the bundle can still be defined, and each L_i given a prescribed order of singularity $-b_i-1$, subject to the conditions $b_i \leq -2, a_i \geq 0$ and $a_i+b_i=m$, the order of homogeneity. In this case the line bundle is written as $\mathcal{O}(a_1,\ldots,a_r;b_1,\ldots,b_r)$ and the result now becomes

$$\chi(\tilde{Z},\mathcal{O}(a_1,\ldots,a_r;b_1,\ldots,b_r)) = \frac{1}{12}(m+1)(m+2)(m+3) - \frac{1}{6}\Sigma_{i=1}^{r} b_i(b_i+1)(3m+5-2b_i),$$

where $\tilde{Z}$ is Z blown up along $L = L_1 \cup \cdots \cup L_r$. Subsequently LeBrun (private communication) has shown that this index calculation can be considerably shortened, avoiding the use of the Hirzebruch–Riemann–Roch theorem and making use instead of knowledge of the index $\chi(Z,\mathcal{O}(m))$ which can be found from the work of Hitchin [2]. This method also works for self-dual, compact Z.

(c) The use of vanishing theorems is more complicated in this case since the alternating sum contains four terms. In the case where the manifold M has negative scalar curvature, it is easy to show both the H^0 and H^3 terms are zero. This leaves H^2 as the awkward term. The Serre dual of $H^2(\tilde{Z},\mathcal{O}(a_1,\ldots,a_r;b_1,\ldots,b_r))$ can be calculated, and is $H^1(\tilde{Z},\mathcal{O}(c_1,\ldots,c_r;d_1,\ldots,d_r))$, where $c_i = -a_i-3 \leq -3, d_i = -b_i-1 \geq -1$. Let $n = -m-4$, so that $c_i+d_i = n$ for $i = 1,\ldots,r$.

It turns out that, using diagram chasing techniques on two Mayer Vietoris sequences, I was able to prove the following relationship between the cohomologies of the blown-up manifold $\tilde{Z}$, and the flat twistor-space Z: if $H^0(Z,\mathcal{O}(n))$ and $H^1(Z,\mathcal{O}(n))$ both vanish, then

$$\dim H^1(\tilde{Z},\mathcal{O}(c_1,\ldots,c_r;d_1,\ldots,d_r)) = r \times \dim H^0(\mathbf{P}^+,\mathcal{O}(n)),$$

and this would enable the final calculation to be made [4].

The question now turns on a vanishing theorem for $H^1(Z,\mathcal{O}(m))$. For the case of M having negative scalar curvature, I was then able to prove the following theorem:

If M is compact, Riemannian, self-dual, Einstein, with negative scalar curvature, then $H^1(Z, \mathcal{O}(m)) = 0$ if $m > 0$, [5].

The method of proof used the Penrose transform to identify $H^1(Z, \mathcal{O}(n-2))$ with certain spinor fields in M, which led to a Bochner style vanishing theorem for the spinor fields. This result was brought to my attention by LeBrun, from some work of his research student (Thornber [7]) on vanishing theorems for quaternionic-Kähler manifolds. His method of approach was completely different, involving a direct attack on the problem in Z.

This information provides an answer when M is a compact, Riemannian, conformally-flat, Einstein manifold with negative scalar curvature, and Z is its twistor space. Let L be the disjoint union of the non-intersecting lines L_i and let $a_i \geq 0, b_i \leq -2$ and $a_i + b_i = m$ for $i = 1, \ldots, r$. If $m < -4$, then we have

$$\dim H^1(\tilde{Z}, \mathcal{O}(a_i, \ldots, a_r; b_i, \ldots, b_r)) = r \times \dim H^0(\mathbf{P}^+, \mathcal{O}(-a-b-4)) - \chi(\tilde{Z}, \mathcal{O}(a, b)).$$

We note that the M are precisely the hyperbolic 4-manifolds. Details of all this work will appear in due course. I wish to express my gratitude to Stephen Huggett, Michael Singer and Paul Tod for all the help, encouragement and guidance given to me in the course of this work, and to Claude LeBrun for pointing out the vanishing theorem and simplification of the index calculation, mentioned above.

References

[1] M. G. Eastwood, L. P. Hughston, *Massless Fields Based on a Line*, in *Advances in Twistor Theory*, edited by L.P.Hughston and R.S.Ward, Pitman, London 1979.

[2] N. J. Hitchin, *Kählerian Twistor Spaces*, Proc. Lond. Math. Soc. **43**, 133-150, 1981.

[3] R. E. Horan, *Holomorphic Euler Characteristics on a Blown-up Twistor Space*, preprint.

[4] R. E. Horan, *The Relationship of Analytic Cohomologies of a Blown-up Twistor space to the Analytic Cohomology of a Flat Twistor Space*, preprint.

[5] R. E. Horan. *Vanishing Theorems*, preprint.

[6] M. A. Singer. *Flat Twistor Spaces, Conformally Flat Manifolds and Four-Dimensional Field Theory*, Comm. Math. Phys. **133**, 75-90, 1990.

[7] M. Thornber. *Vanishing Theorems for Quaternionic Kähler Manifolds*, Ph.D. Dissertation, Stony Brook, SUNY, 1989.

[8] R. O Wells, *Differential Analysis on Complex Manifolds*, Springer Verlag, 1980.

6
Geometric Aspects of Quantum Mechanics

L. P. HUGHSTON Merrill Lynch International Limited,
Ropemaker Place, 25 Ropemaker Street, London EC2Y 9LY, U.K.

1. INTRODUCTION

1.1 Unitary evolution and state reduction

These notes are intended partly as an introduction to certain geometrical aspects of quantum mechanics generally, partly as a distillation of the existing literature on non-linear quantum theory (though no attempt will be made here at a survey), and partly as a vehicle for the expression of some new ideas and results.

We begin with a few remarks indicating briefly the motivation for studying quantum theory from a geometrical point of view, and in particular for the development of a *non-linear geometrical quantum theory*. Quantum mechanical evolution is governed, so it would seem, by two apparently distinct processes. These are, to paraphrase the terminology of Penrose (1989), the so-called 'U-process' (unitary evolution), and the so-called 'R-process' (state vector reduction). The relation of these processes to one another is very mysterious, and in recent years interest in the matter has increased for a variety of reasons to almost feverish levels of intensity. It is now in fact probably a safe bet to say that a proper understanding of the situation cannot be achieved very easily (if at all) within the framework of quantum theory as it is conventionally formulated.

One would like perhaps to think of both processes as representing different aspects, in some sense, of a kind of more general or universal quantum process. On the other hand, the successes of quantum mechanics have been so wide and varied, that one must be wary of making any gross alteration to the structure of the theory.

Here is where we might take a hint from the methodology of twistor theory. The plan will be essentially to reformulate standard quantum mechanics in a new way, bringing out various important geometrical features of the theory that are normally obscured. The new formulation will be in the first place entirely equivalent to the original, but will suggest generalisations that would not otherwise present themselves.

The connection with twistor theory goes perhaps one level deeper, since the particular geometrical characteristics of quantum theory that emerge as being relevant have to do precisely with the complex analytic structure of the quantum mechanical state manifold. It would be nice to think that this is not entirely a coincidence. In any case, part of the problem with the 'conventional' formulation of quantum theory is that this complex analytic structure is thoroughly built in to the formalism, to such an extent in fact that it is difficult to see on the face of it how one could possibly alter it. The scheme offered here remedies that situation.

1.2 Summary

The plan of the paper is as follows. In §2 it will be shown how it can be illuminating to examine ordinary linear quantum mechanics by use of Riemannian geometry. The state manifold is identified as *complex projective space*, and the associated Riemannian structure on the associated real manifold is given by the *Fubini-Study metric*. In this picture we are able to show that the sample path of a U-process is nothing more than a *Killing trajectory*; whereas the transition probabilities of the R-process turn out to be associated in a curious way with the *real geodesics* of the state manifold. Thus in this spirit although we do not achieve a 'unity' among the two processes, we are able nevertheless to develop a language that is sufficiently rich to encourage a useful kind of speculation about their relationship.

In §3 we generalise the state space of quantum mechanics to a fully Riemannian geometry, without symmetry. It turns out (as has been noted by a number of other authors) that many of the essential physical ingredients of quantum mechanics survive if the state space is represented by a general complex manifold, with a compatible Riemannian structure. Thus for the state space we take a complex manifold M with a Kähler metric. A distinctive feature of a such a manifold is the intricate interplay of three related aspects of its geometry: its complex structure, its symplectic structure, and its Riemannian structure. For many applications an 'almost' complex structure suffices.

In §4 we review various physical aspects of this non-linear geometrical quantum theory, and point out a few of its surprising features. The points of the state manifold M are quantum mechanical states; observables are real differentiable functions. The value of a function at a point is the expectation of the observable for that state; the derivative of the function vanishes at an 'eigenstate'; and the value of the function at such a point is the 'eigenvalue'. We shall demonstrate that the *Born approximation* takes a particularly simple form when analysed in this way, thus improving on a related result due to Weinberg (1989b).

When M is viewed as a *real* manifold the geometry determines both a symplectic two-form $\omega(-,-)$ and a Riemannian metric $g(-,-)$. These can be thought of very loosely as associated, respectively, with the U-process and the R-process; alternatively one might say that the symplectic structure is more directly connected with evolutionary aspects of the quantum mechanical system, whereas the Riemannian metric is more directly linked with probabilistic aspects of the theory;—in a Kähler manifold the two structures are of course ultimately equivalent.

Generally there are no Killing vectors: instead, the Schrödinger evolutionary trajectories are integral curves of vector fields that Lie-annihilate the symplectic two-form $\omega(-,-)$, i.e. vector fields of the Hamiltonian type ('canonical' vector fields). The commutator of two observables is given in terms of the Poisson bracket between the corresponding functions, where the bracket operation is determined by the symplectic form $\omega(-,-)$, and where the closure property $d\omega = 0$ ensures the Jacobi identity.

For any observable F we can form a canonical vector field $X_F = (dF)^{\#}$ where the 'sharp' denotes contraction with the inverse symplectic form ω^{-1}. The magnitude of X_F with respect to the metric $g(-,-)$ then turns out to be (a factor of two times) the 'uncertainty' ΔF ; and thus for any two observables a general 'uncertainty relation' $(\Delta F)^2(\Delta G)^2 \geq \frac{1}{4}[F,G]^2$ can be shown to hold, where $[F,G] = -i\omega(dF^{\#}, dG^{\#})$.

The metric also determines a 'covariance' function $\mathrm{Cov}(F,G) = \frac{1}{4}g(dF^{\#}, dG^{\#})$ for any two observables; the associated 'correlation' function $\rho(F,G) = \mathrm{Cov}(F,G)/(\Delta F\,\Delta G)$ satisfies $|\rho| \leq 1$, with equality only if $[F,G] = 0$.

Given the Hamiltonian function H, the Schrödinger equation can be expressed in a remarkably simple way as the dynamical equation $X_H = (dH)^{\#}$, and we are led to the formulation of a *generalised Aharonov-Anandan relation* $g(dH^{\#}, dH^{\#}) = 4(\Delta H)^2$ that relates the local metrical properties of state manifold along the evolutionary trajectory to the uncertainty of the energy.

The content of §5 is devoted to the formulation of some speculative conjectures on the relation of linear theory to its non-linear counterpart. In particular it is argued that in the non-linear theory the Riemannian curvature of M ought to play a special physical role.

This is because the scalar curvature R associated with the metric $g(-,-)$ defines a smooth function on M, which thus has the status of a 'preferred' quantum mechanical observable. In linear quantum mechanics the Ricci scalar is constant, so it is plausible in the general non-linear theory that the gradient of R should be significant. This suggests (since in quantum mechanics only the gradient of the energy is significant) the idea of a natural linear 'Einsteinian identification' between energy and curvature of the form $H = \lambda + \mu R$, where H is the Hamiltonian function, and where λ and μ are constants each with units of energy. One could of course envisage other such relations between energy and curvature, though on the grounds of simplicity the relation $H = \lambda + \mu R$ would appear to be the most compelling .

Notation and conventions. We use units such that the Planck constant $\hbar$ and the speed of light c are both set to unity. In §2 Hilbert space indices are denoted by lower case Greek letters; these also serve as indices for 'homogeneous coordinates' on projective Hilbert space. Upper case Latin letters are reserved for two-component spinors. We use the 'abstract index convention' throughout (cf. Penrose 1968, Penrose & Rindler 1986). Round brackets around a set of indices denote symmetrisation and square brackets denote anti-symmetrisation. In §3 tensor indices for the real Riemannian $2n$-dimensional state manifold M^{2n} are denoted with lower case Latin indices; these are manipulated according to the usual rules of classical Riemannian geometry. We also make some use of the standard index-free notation of symplectic geometry (see Woodhouse 1992, Arnol'd 1989, Abraham & Marsden 1978); in that case the canonical vector field associated with an observable F is denoted X_F ; and the raising of an index by use of the inverse symplectic form is denoted by a sharp (#). For the sake of definiteness and rigour it will be assumed here that the state manifold is finite dimensional;—but almost all of our considerations can be seen to hold in the infinite dimensional case as well, at least formally.

2. GEOMETRIC ANALYSIS OF LINEAR QUANTUM MECHANICS

2.1 Projective geometry

We begin with a recapitulation of ordinary linear quantum mechanics. Our formulation will be entirely equivalent, both mathematically and physically, to the usual one. However, by virtue of the particular geometrical features that receive emphasis here, we shall be led more naturally and expeditiously in the direction of the non-linear generalisation that will be described later.

It turns out to be convenient to use an index notation for Hilbert space operations. Let us write Z^α for coordinates on Hilbert space, with complex conjugate $\bar{Z}_\alpha$ and the usual Hermitian inner product $Z^\alpha \bar{Z}_\alpha$. Thus Z^α corresponds to a Dirac 'ket' vector $|Z\rangle$, and $\bar{Z}_\alpha$ to a 'bra' vector $\langle \bar{Z}|$; the inner product $Z^\alpha \bar{Z}_\alpha$ is equivalent to $\langle \bar{Z}|Z\rangle$. One can think of the coordinate indices as ranging 'abstractly' from 0 to n, where $n+1$ is the dimension of the Hilbert space.

We also use Z^α for homogeneous coordinates on projective Hilbert space, which in the finite dimensional case has the structure of a *complex projective space* equipped with a 'Hermitian correlation', i.e. a complex conjugation operation that maps points to hyperplanes of codimension one.

The complex conjugate of a point P^α is the hyperplane $\bar{P}_\alpha Z^\alpha = 0$. The points on this plane are precisely the states that are 'orthogonal' to the original state P^α. Distinct states P^α and Q^α are joined by a complex projective line represented by the skew tensor $L^{\alpha\beta} = P^{[\alpha}Q^{\beta]}$. The points on this line are the various *complex superpositions* of the original two states. This is a useful thing to know, since the superposition principle in quantum mechanics is very important, and here we have a good representation of it in the language of algebraic geometry. In fact the idea here is to express as much of elementary quantum mechanics as we can in just this sort of way.

The significance of the projective Hilbert space is that the points in it correspond directly to quantum mechanical states, and it is of course the geometry of the actual state space that primarily concerns us. The 'ket' states are points in the projective space; the dual 'bra' states are the projective hyperplanes of codimension one.

2.2 Transition probabilities as cross ratios.

The hyperplanes conjugate to the points P^α and Q^α intersect the joining line $L^{\alpha\beta}$ at a pair of points $\tilde{P}^\alpha$ and $\tilde{Q}^\alpha$ respectively. The projective *cross ratio* between these four points given by

$$\kappa = \frac{P^\alpha \bar{Q}_\alpha \, Q^\beta \bar{P}_\beta}{P^\alpha \bar{P}_\alpha \, Q^\beta \bar{Q}_\beta} \tag{2.1}$$

is the so-called 'transition probability' between P^α and Q^α; in the Dirac notation this is just the familiar expression $\langle \bar{P}|Q\rangle\langle \bar{Q}|P\rangle / \langle \bar{P}|P\rangle\langle \bar{Q}|Q\rangle$.

Viewed as a real manifold the projective line $L^{\alpha\beta}$ is a two-sphere, with $\tilde{P}^\alpha$ and $\tilde{Q}^\alpha$ antipodal to the points P^α and Q^α, thus defining a great circle; and the cross ratio κ is $\frac{1}{2}(1+\cos\theta) = \cos^2(\theta/2)$ where the θ is the angular distance between P^α and Q^α. This construction illustrates the intimate connection between the metrical geometry of the state space and the explicit valuation of the transition probability.

2.3 Fubini-Study metric

Now suppose we let the two states P^α and Q^α approach one another very closely; in the limit the resulting formula for their infinitesimal separation is the natural *Hermitian line element* on complex projective space, given by

$$ds^2 = 8(Z^\alpha \bar{Z}_\alpha)^{-2} Z^{[\alpha} dZ^{\beta]} \bar{Z}_{[\alpha} d\bar{Z}_{\beta]}. \tag{2.2}$$

This is the so-called the 'Fubini-Study metric' (Fubini 1903, Study 1905; cf. Arnol'd 1989, appendix 3; Kobayashi & Nomizu 1969, §IX.6; Page 1987; Anandan & Aharonov 1990; Woodhouse 1992, p. 95). The Fubini-Study metric is symmetric under the induced action of the group of unitary transformations of the Hilbert space—a feature that in fact determines the metric completely, up to an overall constant which here is fixed by the convention that the total length of a geodesic is 2π. More explicitly, formula (2.2) is obtained by setting $P^\alpha = Z^\alpha$ and $Q^\alpha = Z^\alpha + dZ^\alpha$ in expression (2.1) for the cross ratio, while replacing θ with the small angle ds/r_0 in the expression $\frac{1}{2}(1+\cos\theta)$, retaining only terms of the second order in ds, and setting the constant r_0 to unity.

2.4 The projective Schrödinger equation

Let us write H^α_β for the Hamiltonian operator (assumed Hermitian), and $Z^\alpha(t)$ for a smooth curve in Hilbert space. Then in our notation the Schrödinger equation is $dZ^\beta = -\mathrm{i}H^\beta_\gamma Z^\gamma dt$. But we shall be concerned with this equation only inasmuch as it supplies information about the evolution of the actual state of the system—i.e. motion in *projective* Hilbert space. We are interested therefore primarily in the projective Schrödinger equation, given by

$$Z^{[\alpha} dZ^{\beta]} = -\mathrm{i}Z^{[\alpha} H^{\beta]}_\gamma Z^\gamma dt, \tag{2.3}$$

obtained by skew-symmetrising the Schrödinger equation with Z^α. (Here is an example of an operation that is simple enough in our index notation, but is awkward when expressed in the Dirac notation.) Equation (2.3) defines the Schrödinger evolution of a quantum mechanical state in the actual state space. The integral curves generated by the projective Schrödinger equation are Killing trajectories. This does not come as a surprise, for it is well appreciated that the Schrödinger equation represents an 'unfolding of a unitary transformation' on Hilbert space. Here we are stressing the Riemannian geometry associated with the Schrödinger equation, and therefore we think of a solution of the Schrödinger equation as the choice of a Killing vector field on the state manifold; the zeros of this vector field are the 'energy eigenstates'.

2.5 Spin one-half systems

As an illustration suppose we consider the case of an elementary quantum mechanical system consisting of a spin one-half particle in a constant magnetic field. In that case the Hilbert space is two-dimensional, and the actual state space is the one-dimensional complex projective space CP^1. The two energy eigenstates correspond to the north and south poles of the sphere, and the other points on the sphere correspond to complex superpositions of these two energy eigenstates. The Schrödinger trajectories are the integral curves of the Killing vector field defined by the Hamiltonian. In this case the relevant Killing vector field is determined by the requirement that the energy eigenstates are stationary points; it follows therefore that the evolutionary trajectories are generated by *rotations about the north-south axis*, and correspond to 'latitudinal circles'. If a state is initially at a certain latitude, then it stays at that latitude.

It is instructive to examine this particular set-up rather more closely by use of the standard two-component spinor calculus (see e.g. Pirani 1965, Penrose 1968, Penrose & Rindler 1984, 1986). For coordinates on the Hilbert space we write Z^A, with complex conjugate $\bar{Z}^A$. Then by use of the elementary spinor identity $2X^{[A}Y^{B]} = \varepsilon^{AB}X_C Y^C$ where $\varepsilon^{AB} = -\varepsilon^{BA}$ we are able to obtain $ds^2 = 4Z_A dZ^A \bar{Z}_B d\bar{Z}^B / (\bar{Z}_C Z^C)^2$ for the metric, and $Z_C dZ^C = \mathrm{i}H_{AB}Z^A Z^B dt$ for the projective Schrödinger equation. The Hamiltonian is real and symmetric, and is therefore necessarily of the form $H_{AB} = 2E\,P_{(A}\bar{P}_{B)}/(P_C P^C)$ where E is a coupling constant; the points P^A and $\bar{P}^A$ are opposite each other on the sphere, and can be thought of as determining a direction in an ambient real 3-space; the normalization has been arranged so that P^A (*resp.* $\bar{P}^A$) is an eigenstate of the Hamiltonian with eigenvalue E (*resp.* $-E$). For the metric we deduce that

$$\left(\frac{ds}{dt}\right)^2 = 16E^2\,\mathbf{P}[Z^A \to P^A]\mathbf{P}[Z^A \to \bar{P}^A] \tag{2.4}$$

where $\mathbf{P}[Z^A \to P^A]$ is the transition probability from Z^A to P^A, given by the cross ratio $(Z_A\bar{P}^A)(\bar{Z}_B P^B)/(Z_A\bar{Z}^A)(P_B\bar{P}^B) = \frac{1}{2}(1+\cos\theta)$, and where the probability $\mathbf{P}[Z^A \to \bar{P}^A]$ is given analogously by $(Z_A P^A)(\bar{Z}_B\bar{P}^B)/(Z_A\bar{Z}^A)(P_B\bar{P}^B) = \frac{1}{2}(1-\cos\theta)$.

Combining these expressions we then get $ds/dt = 2E\sin\theta$ for the *rate of evolution through state space*; and for the state vector itself we have

$$Z^A = \sin\theta\exp(\mathrm{i}Et)\,P^A + \cos\theta\,\exp(-\mathrm{i}Et)\,\bar{P}^A. \tag{2.5}$$

The latitude is fixed by the angle θ, and the angular rate of revolution about the axis is determined by the difference in the energy levels.

2.6 State vector reduction

A significant exception to the dynamics outlined above occurs when a measurement of the energy is made. In that case the state of the system leaves the Killing trajectory, and moves to one or the other of the poles. The motion here can for practical purposes be regarded as instantaneous, though the precise nature of and mechanism for this particular aspect of quantum mechanical evolution, which we are calling the 'R-process' (cf. Penrose 1989, von Neumann 1955), is the subject of controversy and differing interpretations (see e.g. Bell 1987, Wheeler & Zurek 1983, Cushing & McMullin 1989). Some authors have even maintained that there is no such thing as the R-process.

In any case it is interesting to note that the geometry of the situation is such that the probability of movement to a specified pole is a simple function of the *geodesic distance* between the initial point and the given pole. Consider more generally the case of the evolution of a state in CP^n, and suppose at a certain time the system is in the state $Z^\alpha(t)$ when a measurement is made to decide if the system is in the state P^α.

To see what happens in that situation we construct the complex projective line joining $Z^\alpha(t)$ and the target point P^α, and intersect that line with the plane conjugate to P^α, calling the intersection point $\tilde{P}^\alpha$. On measurement the system moves either to the state P^α with probability $\frac{1}{2}(1+\cos\theta)$ or to the state $\tilde{P}^\alpha$ with probability $\frac{1}{2}(1-\cos\theta)$, where θ is the geodesic distance between the initial state $Z^\alpha(t)$ and the P^α.

These remarks are consistent with our earlier observations about transition probabilities. But note that even in the case of more complicated measurements (for example, for a spin s system, with $2s+1$ distinct energy eigenstates), the probability of an R-process transition from a given evolutionary state to any particular eigenstate of the measurement observable is always determined entirely by the geodesic distances between the initial and final states.

2.7 Energy uncertainty

The spin one-half example for motion in a magnetic field described above can be used to illustrate another important principle. Observe that the angular rate of rotation of the state of the system about the north-south axis is the same for all states, and fixed by the difference of the two energy levels. However, the actual speed of evolution depends on the latitude, and is clearly higher for states with a greater energy uncertainty, i.e. as one approaches the equator.

In fact it is a general result in elementary quantum mechanics that speed of evolution in the state manifold is given by the remarkable relation $ds/dt = 2(\Delta H)$ where ΔH is the energy uncertainty (Anandan & Aharonov 1990).

This can be deduced very directly by insertion of the projective Schrödinger equation $Z^{[\alpha}dZ^{\beta]} = -iZ^{[\alpha}H^{\beta]}_{\gamma}Z^{\gamma}dt$ into the metric $ds^2 = 8(Z^{\alpha}\bar{Z}_{\alpha})^{-2}Z^{[\alpha}dZ^{\beta]}\bar{Z}_{[\alpha}d\bar{Z}_{\beta]}$, followed by simplification of the result by use of the standard formulae

$$\langle H\rangle = \frac{\bar{Z}_{\alpha}H^{\alpha}_{\beta}Z^{\beta}}{\bar{Z}_{\gamma}Z^{\gamma}} \tag{2.6}$$

and

$$(\Delta H)^2 = \langle H^2\rangle - \langle H\rangle^2 = \frac{\bar{Z}_{\alpha}H^{\alpha}_{\beta}H^{\beta}_{\gamma}Z^{\gamma}}{\bar{Z}_{\alpha}Z^{\alpha}} - \left(\frac{\bar{Z}_{\alpha}H^{\alpha}_{\beta}Z^{\beta}}{\bar{Z}_{\alpha}Z^{\alpha}}\right)^2 \tag{2.7}$$

for the expectation and variance of the energy. As we shall see later, this interesting connection between the uncertainty in the energy and the rate of state evolution is a feature of the linear theory that extends to the non-linear theory, where the relation $ds/dt = 2(\Delta H)$ remains fully valid. In the spin one-half example above we have $\Delta H = E\sin\theta$; as one might expect, the speed of evolution and the energy uncertainty are minimised on the poles and maximised on the equator.

3. ANALYTICAL PROPERTIES OF THE STATE MANIFOLD

3.1 The need for a generalised state space

The two broadly unsatisfactory aspects of linear quantum mechanics are (a) that the theory admits superpositions of macroscopically distinct states, and (b) that the evolution of a quantum mechanical system suffers a discontinuity when a measurement is made. The first problem is exemplified by the famous 'cat paradox' (Schrödinger 1935), and the second problem is well-represented by considerations associated with the EPR paradox (Einstein *et al* 1935). It would appear to be simply a matter of fact that 'macro-superpositions' are not ordinarily observed (cf. Leggett 1987); and while it may be possible to account for this in part by the introduction (say) of a suitable system of superselection sectors, it remains that elementary quantum mechanics as it stands carries no such prescription.

As for discontinuities, we can sympathise with Schrödinger (cf. Bell 1987), who said, 'If we have to go on with these damned quantum jumps, then I'm sorry that I ever got involved.' Even now it can be an awkward moment in the teaching of basic quantum mechanics when the idea of state reduction is introduced—with no satisfactory explanation.

Let us assume provisionally that these are problems with quantum mechanics itself, and not merely with its 'interpretation'. It is tempting then to speculate, as others have from time to time, that the resolution of the matter may involve in some way an interplay of quantum mechanics and gravitation (cf. Penrose 1985a,b, 1987, 1989, 1990, 1991). By this we do not necessarily mean 'quantum gravity' as usually understood, whereby one attempts (say) by various techniques of varying degrees of sophistication to 'quantise gravity', whatever that might ultimately mean. We shall suppose that it is the mathematical structure of basic quantum mechanics in its own right that needs to be reconsidered, even in the non-relativistic limit.

But this does not mean simply making adjustments to the dynamical equations of the standard theory, and leaving the matter at that. What is being aimed at here, rather, is the notion that the new approach should bear a relation to ordinary quantum mechanics more like the relation between Einsteinian and Newtonian gravitation: in other words the state manifold needs to be 'curved' in some suitably non-trivial sense, and the curvature should in some manner 'assimilate' the Hamiltonian, in the same sense that the curvature tensor in Einstein's theory assimilates the energy tensor. Such a thought must be regarded as highly speculative, and any such approach must be viewed with caution. But it is interesting nevertheless that the geometrical framework outlined above suggests a way forward that actually exhibits many of these features, and thus probably warrants further investigation.

3.2 Kähler geometry

For our general state space we shall take a complex manifold M^n of complex dimension n (real dimension $2n$). We require that M^n should have a compatible Riemannian structure. This is a non-trivial requirement and can be viewed in various ways, so it is appropriate to go into some detail.

The essential mathematical tool here is Kähler geometry. A Kähler manifold can either be regarded as a real Riemannian manifold that is endowed also with a complex structure, or as a complex manifold that is also endowed with a Riemannian structure. The two points of view are to some extent complementary, and it is an advantage to be able to switch back and forth.

From the real Riemannian point of view the complex structure appears in the form of a special globally defined tensor field that satisfies various conditions. From the complex point of view the Riemannian structure appears in the form of a Hermitian metric tensor that satisfies a certain rather stringent closure condition.

This elaborate interpenetration of the real and the complex is strikingly reminiscent of an analogous situation in quantum mechanics, where on the one hand we have complex wave functions, complex superpositions of states, a complex wave equation as the fundamental dynamical law, and so on;—yet these all somehow conspire to produce expressions for real probabilities, real energy levels, real angular momentum values, real scattering angles, and so on—and more generally a physical description based in *real space-time* for the analysis of *real observables*.

As Norbert Wiener (1958) put it, 'The large part of the mystery of quantum theory is why, from this whole complex apparatus, we should get a real probability.' But it should be clear from the arguments being put forth here that the parallelism between these two lines of thought is more than mere analogy—indeed, there is a case for suggesting that it is the special two-fold nature of Kähler geometry that lies at the heart of the relation between the real and the complex in quantum mechanics.

3.3 Complex manifolds

To develop matters further it will be convenient first if we review briefly and informally certain aspects of complex manifold theory, at the same time setting the notation and placing particular emphasis on those features relevant to our study. A complex manifold M^n can be characterised as a real differentiable manifold M^{2n} with a globally defined tensor field $J_a^{\ b}$ called an 'almost complex structure' (the indices a, b, and so on, range 'abstractly' from 1 to $2n$) which is required to satisfy $J_a^{\ b} J_b^{\ c} = -\delta_a^{\ c}$. In that case M^{2n} is said to be an 'almost complex manifold'. For a (fully) complex manifold we also require the satisfaction of an integrability condition (Newlander & Nirenberg 1957). This is given by the vanishing of the so-called Nijenhuis torsion tensor, defined by

$$N_{ab}^{c} = J_d^{\ c} \nabla_{[a} J_{b]}^{\ d} - J_{[a}^{\ d} \nabla_{|d|} J_{b]}^{\ c}. \tag{3.1}$$

It is straightforward to verify that N_{ab}^{c} is in fact independent of the choice of symmetric connection ∇_a on M^{2n}; see for example Penrose & Rindler (1984) pp. 204-207. The interpretation of the torsion-free condition can be explained as follows. A complex vector field V^a on M^{2n} is said to be of 'unprimed' (*resp.* 'primed') type with respect to the almost complex structure $J_a^{\ b}$ if it satisfies $J_b^{\ a} V^b = -\mathrm{i}V^a$ (*resp.* $+\mathrm{i}V^a$). Unprimed and primed vector fields are also called vector fields of type (1,0) and type (0,1) respectively. The torsion-free condition is necessary and sufficient to ensure that the commutator (Lie bracket) of any two complex vector fields of the same type is itself also of that type;—in that case we say that the 'distribution' spanned by vector fields of the same type is 'integrable'. A covariant vector field (one-form) W_a is of unprimed (*resp.* primed) type if it is annihilated by the operator $\delta_b^{\ a} - \mathrm{i}J_b^{\ a}$ (*resp.* $\delta_b^{\ a} + \mathrm{i}J_b^{\ a}$). Equivalently, a one-form is of unprimed (*resp.* primed) type if its inner product with any vector field of primed (*resp.* unprimed) type vanishes.

A smooth complex function is said to be *holomorphic* in a given neighbourhood if and only if its gradient there with respect to any vector field of primed type vanishes—the Cauchy-Riemann condition. The gradient of a holomorphic function is therefore a one-form of the unprimed type. An unprimed vector field V^a is said to be a *holomorphic vector field* if locally for any holomorphic function f the gradient $V^a\nabla_a f$ is also holomorphic.

3.4 Metric compatible complex structures

A Riemannian metric g_{ab} is said to be 'compatible' with the complex structure J_a^b if $J_a^c J_b^d g_{cd} = g_{ab}$, and $\nabla_a J_b^c = 0$, where ∇_a now denotes the usual torsion-free symmetric connection (the Levi-Civita connection) associated with g_{ab}. Taken together these two conditions are quite strong—and when they are satisfied the implications for the manifold are profound, for this is precisely the circumstance where both complex analysis and Riemannian geometry are applicable. The manifold in such a set-up lives a kind of 'double life'; for in its guise as a complex space M^n its analytic properties are brought to the forefront—whereas when we take note of its real 'alter ego' M^{2n} the metrical geometry is more prominent.

It will be instructive therefore for us to examine the two metric compatibility conditions rather more closely. The first condition, $J_a^c J_b^d g_{cd} = g_{ab}$, says that g_{ab} is a 'Hermitian metric'. On account of the relation $J_a^b J_b^c = -\delta_a^c$ the Hermitian condition can be expressed in another way—namely, $\Omega_{ab} = -\Omega_{ba}$, where $\Omega_{ab} = J_a^c g_{bc}$. This means that the raising and lowering of indices by use of the metric 'anti-commutes' with the almost complex structure; i.e. if we use the metric to lower the index of a type (1,0) vector field, then the resulting one-form will be of type $(0,1)$. It follows that a complex vector field of a given definite type is necessarily *null* with respect to the Hermitian metric. The skew tensor Ω_{ab} is called the 'fundamental two-form'.

The condition $\nabla_a J_b^c = 0$ says that J_a^b should be parallel or 'covariantly constant' with respect to the connection associated with g_{ab}. This means that the 'type' of an index is unaffected by covariant differentiation of the tensor on which it appears; thus if W_b is of type (1,0) on the index b, then the tensor field $\nabla_a W_b$ will also be of type (1,0) on that index. Clearly we have $\nabla_a J_b^c = 0$ if and only if $\nabla_a \Omega_{bc} = 0$. However, if J_a^b is an integrable almost complex structure ($N_{ab}^c = 0$) a sufficient condition to ensure a parallel complex structure $\nabla_a J_b^c = 0$ is merely that $\nabla_{[a}\Omega_{bc]} = 0$, i.e. closure of the fundamental two-form. This follows from the identity

$$4\nabla_a \Omega_{bc} = 6\nabla_{[a}\Omega_{pq]} J_b^p J_c^q - 6\nabla_{[a}\Omega_{bc]} + N_{bc}^d \Omega_{ad} = 0 \tag{3.2}$$

which is valid in the general situation where J_a^b is an almost complex structure, g_{ab} is a Hermitian metric, and Ω_{ab} is the associated fundamental two-form. From (3.2) it follows that if Ω_{ab} is closed, and if the almost complex structure is integrable, then Ω_{ab} is in fact covariantly constant. Thus a complex structure is compatible with a Riemannian structure if the metric is Hermitian and the associated fundamental two-form is closed. A complex manifold in possession of such a compatible Riemannian structure is called a 'Kähler manifold'. For further general information on Kähler manifolds see for example Morrow & Kodaira (1971), Kobayashi & Nomizu (1969).

3.5 Identification of the quantum mechanical symplectic structure

From the foregoing discussion it should be clear that a Kähler manifold is also in possession of another geometrical feature, namely a *symplectic structure*. A manifold is said to have a symplectic structure if there exists a globally non-degenerate closed two-form, and here it is the tensor

$$\omega_{ab} = \tfrac{1}{2}\Omega_{ab} \tag{3.3}$$

that specifically plays that role. It is quite important to note that even in ordinary linear quantum mechanics the state space is in possession of a canonical symplectic structure (cf. Mielnik 1974, Abraham & Marsden 1978, Kibble 1979, Heslot 1985, Page 1987, Weinberg 1989a,b, Cirelli *et al.* 1990, Gibbons 1992). In fact, we shall see that the Schrödinger trajectories of linear quantum mechanics can be viewed consistently as 'classical' dynamical trajectories on the state space; i.e. as the integral curves of the classical dynamical system determined by the natural symplectic structure on the state manifold.

This is a point that has been observed by a number of authors, and is undoubtedly of great significance. Indeed, it may be instructive to regard 'quantisation' as a sort of procedure that takes one from the consideration of a given symplectic manifold (the original classical phase space) to another (the quantum mechanical state space)—where in each case the dynamical path is determined by a canonical vector field associated with the relevant symplectic structure.

Note, incidentally, that the quantum mechanical symplectic structure ω_{ab} differs by a factor of one-half from the 'natural' fundamental two-form Ω_{ab}. We might have adopted a notation that took advantage of this situation (e.g. as in Woodhouse 1992); but it serves a useful purpose here to keep them distinct. By the same token there are gains also in going over to the standard notation of Kähler geometry (by use of primed and unprimed Hermitian indices), but in doing so one loses touch to some extent with the real structure of the manifold, and also the conventions as regards the factor of one-half are not so easy to keep under control. A Kähler geometry clearly determines a one parameter family of symplectic structures on M^{2n} given by $k\Omega_{ab}$ where k is a constant. Of these Ω_{ab} itself is singled out by the property that its inverse is the same as the tensor obtained by raising its indices with the metric; with that convention the correct quantum mechanical symplectic structure turns out to be given by $\omega_{ab} = \frac{1}{2}\Omega_{ab}$, as indicated above.

4. NON-LINEAR QUANTUM MECHANICS

4.1 Eigenstates and eigenvalues

Let us suppose that we have our complex state manifold M^n together with a compatible Riemannian metric g_{ab} and the associated symplectic structure $\omega_{ab} = \frac{1}{2}\Omega_{ab}$. It is useful to have in mind the case when M^n is CP^n and g_{ab} is the Fubini-Study metric, but the analysis here will be general, with no precise specification of the properties of the state manifold. Our programme in the first instance will be to attempt to reconstruct in this environment as many of the familiar physical ideas of quantum mechanics as we can. Each point of the manifold is to be regarded as a quantum mechanical state. Observables correspond to (globally defined) real differentiable functions on the manifold; these will generally be taken to have continuous second derivatives. The idea here is that the value of an observable at a point is the expectation of that observable in the state corresponding to the given point. The eigenstates of an observable F are those points at which its gradient $\nabla_a F$ vanishes; the eigenvalue corresponding to such a state is then the value of the observable at such a point, and equals the expectation. The eigenvalues of an observable are clearly real; and the expectation of a observable always takes on a value between its highest and lowest eigenvalues.

Now let F_0 be an observable with an eigenvalue α_0 at a point x^a in the state manifold, and consider the effect of perturbing F_0 with a small term of the form εF_1 (with $\varepsilon << 1$). In the general situation we expect the perturbed observable to have an perturbed eigenvalue of the form $\alpha_0 + \varepsilon\alpha_1$ at some nearby point $x^a + \varepsilon\eta^a$.

It follows immediately that $\alpha_1 = F_1(x^a)$ to first order in ε. This says simply that the first order correction ('shift') to the eigenvalue is given by the expectation (taken at the original state) of the perturbing term in the observable. In the linear theory this result is known as the 'Born approximation', and it is useful to see that all the relevant ideas carry through essentially without qualification to the nonlinear theory; in fact, the calculations are rather simpler here, in the fully non-linear theory, than in the linear theory in its more usual formulation.

4.2 Commutation relations

For any observable F we can form a so-called canonical (or 'Hamiltonian') vector field $V^a = \varpi^{ab}\nabla_b F$ by transvecting the gradient of F with the inverse ϖ^{ab} of the symplectic form ω_{ab}, defined by $\omega_{ab}\varpi^{bc} = \delta_a^c$. Note, as we remarked earlier, that the inverse of the fundamental two-form Ω_{ab} is obtained simply by raising its indices by use of the metric: by the Hermitian condition we have $\Omega_{ac}\Omega^{bc} = \delta_a^b$ where $\Omega^{ab} = g^{ac}J_c^b$. The symplectic structure ω_{ab} and its inverse ϖ^{ab} are then given by $\omega_{ab} = \frac{1}{2}\Omega_{ab}$ and $\varpi^{ab} = 2\Omega^{ab}$. The factors of two here are slightly awkward—but that is not *our* fault.

We say that the canonical vector field V^a constructed in this way is *generated* by the observable F. The commutator $[F,G]$ of two observables F and G is defined by the formula $[F,G] = \mathrm{i}\{F,G\}$ where $\{F,G\}$ is the Poisson bracket $\{F,G\} = \varpi^{ab}\nabla_a F\nabla_b G$. The real vector space of all observables (smooth functions) has the structure of a Lie algebra under the Poisson bracket operation; and by a standard argument one deduces the Jacobi identity $\{P,\{Q,R\}\} + \{Q,\{R,P\}\} + \{R,\{P,Q\}\} = 0$ from the fact that ω_{ab} is closed.

As an alternative useful expression for the Poisson bracket, incidentally, we can write $\{F,G\} = \omega(X_F, X_G)$, where X_F denotes the canonical vector field associated with the observable F. Here for any two vector fields U^a, V^a, the so-called Lagrange bracket $\omega(U,V)$ is defined by $\omega(U,V) = \omega_{ab}U^aV^b$.

Note that if a point is an eigenstate of either of the observables F or G then the commutator $[F,G]$ necessarily has a vanishing expectation at that point, though the gradient of the commutator need not vanish there. However, a necessary and sufficient condition for the commutator of two observables to have a vanishing gradient is that the Lie bracket of the associated canonical vector fields vanish. This follows from the commutator identity $X_{\{F,G\}} = -[X_F, X_G]$. The Lie bracket $W = [U,V]$ between two vector fields is given as usual by $W^a = U^b\nabla_b V^a - V^b\nabla_b U^a$.

4.3 Uncertainty relations

Now we are in a position to introduce a geometrical characterisation of an important probabilistic notion. For any observable F the value of the magnitude of its gradient is the variance (squared uncertainty) in F at the point of valuation:

$$(\Delta F)^2 = g^{ab}\nabla_a F\nabla_b F. \tag{4.1}$$

It follows that the uncertainty can also be expressed in terms of the magnitude of the canonical vector field associated with F; i.e. by the relation $(\Delta F)^2 = \frac{1}{4}g_{ab}V^aV^b$, where $V^a = \varpi^{ab}\nabla_b F$. Note that since g_{ab} is positive definite, the variance vanishes at and only at an eigenstate.

For any pair of vector fields on a Riemannian manifold we have the elementary Schwarz inequality $(g_{ab}X^aX^b)(g_{ab}Y^aY^b) \geq (g_{ab}X^aY^b)^2$. On a Hermitian manifold however there is a *stronger* inequality that holds, which we shall call the *Hermitian* Schwarz inequality, given by

$$(g_{ab}X^aX^b)(g_{ab}Y^aY^b) \geq (g_{ab}X^aY^b)^2 + (\Omega_{ab}X^aY^b)^2 \tag{4.2}$$

We get a more powerful inequality in the Hermitian case since the metric is more 'restricted' in its character than in the general Riemannian situation. Note that it is the Lagrange bracket between the two vector fields that determines the extent to which the inequality is strengthened beyond the merely Riemannian situation. The surprise that results from this is that we are immediately lead for any pair of real observables F and G without any further calculation, to the following *generalized Heisenberg relation*:

$$(\Delta F)^2(\Delta G)^2 \geq \tfrac{1}{4}|[F,G]|^2 \,. \tag{4.3}$$

This construction also of course provides a kind of a geometrized derivation of the uncertainty relations in the standard theory which may have some advantages over the customary approach.

4.4 Quantum correlations

The metric also determines a 'covariance' function defined by $\mathrm{Cov}(F,G) = \tfrac{1}{4}g(dF^\#, dG^\#)$ for any two observables; here the operator d signifies exterior differentiation and the sharp sign denotes raising an index by use of the inverse of the symplectic form, so $X_F = (dF)^\#$. The covariance between any two observables is apart from a factor of one-quarter simply the inner product between the associated canonical vector fields.

Reverting to the index notation it is easily verified that we also have the equivalent relation $\mathrm{Cov}(F,G) = g^{ab}\nabla_a F \nabla_b G$. Note that $\mathrm{Cov}(F,F) = \mathrm{Var}(F) = (\Delta F)^2$. The associated correlation function, which is defined according to the usual conventions by the formula $\rho(F,G) = \mathrm{Cov}(F,G)/(\Delta F\,\Delta G)$, clearly satisfies $|\rho| \leq 1$ on account of the cosine rule. In fact by use of the fundamental Hermitian inequality (4.2) we can deduce the stronger relation

$$|\rho| \leq \sqrt{1 - \left(\frac{|[F,G]|}{2\Delta F\,\Delta G}\right)^2}\,, \tag{4.4}$$

a formula that is valid also in the linear theory (cf. Bohm 1951); this shows, as one would suppose, that when two observables have a non-vanishing Poisson bracket their correlation is necessarily imperfect.

4.5 Schrödinger trajectories

The special canonical vector field $X_H = (dH)^{\#}$ generated by the physical Hamiltonian has as its integral curves the Schrödinger evolutionary trajectories. Thus once the Hamiltonian function has been specified the evolutionary trajectory is determined by the usual rules of *classical* mechanics. This feature is *a fortiori* also present in the linear theory.

We see therefore that the U-process can be regarded in an entirely natural way as a classical dynamical system on a compact complex manifold. (The manifold in question in the case of ordinary linear quantum mechanics is CP^n with the Fubini-Study metric.) From the expression for the variance above we are able to go further and deduce immediately a *generalised Aharonov-Anandan relation,*

$$g(dH^{\#}, dH^{\#}) = 4(\Delta H)^2 \tag{4.5}$$

that relates the speed of evolution (the magnitude of the Hamiltonian vector field $dH^{\#}$) to the uncertainty in the energy ΔH; and it is interesting to note that in our framework this important relation is essentially an identity.

The symplectic two-form admits a further elegant dynamical interpretation: let β be the integral of ω over a 2-surface spanning a loop γ in M. If γ is a closed evolutionary path, β is the so-called 'geometric phase' (Berry 1984, Simon 1983, Aharonov & Anandan 1987, 1990). On the other hand if γ represents a given loop in M that subsequently evolves, then β is a quantum mechanical analogue of the Poincaré-Cartan integral invariant (cf. Arnol'd 1989). These constructions are both applicable in a non-trivial way in the linear theory, so it is significant to note that they go over readily to the non-linear theory as well.

5. FURTHER DEVELOPMENTS

5.1 The role of the curvature of the state space

It was observed at the outset of §3 that one aspect of the motivation behind this analysis was the thought that with a 'curved' state manifold it ought to be possible to 'assimilate' the Hamiltonian in some sense into the structure of the manifold, by identifying it with some geometric object. The plausibility of this point of view is suggested, in broad terms at any rate, by general relativity—and is certainly attractive from a mathematical angle, however indirect the physical reasoning behind it might be. It is interesting therefore to see that in fact a natural candidate for an identification of this sort does indeed present itself—namely, the scalar curvature.

The scalar curvature of the state manifold is an observable, and clearly has a 'preferred' status. But there is only one observable in quantum mechanics that necessarily exists for any quantum mechanical system, and therefore has a corresponding status as a *physically* 'preferred' observable—this is of course the Hamiltonian. Since the Ricci curvature of the state manifold in ordinary quantum mechanics is constant this suggests in the non-linear situation an identification of the form

$$H = \lambda + \mu R \tag{5.1}$$

where H is the Hamiltonian function, and where λ and μ are constants each with units of mass.

In fact if we set $\lambda = -\mu R_0$ where R_0 is the curvature of the state manifold of linear quantum mechanics, then we are ensured that in the limit as $R \to R_0$ the Hamiltonian will be extremely small; and by the same token we want μ to be quite large, compared with H, for everyday quantum mechanical circumstances, to ensure that the curvature does not deviate too far from the Fubini-Study curvature.

As matters stand we are not of course in a position to say what the value of μ should be, though one can speculate that if the non-linearities studied here arise in some way in connection with the relation between quantum mechanics and gravitation, then it would be natural to consider μ to be (say) on the order of the Planck mass.

It may be that one should view equation (5.1) in rather the same spirit as Einstein's gravitational equation. The idea is that one specifies the general form of the Hamiltonian function by consideration of the general makeup of the particular physical system under consideration (rather in the way one specifies the structure of an energy tensor).

Since a Kähler metric is determined locally by a single real function, the Kähler potential, equation (5.1) can be interpreted as a non-linear relation between that potential and the Hamiltonian, which given suitable boundary conditions determines the metric in terms of the Hamiltonian.

This then feeds back non-linearly into the Schrödinger equation via the symplectic structure. It would be interesting to have some exact solutions for simple physical systems.

5.2 Relation to theories of the Weinberg type

Now let us consider in more detail the relation of the non-linear theory to the linear theory. In the linear case the manifold is CP^n and the Riemannian structure is given by the Fubini-Study metric. This metric is symmetric under the induced action of the unitary group, and the symmetries are represented by a family of $n(n+2)$ real Killing vectors. The primed (*resp.* unprimed) parts of these Killing vectors are holomorphic (*resp.* anti-holomorphic) vector fields; and conversely the $n(n+2)$ independent real Killing vectors are the real parts of the $n(n+2)$ independent global holomorphic vector fields on CP^n. These Killing vector fields are also canonical vector fields with respect to the symplectic structure, and the standard linear observables of ordinary quantum mechanics are the generators of these canonical vector fields. In the general non-linear situation there are no Killing vectors; observables are introduced in the form of a broader concept, i.e. as generators of canonical vector fields.

But there is an intermediate case of some interest in its own right; namely, the case where we look at the manifold CP^n and equip it with the Fubini-Study metric (as in ordinary quantum mechanics), but consider as observables the more general category of generators of canonical vector fields.

In this case the Schrödinger equation is $X_H = (dH)^\#$ where H is a real function, but not necessarily the generator of a Killing vector of the Fubini-Study metric. This special class of non-linear theory turns out to be precisely equivalent to a type of non-linear theory studied by Weinberg (1989a,b,c) and others.

The essence of the Weinberg-type theories is that the 'Killing flow' of linear quantum mechanics is replaced by a general Hamiltonian flow. In the notation of §2 the projective Schrödinger equation $Z^{[\alpha}dZ^{\beta]} = -\mathrm{i}Z^{[\alpha}H^{\beta]}_{\gamma}Z^{\gamma}dt$ of the linear theory gets displaced in Weinberg's approach by the non-linear dynamical equation

$$Z^{[\alpha}dZ^{\beta]} = -\mathrm{i}(Z^{[\alpha}\partial^{\beta]}h)\,dt \tag{5.2}$$

where $\partial^{\beta} = \partial/\partial\bar{Z}_{\beta}$ and the real function $h(Z^{\alpha},\bar{Z}_{\beta})$ is required to be homogeneous of degree one in each variable; the Hamiltonian function H in our formulation, which is well-defined directly as a function on the projective space, is given by $H = h(Z^{\alpha},\bar{Z}_{\beta})/(Z^{\gamma}\bar{Z}_{\gamma})$. The linear theory is then recovered in the case $h(Z^{\alpha},\bar{Z}_{\beta}) = \bar{Z}_{\beta}H^{\beta}_{\alpha}Z^{\alpha}$. The non-linearities introduced in the Weinberg-type theory are of a very special sort, and in fact relatively tractable, since the metrical structure and the complex analytic structure of linear quantum mechanics are both retained.

There is still another intermediate situation of some importance, of greater generality than Weinberg's theory, but still relatively tractable; this is where we stay within the CP^n framework, but consider a compatible Riemannian geometry more general than the

Fubini-Study metric—this case is particularly interesting since it is sufficiently general to admit the relation $H = \mu(R - R_0)$ between the curvature and the Hamiltonian, as discussed above, while retaining the simplicity associated with complex projective space as the underlying manifold.

It would be very interesting, incidentally, to pursue the probabilistic foundations of quantum mechanics within the geometrical framework outlined here, and in particular it would be useful I think to have further clarification of the role of density matrices, from a geometrical point of view, both in the linear and non-linear theories (cf. Gisin 1989, Peres 1989). Also it would be useful to understand the precise sense in which quantum mechanics requires a complex structure for the state manifold, rather than merely an almost complex structure. Clearly for much of the apparatus the latter suffices, but perhaps for the superposition principle we require the additional integrability conditions for a complex structure. In that case it may be that the collapse of the wave function is in some sense associated with a kind of breakdown or shift in the complex structure of the state manifold.

Acknowledgements. The author is grateful to J. Anandan, L. Dabrowski, R. Jozsa, C. R. LeBrun, L. J. Mason, R. Penrose, D. C. Robinson, R. F. Streater, F. J. Tipler, W. Triggs, H. Urbantke, K. Wanelik, and N. M. J. Woodhouse for helpful discussions and comments.

References

Abraham, R., & Marsden, J. E. (1978) **Foundations of Mechanics**, Addison-Wesley.

Aharonov, Y., & Anandan, J. (1987) *Phys. Rev. Lett.* 58, 1593.

Anandan, J. & Aharonov, Y. (1990) *Phys. Rev. Lett.* 65, 1697.

Arnol'd, V. I. (1989) **Mathematical Methods of Classical Mechanics**, Springer Verlag.

Bell, J. S. (1987) **Speakable and Unspeakable in Quantum Mechanics**, CUP.

Berry, M. V. (1984) *Proc. Roy. Soc. Lond.* A 392, 45.

Bohm, D. (1951) **Quantum Theory** (first edition) Prentice-Hall.

Cirelli, R., Mania, A. & Pizzocchero, L. (1990) *J. Math Phys.* 31, No. 12, 2891-2903.

Cushing, J.T. & McMullin, E., eds., (1989) **Philosophical Consequences of Quantum Theory—reflections on Bell's theorem**, University of Notre Dame Press.

Einstein, A. Padolsky, P., & Rosen, N. (1935) *Phys. Rev.* 47, 777.

Fubini, G. (1903) *Atti. Instit. Veneto* 6, 501-513.

Gibbons, G. W. (1992) *J. Geom. Phys.* 8, 147-162.

Gisin, N. (1989) *Helv. Phys. Acta.* 62, 363-371.

Heslot, A. (1985) *Phys. Rev.* D 31, 1341.

Kibble, T. W. B. (1979) *Commun. Math. Phys.* 65, 189.

Kobayashi, S. & Nomizu, K. (1969) **Foundations of Differential Geometry**, Wiley.

Leggett, A. J. (1987) *Reflections on the quantum measurement paradox*, in **Quantum Implications—essays in honour of David Bohm** (B. J. Hiley, F. D. Peat, eds.), Routledge.

Mielnik, B. (1974) *Comm. Math. Phys.* 37, 221.

Morrow, J. & Kodaira, K. (1971) **Complex Manifolds**, Holt, Rinehart & Winston.

Newlander, A., & Nirenberg, L. (1957) *Ann. of Math.* 65, 391-404.

Neumann, J. v. (1955) **Mathematical Foundations of Quantum Mechanics**, PUP.

Page, D. A. (1987) *Phys. Rev.* A 36, 3479.

Penrose, R. (1968) *Structure of Space-time*, in **Battelle Rencontres, 1967 Lectures in Mathematics and Physics**, (C. M. DeWitt & J. A. Wheeler, eds.) Benjamin.

Penrose, R. (1985) *Objective State-Vector Reduction?*, in *Twistor Newsletter* 19, 1-4.

Penrose, R. (1985) *Quantum gravity and state-vector reductions*, in **Quantum Concepts in Space and Time** (C. J. Isham & R. Penrose, eds.) OUP.

Penrose, R. (1987) *Quantum physics and conscious thought*, in **Quantum Implications—essays in honour of David Bohm** (B. J. Hiley, F. D. Peat, eds.) Routledge.

Penrose, R. (1989) **The Emperor's New Mind**, OUP.

Penrose, R. (1990) *Twistor theory after 25 Years—its physical status and prospects*, pp 1-29 in **Twistors in Mathematics and Physics** (T. N. Bailey & R. J. Baston, eds.) CUP.

Penrose, R. (1991) *The Modern Physicist's View of Nature*, Herbert Spencer Lecture, Oxford.

Penrose, R. & Rindler, W. 1984, 1986 **Spinors and Space-Time** (Vols. I, II) CUP.

Peres, A. (1989) *Phys. Rev. Lett.* 63, 1114.

Pirani, F. A. E. (1965) *Introduction to gravitational radiation theory*, in **Lectures on General Relativity** (A. Trautman, F.A.E. Pirani, & H. Bondi), Prentice Hall.

Pizzocchero, L. (1990) *La Mecchanica di Schrödinger nell'approccio Geometrico*, Tesi di dottorato, University of Milan.

Schrödinger, E. (1935) *Die Naturwissenschaften*, 23, 844 (reprinted in J. A. Wheeler & W. O. Zurek, eds., 1983, **Quantum Theory and Measurement**, PUP).

Simon, B. (1983) *Phys. Rev. Lett.* 51, 2167.

Study, E. (1905) *Math Ann.* 60, 321-377.

Weinberg, S. (1989a) *Phys. Rev. Lett.* 62, 485

Weinberg, S. (1989b) *Ann. Phys.* 194, 336.

Weinberg, S. (1989c) *Phys. Rev. Lett.* 63, 1115.

Wheeler, J. A. & Zurek, W. O., eds. (1983) **Quantum Theory and Measurement**, PUP.

Wiener, N. (1958) **Non-linear Problems in Random Theory**, MIT.

Woodhouse, N. M. J. (1992) **Geometric Quantization** (second edition) OUP.

7
Anti-Self-Dual Riemannian 4-Manifolds

Claude LeBrun Mathematics Department, State University of New York at Stony Brook, Stony Brook, New York

Dimension four is utterly unique from a geometric stand-point. In particular, the orthogonal group $SO(n)$ is non-simple only for $n = 4$; and, as a consequence, the Weyl tensor of an oriented Riemannian 4-manifold invariantly splits as a sum $W = W_+ + W_-$ of its so-called self-dual and anti-self-dual parts. This decomposition is also *conformally* invariant, and so, letting $[g] = \{ug \mid u : M \overset{C^\infty}{\to} \mathbb{R}^+\}$ denote the conformal class determined by a smooth Riemannian metric g, we may say that an oriented conformal-Riemannian 4-manifold $(M, [g])$ is *anti-self-dual* if $W_+ = 0$, or *self-dual* if $W_- = 0$. Reversing the orientation of M just interchanges these conditions, so it will suffice to discuss the anti-self-dual case; the self-dual case then follows by reading the article in a mirror!

Which smooth oriented compact 4-manifolds M admit anti-self-dual metrics? Certainly not all; for example, the existence of anti-self-dual metrics is obstructed when the signature $\tau(M) = b_+(M) - b_-(M)$ is positive, as

$$\tau(M) = \frac{1}{3}\int_M p_1(M) = \frac{1}{12\pi^2}\int_M (\|W_+\|^2 - \|W_-\|^2)d\mathrm{vol}_g$$

for any Riemannian metric g. Nonetheless, Taubes [20] has proved the following remarkable result:

Theorem 1 (Taubes) *Let M be a smooth compact oriented 4-manifold. Then the connected sum $M\#m\overline{\mathbb{C}P}_2$ of the given manifold with m reverse-oriented copies of the complex projective plane admits anti-self-dual metrics provided that $m \gg 0$ is sufficiently large.*

Supported in part by NSF grant DMS-9204093.

Here $\overline{\mathbb{C}P}_2$ is the smooth manifold underlying $\mathbb{C}P_2$ equipped with the orientation which is *not* compatible with any complex structure; and the connected-sum operation $\#$ is performed on two oriented manifolds by removing a small ball from each and then identifying the resulting boundary spheres by a reflection.

The depth of Taubes' result is indicated by the following:

Corollary 1 *Let Γ be any finitely-presented group. Then there exists a compact complex manifold Z with $\pi_1(Z) \cong \Gamma$.*

Proof. By performing elementary surgeries on S^4, one can construct a compact oriented 4-manifold M with $\pi_1(M) \cong \Gamma$; indeed, one just needs to replace an arbitrary $S^0 \times D^4$ with a $D^1 \times S^3$ for each generator, and then replace a suitable $S^1 \times D^3$ with a $D^2 \times S^2$ for each relation. Now Theorem 1 tells us that that there is an m such that $\tilde{M} = M\#m\overline{\mathbb{C}P}_2$ admits an anti-self-dual conformal metric $[g]$, and $\pi_1(\tilde{M}) = \pi_1(M) = \Gamma$ by the Seifert-van Kampen Theorem. By the twistor correspondence (see §1 below), there is then a complex 3-manifold Z, called the *twistor space* of $(\tilde{M}, [g])$, which fibers over $\tilde{M}$ with fiber S^2. Since the long exact homotopy sequence of this fibration reads

$$\begin{array}{ccccccccc} \cdots & \rightarrow & \pi_1(S^2) & \rightarrow & \pi_1(Z) & \rightarrow & \pi_1(\tilde{M}) & \rightarrow & 0, \\ & & \| & & & & \| & & \\ & & 0 & & & & \Gamma & & \end{array}$$

the compact complex manifold Z therefore has fundamental group Γ. ∎

This corollary should be understood in the context of recent results [3] which show that the fundamental groups of compact Kähler manifolds— e.g. smooth projective varieties— are highly restricted. The upshot is that most compact complex manifolds are profoundly different from anything ever seen in algebraic geometry.

Taubes' proof begins by arguing that, for m sufficiently large, $M\#m\overline{\mathbb{C}P}_2$ admits Riemannian metrics for which W_+ is small; the point is that the standard metric on $\overline{\mathbb{C}P}_2$ is anti-self-dual, and one is free to cut out pieces of M where W_+ is large and replace them with $\overline{\mathbb{C}P}_2$'s. He then proves, by the methods of non-linear elliptic PDE, that any metric with sufficiently small W_+ (in a cunningly defined norm) is actually near an anti-self-dual metric.

As we have seen, Taubes' result is very powerful indeed. Nonetheless, it gives us no reasonable estimate of how large m needs to be for a given M. For example, W_+ is uniformly large for the standard metric on $M = \mathbb{C}P_2$, and one might therefore think that only astronomically large m would work for this manifold. By a direct construction, however, we will see below that reasonably simple M are sated by rather small values of m— and, for example, $m = 14$ suffices when $M = \mathbb{C}P_2$.

The construction described herein proceeds in exactly the opposite manner from the proof of Corollary 1; namely, we will first construct the twistor space Z, and only then infer the existence of the sought-after self-dual metric on the 4-manifold in question. In fact, by carefully exploiting the complex analytic aspects of this approach, we will simultaneously solve an interesting problem in Kähler geometry.

A Kähler manifold of complex dimension 2 is anti-self-dual with respect to the standard orientation iff its scalar curvature is identically zero [6]; a compact manifold of this type is called a *scalar-flat Kähler surface.* It turns out [2, 18] that a compact anti-self-dual 4-manifold with even first Betti number is (conformally) a scalar-flat Kähler surface iff its twistor space has a smooth cross-section which is also a complex hypersurface. Thus, by making sure that enough of the constructed twistor spaces carry suitable divisors, one is able to give an essentially complete classification [9] of scalar-flat Kähler surfaces.

Acknowledgements. The author would like to thank Michael Singer, Jongsu Kim, and Massimiliano Pontecorvo for their help in working out many of the ideas described here. He would also like to express his profound gratitude to Dominic Joyce [7] for a remark which inspired his discovery of the key examples described here.

1 The Twistor Correspondence

While there are any number of good reasons to study anti-self-dual manifolds, surely one of the most compelling is given by the Penrose twistor correspondence [17], which associates a complex 3-manifold (Z, J) with every anti-self-dual manifold $(M, [g])$. This is done in such a way that the conformal metric $[g]$ is completely encoded by the complex structure, and the class of complex 3-manifolds (Z, J) arising in this way can be completely characterized. Penrose's original description of (Z, J) actually needs the conformal metric $[g]$ to be real-analytic, and entails certain global difficulties. The Atiyah-Hitchin-Singer construction [1], which we will now sketch, avoids these global difficulties, and only requires that the metric be sufficiently differentiable[1].

If (M, g) is *any* oriented Riemannian 4-manifold, we can consider the 2-sphere bundle $\wp : Z \to M$ defined by $Z = F/U(2)$, where $F \to M$ is the principal $SO(4)$-bundle of oriented orthonormal frames on M. The smooth 6-manifold Z then automatically carries a natural almost-complex structure $J : TZ \to TZ$, $J^2 = -1$, which respects the g-induced splitting of TZ into vertical and horizontal tangent spaces. Indeed, observe that there is a canonical identification

$$Z \cong \{\jmath \in \mathrm{End}(TM) \mid \jmath^2 = -1, \jmath^* g = g, \ \mathrm{Pfaffian}(\jmath) > 0\} \ ,$$

so that, thinking of the horizontal sub-bundle of TZ as the pull-back of TM to Z, one may let J act on the horizontal sub-bundle at $\jmath \in \wp^{-1}(x)$ by $\jmath : T_x M \to T_x M$. In the vertical subspace, on the other hand, J will simply act as the standard complex structure on S^2, namely rotation by $+90°$. Provided that we give the fibers the correct orientation in defining this almost-complex structure J, the entire construction turns out to be conformally invariant, and so associates an almost-complex manifold with each conformal Riemannian 4-manifold. However, the almost-complex manifold (Z, J) will not, in general, be a complex manifold. Instead, the relevant integrability [16] condition $\bar{\partial}^2 = 0$ amounts to the vanishing of a conformally-invariant piece of the Riemannian curvature ∇^2, and so, with a little thought, turns out to simply read $W_+ = 0$. Thus every anti-self-dual manifold $(M, [g])$ has an associated complex 3-manifold (Z, J) called its *twistor space*.

The class of complex 3-manifolds which arise by the twistor construction can be completely characterized in terms of their complex geometry, as we now describe. First, one needs a free anti-holomorphic involution $\sigma : Z \overset{\overline{\circlearrowleft}}{\to} Z$, $\sigma^2 = id_Z$, which in the above construction arises as the fiber-wise antipodal map; such an involution is called a *real structure*. Secondly, there should

[1]In fact, combining the two points of view leads to a simple and elegant proof of the fact that any anti-self-dual metric is actually real-analytic with respect to a suitably chosen coordinate atlas.

exist at least one σ-invariant rational curve $\mathbb{C}P_1 \subset Z$ with normal bundle $\mathcal{O}(1) \oplus \mathcal{O}(1)$; in the above construction, any fiber of $\wp : Z \to M$ would fit the bill. A curve of the latter type is called a *real twistor line*, and, using a theorem of Kodaira [11], it is not difficult to show that the real twistor lines locally foliate Z. The 4-manifold M of such real twistor lines now comes equipped with a canonical self-dual conformal metric for which the complex null vectors are precisely the images of those holomorphic sections of the the normal bundle of a leaf which have a zero somewhere. If, moreover, the real twistor lines *globally* foliate Z, then Z is precisely the twistor space of the leaf space M with its induced conformal structure $[g]$.

We will now describe some examples [12] of anti-self-dual 4-manifolds, beginning with some families of anti-self-dual conformal metrics on the n-fold connected sums $\overline{\mathbb{C}P}_2 \# \cdots \# \overline{\mathbb{C}P}_2$. Choose n points $p_1, \ldots, p_n$ in hyperbolic 3-space $\mathcal{H}^3$, and set $X = \mathcal{H}^3 - \{p_1, \ldots, p_n\}$. Define $V : X \to \mathbb{R}^+$ by

$$V = 1 + \sum_{j=1}^{n} \frac{1}{e^{2r_j} - 1} ,$$

where r_j denotes the hyperbolic distance from p_j. Then $d \star dV = 0$, and $[\frac{1}{2\pi} \star dV]$ is an integral deRahm class in $H^2(X)$. Hence there exists a circle bundle $P \to (\mathcal{H}^3 - \{p_1, \ldots, p_n\})$ with connection 1-form θ whose curvature is $\star dV$. Now let r denote the hyperbolic distance from any reference point. The metric

$$g = (sech^2\, r)\,(Vh + V^{-1}\theta^{\otimes 2}) \tag{1.1}$$

is then anti-self-dual, and, because of our choice of conformal gauge, may be smoothly compactified by adding a 2-sphere and n points $\hat{p}_1, \ldots, \hat{p}_n$. Indeed, let D^3 denote the closed unit ball in $\mathbb{R}^3$, and identify the interior of D^3 with $\mathcal{H}^3$ via the Poincaré conformal model. Then $M = P \cup S^2 \cup \{\hat{p}_1, \ldots, \hat{p}_n\}$ can be made into a smooth 4-manifold with circle-action in such a manner that $S^2 \cup \{\hat{p}_1, \ldots, \hat{p}_n\}$ is the fixed point set and D^3 is the orbit space. The projection to D^3 is thus as follows:

$$\begin{array}{ccccccc} M & = & P & \cup & S^2 & \cup & \{\hat{p}_1, \ldots, \hat{p}_n\} \\ \downarrow & & \downarrow & & \downarrow & & \downarrow \\ D^3 & = & X & \cup & \partial D^3 & \cup & \{p_1, \ldots, p_n\} \end{array}$$

The metric g of equation (1.1) then extends to M so as to yield a compact antiself-dual 4-manifold diffeomorphic to $\overline{\mathbb{C}P}_2 \# \cdots \# \overline{\mathbb{C}P}_2$. When $n = 0, 1,$

this construction produces the standard conformal structures of the symmetric metrics on S^4 and $\overline{\mathbb{C}P}_2$, respectively. When $n = 2$, we instead get the metrics of Poon [19].

To explicitly describe the twistor spaces of the above manifolds, we let $\mathcal{O}(k, \ell)$ denote the unique holomorphic line-bundle over $\mathbb{C}P_1 \times \mathbb{C}P_1$ with degree k on the first factor and degree ℓ on the second. The data points $p_1, \ldots, p_n \in \mathcal{H}^3 \subset \mathbb{R}^4$ determine n sections

$$\mathcal{P}_1, \ldots, \mathcal{P}_n \in \Gamma(\mathbb{C}P_1 \times \mathbb{C}P_1, \mathcal{O}(1,1)) \cong \mathbb{C}^4.$$

Let $\mathcal{B}$ denote the total space of the $\mathbb{C}P_2$-bundle

$$\mathcal{B} := \mathbb{P}(\mathcal{O}(n-1,1) \oplus \mathcal{O}(1,n-1) \oplus \mathcal{O}) \xrightarrow{\pi} \mathbb{C}P_1 \times \mathbb{C}P_1 ,$$

and define an algebraic variety $\tilde{Z} \subset \mathcal{B}$ by the equation

$$xy = t^2 \prod_{j=1}^{n} \mathcal{P}_j ,$$

where $x \in \mathcal{O}(n-1,1)$, $y \in \mathcal{O}(1,n-1)$, and $t \in \mathcal{O} := \mathcal{O}(0,0)$. The twistor space Z of the constructed metric is then obtained from $\tilde{Z}$ by making small resolutions of the singular points and blowing down the surfaces $x = t = 0$ and $y = t = 0$ to $\mathbb{C}P_1$'s. Notice that these twistor spaces are all *Moishezon*, meaning that they are bimeromorphic to complex projective varieties.

The above construction can also be carried out [13, 8] if $\mathcal{H}^3$ is replaced by another sufficiently nice 3-manifold (Y, h) of constant curvature -1. For example, this can be done for $Y = \mathbb{R} \times \Sigma$ and $h = dt^2 + (\cosh^2 t)h_\Sigma$, where h_Σ is the curvature -1 metric on a compact Riemann surface Σ of genus $g \geq 2$ and $t \in \mathbb{R}$. Let $p_1, \ldots, p_n \in Y$ be given, let G_j be the corresponding hyperbolic Green's functions; and let

$$V = 1 + \sum_{j=1}^{n} G_j$$

on $X = Y - \{p_1, \ldots, p_n\}$. If there is a circle-bundle P with connection 1-form θ whose curvature is $\star dV$, then the completion of (P, g), where

$$g = (\text{sech}^2 t)(Vh + V^{-1}\theta^{\otimes 2}) ,$$

is again a compact anti-self-dual manifold; indeed, for the particular Y and the particular conformal factor given, it is even a compact scalar-flat Kähler manifold! The only catch lies in showing that $[\frac{1}{2\pi}\star dV]$ is an integral cohomology class. In our example $Y = \Sigma \times \mathbb{R}$, the integrality condition is non-trivial, but there are configurations which verify it for all $n \neq 1$, and for these values one can thereby construct, for example, scalar-flat Kähler metrics on n-fold blow-ups of $\mathbb{C}P_1 \times \Sigma$.

The twistor spaces spaces of these manifolds are more difficult to describe explicitly than those of the previous examples. However, the *universal cover* of any of these twistor spaces becomes a holomorphic conic bundle over an open dense subset of $\mathbb{C}P_1 \times \mathbb{C}P_1$ once one blows up a pair of curves for each boundary component of Y.

2 Smoothing Singular Twistor Spaces

We will now construct new families of anti-self-dual 4-manifolds by smoothing singular quotients of the explicit solutions described in §1.

Theorem 2 *[9] Let (N, J_N, g_N) be a non-minimal compact complex surface with scalar-flat Kähler metric, and let $\Phi : N \to N$, $\Phi^2 = 1$, be a holomorphic isometry with only isolated fixed points. Let (M, J_M) be obtained from N/Φ by replacing each singular point with a $\mathbb{C}P_1$ of self-intersection -2. Then there exist scalar-flat Kähler metrics g_M on (M, J_M).*

Strategy of Proof. We apply an orbifold modification of the Donaldson-Friedman construction [4], combining observations of [15] and [10]. Let Z_N denote the twistor space of (N, g_N), and let $L_1, \ldots, L_k$ be the twistor lines of the fixed points of Φ. Let $\tilde{Z}_N$ be the blow-up of Z_N along these lines, and let $Q_1, \ldots, Q_k$ be the exceptional divisors in $\tilde{Z}_N$ corresponding to $L_1, \ldots, L_k$; thus $Q_j \cong \mathbb{C}P_1 \times \mathbb{C}P_1$, $j = 1, \ldots k$, and each of these 2-quadrics has normal bundle $\mathcal{O}(1, -1)$. Since the derivative of Φ at its k isolated fixed points must be -1, the induced biholomorphism $\tilde{\Phi} : \tilde{Z}_N \to \tilde{Z}_N$ fixes each Q_j and acts on its normal bundle by -1. The quotient $Z_- := \tilde{Z}_N/\hat{\Phi}$ can thus be given the structure of a compact complex manifold in a unique way that the quotient map $\tilde{Z}_N/ \to Z_-$ becomes a branched covering, with $\coprod Q_j$ as ramification locus. By an abuse of notation, we will also use Q_j to denote the image of this quadric in Z_-, where it sits embedded with normal bundle $\mathcal{O}(2, -2)$.

The twistor space Z_N contains a hypersurface D_N corresponding to the complex structure J_N, as well as a disjoint hypersurface $\bar{D}_N := \sigma(D_N)$ corresponding to the conjugate complex structure $-J_N$; indeed, D_N and $\bar{D}_N$ are respectively isomorphic to $(N, \pm J_N)$ as complex surfaces. As the action of Φ sends each such surface to itself, there are disjoint hypersurfaces D_- and $\bar{D}_-$ in Z_- obtained by first taking the the proper transforms in $\tilde{Z}_N$ of D_N and $\bar{D}_N$ and then projecting these hypersurfaces to Z_-. Notice that D_- is exactly a copy of (M, J_M), whereas $\bar{D}_-$ is a copy of $(M, -J_M)$.

Next let Z_+ consist of k disjoint copies of the complex 3-fold $\tilde{Z}_{EH}$ obtained from the orbifold twistor space of the conformally compactified Eguchi-Hanson metric by blowing up the twistor line of infinity. Each $\tilde{Z}_{EH}$ contains a quadric $Q = \mathbb{C}P_1 \times \mathbb{C}P_1$ corresponding to the line at infinity, as well as two Hirzebruch surfaces, $D_{EH} \cong \bar{D}_{EH} \cong \mathbb{P}(\mathcal{O}(1) \oplus \mathcal{O}) \to \mathbb{C}P_1$, which meet Q in two disjoint lines $\mathbb{C}P_1 \times \{0, \infty\}$. Define $D_+ \subset Z_+$ (respectively, $\bar{D}_+ \subset Z_+$) to be the disjoint union of the k obvious copies of D_{EH} (respectively, $\bar{D}_{EH}$).

Now let $Z_0 = Z_- \cup_{Q_j} Z_+$ be obtained from $Z_- \coprod Z_+$ by identifying each quadric $Q_j \subset Z_-$ with a quadric Q in a different copy of $\tilde{Z}_{EH}$. These identifications are to be made, moreover, in a way such that the real structures of $Z_\pm$ agree on Q_j, and such that $D_- \cap Q_j = D_+ \cap Q_j$. We now set $D_0 = D_- \cup D_+$ and $\bar{D}_0 = \bar{D}_- \cup \bar{D}_+$. The result is that Z_0, D_0 and $\bar{D}_0$ are complex spaces with normal crossing singularities. One now constructs twistor spaces Z_t containing hypersurfaces $D_t \cong (M, J_M)$ by simultaneously smoothing the singularities of Z_0 and $D_0 \subset Z_0$. This can be done in such a way that the smoothing of $D_0 \subset Z_0$ is just the trivial one (obtained from $D_- \times \mathbb{C}$ by blowing up k lines in $D_- \times \{0\}$) and so that the whole family admits a real structure. The fact that Z_t is then a twistor space for small real values of t follows by an argument identical to that of [4]. ∎

Theorem 3 *If $\mathbb{C}P_1 \times \mathbb{C}P_1$ is blown up at 13 suitably chosen points, the resulting complex surface admits scalar-flat Kähler metrics.*

Proof. The strategy is to apply Theorem 2 when N is a reverse-oriented scalar-flat Kähler metric on a two-fold blow-up of $\mathbb{C}P_1 \times \Sigma$, where Σ is a compact complex curve of genus 2. We therefore begin by constructing such a Kähler manifold with a suitable involution Φ. We will do this do this by means of the construction of page 7.

Let Σ be a of genus 2, let $\phi : \Sigma \to \Sigma$ be the Weierstraß involution which realizes Σ as a 2-sheeted branched cover over $\mathbb{C}P_1$, and let $q \in \Sigma$ be one of the 6 fixed points of ϕ. Let h_Σ be the curvature -1 Hermitian metric on Σ, and notice that ϕ is an isometry of h_Σ.

As before, we equip the 3-manifold $Y = \mathbb{R} \times \Sigma$ with the hyperbolic metric $h = dt^2 + (\cosh^2 t)h_\Sigma$. Let $p_\pm = (q, \pm\frac{1}{2}) \in Y$, let $G_\pm$ be the hyperbolic Green's functions of $p_\pm \in Y$, and set $V = 1 + G_+ + G_-$. Observe that $\star dV$ vanishes on $t = 0$ by symmetry. Let P be the unique principal S^1-bundle on $Y - \{p_\pm\}$ with connection 1-form θ such that

$$\star dV = d\theta$$

and such that the restriction of (P, θ) to $t = 0$ is the trivial bundle-with-connection $\Sigma \times S^1$. We then endow P with the Riemannian metric

$$g = (\text{sech}^2 t)[Vh + V^{-1}\theta^2].$$

The metric space completion of (P, g) is then a smooth compact Riemannian 4-manifold (N, g_N) of scalar-curvature zero, and admits a complex structure J_N with respect to which g_N is Kähler. Moreover [13], the complex surface (N, J_N) is biholomorphic to $\mathbb{C}P_1 \times \Sigma$ blown up at two points in the fiber over $q \in \Sigma$.

Consider the map $\psi : \Sigma \times (-1,1) \to \Sigma \times (-1,1)$ given by $(\zeta, t) \mapsto (\phi(\zeta), -t)$. As $\psi^* V = V$ and ψ is an orientation-*reversing* isometry of Y, it follows that $\psi^* P \cong \bar{P}$ as a principal bundle-with-connection, where $\bar{P}$ denotes P equipped with the inverse S^1 action. There is therefore a unique isometry Φ of P which covers ψ and restricts to the hypersurface $t = 0$ as $\phi \times r : \Sigma \times S^1 \to \Sigma \times S^1$, where $r : S^1 \to S^1$ is the reflection $e^{i\vartheta} \to e^{-i\vartheta}$. This extends to the Riemannian completion N as an involution of the desired type.

One may now apply Theorem 2. All that remains is to understand the structure of the resulting (M, J_M). In order to do this, first blow up N at the 12 fixed points of Φ, and notice that we have the following arrangement of curves in the blow-up $\tilde{N}$:

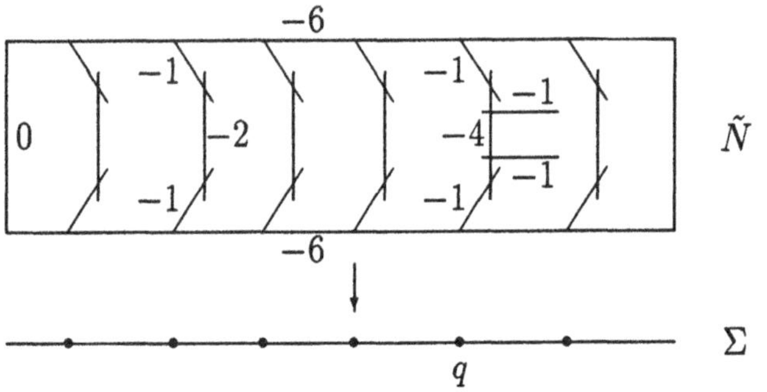

But $\tilde{N}$ is a 2-fold branched cover of M, with ramification locus equal to the the 12 (-1)-curves introduced by the blowing up the fixed points. Descending to M will thus double the self-intersection of these branch curves, while halving the self-intersection of any curve on which Φ acts non-trivially. The corresponding picture of M is therefore as follows:

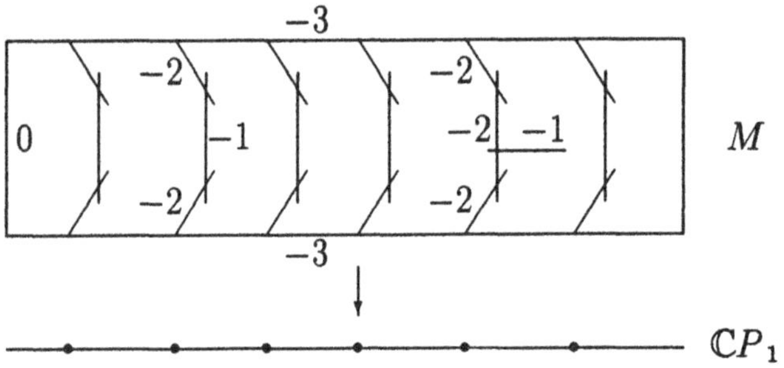

Contracting the 13 exceptional curves in a judicious order,

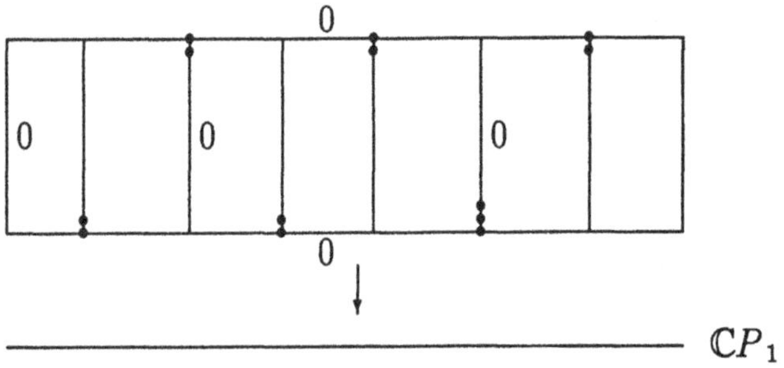

we get $\mathbb{C}P_1 \times \mathbb{C}P_1$ as our minimal model. Invoking Theorem 2, we thus conclude that the above iterated blow-up of $\mathbb{C}P_1 \times \mathbb{C}P_1$ admits scalar-flat

Kähler metrics. Moreover, it also follows from [14] that any sufficiently small deformation

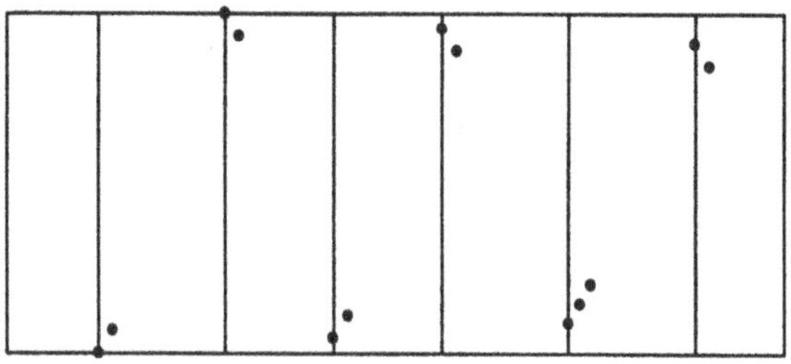

of M also admits scalar-flat Kähler metrics. ∎

Theorem 4 *If $\mathbb{C}P_2$ is blown up at $m \geq 14$ suitably chosen points, the resulting complex surface admits scalar-flat Kähler metrics.*

Proof. For $m = 14$, we just need quote the previous result, since $\mathbb{C}P_2$ blown up at two distinct points is biholomorphic to $\mathbb{C}P_1 \times \mathbb{C}P_1$ blown up once. One can now increase m by k merely by choosing $2k$ rather than 2 centers for our Green's functions. ∎

Theorem 5 *If $n \geq 14m$, $m\mathbb{C}P_2\#n\overline{\mathbb{C}P}_2$ admits anti-self-dual metrics.*

Proof. By [4, 5], connected sums of the anti-self-dual manifolds constructed in Proposition 4 also admit anti-self-dual metrics. ∎

This also gives us a simple proof of a weak version of Theorem 1:

Theorem 6 *Let M be a simply-connected, compact, oriented, smooth 4-manifold. Then a connected sum of M with sufficiently many $\mathbb{C}P_2$'s and $\overline{\mathbb{C}P}_2$'s admits anti-self-dual metrics.*

Proof. In [21], Wall proved that if M_1 and M_2 are simply-connected, compact, oriented, smooth 4-manifolds with isomorphic intersection-forms, then there is an integer k and a diffeomorphism $M_1\#k(S^2\times S^2) = M_2\#k(S^2\times S^2)$. If the intersection-form of M_1 is odd (i.e., M_1 is not spin) and indefinite, then it is isomorphic to the standard form realized by $p\mathbb{C}P_2\#q\overline{\mathbb{C}P}_2$. Now whatever the intersection-form of M, at least one of $M\#\mathbb{C}P_2$ or $M\#\overline{\mathbb{C}P}_2$ will have indefinite odd intersection form. Calling this M_1, and applying Wall's theorem, we have, for some p and q, a diffeomorphism $M_1\#k(S^2 \times S^2) = p\mathbb{C}P_2\#q\overline{\mathbb{C}P}_2\#k(S^2 \times S^2)$.

On the other hand, $S^2 \times S^2\#\overline{\mathbb{C}P}_2$ is diffeomorphic to $\mathbb{C}P_2\#2(\overline{\mathbb{C}P}_2)$, and applying this to the $S^2 \times S^2$-summands in the above, we obtain a diffeomorphism

$$M\#r\mathbb{C}P_2\#s\overline{\mathbb{C}P}_2 \cong m\mathbb{C}P_2\#n\overline{\mathbb{C}P}_2$$

for some m, n, r, s. By increasing n and s if necessary, the result follows from Corollary 5. ∎

These techniques seem particularly well-suited to the scalar-flat Kähler case. By an argument similar to the proof of Theorem 3, one can construct scalar-flat metrics on blow-ups of $E\times\mathbb{C}P_1$ for any elliptic curve $E \approx S^1\times S^1$. Combined with Theorem 3 and our explicit examples, this shows that blow-ups of $\Sigma \times \mathbb{C}P_1$ admit scalar-flat metrics for each compact complex curve Σ; and with a little extra thought, one can even specify most of the blown up points arbitrarily. In conjunction with surface classification theory, this allows one to prove the following result [9], originally conjectured in [14]:

Theorem 7 *Let (M, J) be a compact complex 2-manifold which admits a Kähler metric for which the integral of the scalar curvature is non-negative. Then precisely one of the following holds:*

- *(M, J) admits a Ricci-flat Kähler metric; or*
- *any blow-up of (M, J) has blow-ups $(\tilde{M}, \tilde{J})$ which admit scalar-flat Kähler metrics.*

References

[1] M. Atiyah, N. Hitchin and I. Singer, "Self-Duality in Four Dimensional Riemannian Geometry," **Proc. R. Soc. Lond. A 362** (1978) 425–461.

[2] C. Boyer, "Conformal Duality and Compact Complex Surfaces," **Math. Ann. 274** (1986) 517–526.

[3] J.A. Carlson and D. Toledo, "Harmonic Mappings of Kähler Manifolds to Locally Symmetric Spaces," **Publ. Math. IHES 69** (1989) 173–201.

[4] S.K. Donaldson and R.D. Friedman, "Connected Sums of Self-Dual Manifolds and Deformations of Singular Spaces," **Nonlinearity 2** (1989) 197–239.

[5] A. Floer, "Self-Dual Conformal Structures on $\ell\mathbb{C}P^2$," **J. Diff. Geom. 33** (1991) 551–573.

[6] P. Gauduchon, "Surfaces Kählériennes dont la Courbure Vérifie Certaines Conditions de Positivité," in **Géometrie Riemannienne en Dimension 4**. *Séminaire A. Besse, 1978/1979*, (Bérard-Bergery, Berger, and Houzel, eds.), CEDIC/Fernand Nathan, 1981.

[7] D. Joyce, remark concerning $4\mathbb{C}P_2$, private communication.

[8] J.-S. Kim, **On a class of 4-dimensional minimum-energy metrics and hyperbolic geometry**, Ph.D. Thesis, SUNY Stony Brook, 1991.

[9] J.-S. Kim, C.R. LeBrun, and M. Pontecorvo, "Scalar-Flat Kähler Surfaces of All Genera," preprint, 1993.

[10] J.-S. Kim and M. Pontecorvo, "A New Method of Constructing Scalar-Flat Kähler Metrics," preprint, 1993.

[11] K. Kodaira, "A Theorem of Completeness of Characteristic Systems for Analytic Families of Compact Submanifolds of Complex Manifolds," **Ann. Math. 75** (1962) 146–162.

[12] C.R. LeBrun, "Explicit Self-Dual Metrics on $\mathbb{C}P_2\# \cdots \#\mathbb{C}P_2$," **J. Diff. Geom. 34** (1991) 223–253.

[13] C.R. LeBrun, "Scalar-Flat Kähler Metrics on Blown-Up Ruled Surfaces," **J. reine angew. Math.** 420 (1991) 161–177.

[14] C.R. LeBrun and M.A. Singer, "Existence and Deformation Theory for Scalar-Flat Kähler Metrics on Compact Complex Surfaces," **Inv. Math. 112** (1993) 273–313.

[15] C.R. LeBrun and M.A. Singer, "A Kummer-type Construction of Self-Dual 4-Manifolds," preprint, 1993.

[16] A. Newlander and L. Nirenberg, "Complex Analytic Coordinates in Almost-Complex Manifolds," **Ann. Math. 65** (1957) 391–404.

[17] R. Penrose, "Non-linear Gravitons and Curved Twistor Theory," **Gen. Rel. Grav. 7** (1976) 31–52.

[18] M. Pontecorvo, "On Twistor Spaces of Anti-Self-Dual Hermitian Surfaces," **Trans. Am. Math. Soc. 331** (1992) 653–661.

[19] Y.S. Poon, "Compact Self-Dual Manifolds with Positive Scalar Curvature," **J. Diff. Geom. 24** (1986) 97–132.

[20] C. H. Taubes, "The Existence of Anti-Self-Dual Metrics," **J. Diff. Geom. 36** (1992)163–253.

[21] C.T.C. Wall, On Simply Connected 4-manifolds, **J. Lond. Math. Soc. 39** (1964) 141–149.

8
Generalized Twistor Correspondences, d-Bar Problems, and the KP Equations

L.J. Mason
Mathematical Institute,
24-29 St Giles,
Oxford, OX1 3LB, U.K.

1 Introduction

Like other programmes in mathematical physics, one of the most impressive features of twistor theory is the diversity and depth of its applications in pure and applied mathematics. In this conference and its proceedings we have applications to self-dual 4-manifolds, integral geometry, differential geometry, representation theory and this contribution is concerned with the connections between twistor theory and nonlinear integrable equations. However, there are indications that there might be significant feedback to the twistor programme itself from this last application and this is also explored in the following.

This article is concerned with the overview that twistor theory provides over the theory of nonlinear integrable equations. This overview arises from the observation of Ward [11] and others [4, 6, 7] that most of the familiar examples of integrable equations are symmetry reductions of the self-duality equations. This means that most integrable systems admit a twistor correspondence—that obtained by taking the symmetry reduction of one of the standard ones for the self-duality equations. However, there is a family of equations that have not yet been expressed as symmetry reductions of the self-duality equations: the Kadomtsev-Petviashvilii (KP) equation and its relatives. The purpose of this article is to argue firstly that these equations will not be expressible as reductions of the self-duality equations but on the other hand to present a generalized twistor correspondence that should, with further development, extend the applications of twistor theory to these anomalous integrable systems. Thus, despite the fact that these equations are not directly related to the self-duality equations, a generalization of the standard twistor construction still applies. A secondary purpose of this article is to discuss the possibility that these generalizations might be a first step towards the twistor description of the non-integrable equations of basic physics that must be found for the twistor programme to succeed.

2 Linear systems and twistor correspondences

The main obstacle to the twistor programme is the lack of a twistor description of the full Einstein and Yang-Mills equations in Lorentz signature. These equations are certainly far from being integrable. However, the links between twistor theory and integrable systems highlight the pivotal role played by the Lax pair (or linear system) in twistor correspondences and the full Yang-Mills and the Einstein vacuum equations do in fact have linear systems. (Recall that a linear system or Lax pair for a set of nonlinear equations is an overdetermined set of linear equations whose coefficients contain the dependent variables of the nonlinear equation and whose consistency conditions are the original nonlinear equation itself.) In the case of the Einstein vacuum equations, the linear system is the Rarita Schwinger equation for the potential of a helicity-3/2 field (see Penrose's article in this volume for a discussion of this) and in the case of the Yang-Mills equations, the linear system is just the linearized equations off the given background. Furthermore, there appear to be links between the linear system for the Einstein vacuum equations and twistors (see the article by Penrose in this proceedings and our joint article in $\mathbb{TN}$ 27).

Thus one might hope that there exist twistor correspondences for these non-integrable equations somehow arising from these linear systems. One might expect that the non-integrability would not be an obstruction to the existence of a twistor correspondence, but would manifest itself in the fact that the twistor correspondence no longer provides a linearization of the equations (i.e. a solution procedure that can be implemented by the solution of sets of linear equations).

There are, however, significant differences between the structure of the linear system of the self-duality equations and the linear system for the full Yang-Mills and Einstein systems. Some of the differences are as follows.

1. The linear system for the self-duality equations can be represented by a pair of vector fields (after perhaps introducing some extra degrees of freedom by going to the total space of the Yang-Mills vector bundle over the spin bundle in the Yang-Mills case) whereas the linear system for the full equations consists of genuine partial differential equations in the sense that it is impossible to reformulate them as O.D.E.'s even on a larger (but finite dimensional) space.

2. The linear system of most integrable systems involves an additional parameter (the spectral parameter) over and above the independent

variables of the original equation whereas that for the full Yang-Mills and vacuum equations are straightforward background coupled differential equations with no additional parameters involved.

3. For all integrable systems (that I am familiar with) the Lax pair is sufficiently overdetermined that it propagates initial data from a codimension-2 submanifold whereas the linear system for the full Einstein and Yang-Mills equations propagates only from a codimension-1 hypersurface as the full equations do.

These differences make it very difficult to see what the analogue of a twistor construction should be for the full Einstein or Yang-Mills equations. The first item is the most crucial: if it is possible to reformulate the linear system as a pair of vector fields on some larger space, then it is straightforward to write down a candidate for the twistor correspondence. This can be seen as follows.

Suppose that the differential equations are defined on a space X. Thus, for SDYM, X is Minkowski space with standard coordinates $x^{AA'}$. Suppose also that the linear system can be reformulated as a distribution D (usually 2-dimensional) on a space Y that fibres over X, $p : Y \mapsto X$. The fibre of p at $x \in X$ is, roughly speaking, the Cartesian product of the space coordinatized by the spectral parameter (the projective spin space $\mathbb{PS}$ in the SDYM case with homogeneous coordinates π_A) and an additional space E_x that is required to reduce the operators of the linear system to vector fields. In the SDYM case E_x can be taken to be the fibre of the Yang-Mills vector bundle E at x. The Lax pair for SDYM, $\pi^A D_{AA'}$ (here $D_{AA'} = \partial_{AA'} - A_{AA'}$ is the horizontal lift of $\partial_{AA'}$ to E) spans a subbundle of the horizontal distribution on E pulled back to the spin bundle. The introduction of the additional space E_x is sometimes referred to as prolongation. In the case of the n-KdV equations, one of the operators of the Lax pair is an nth-order scalar differential operator in 1-variable and E_x is the fibre of the $(n-1)$'th jet bundle of $\mathbb{R}$. On the total space of the jet bundle the operator can be represented as a vector field. In the case of self-dual conformal structures, Y is just the projective spin bundle.

In this situation, one can directly define the twistor space Z to be the quotient of Y by the distribution D. This then leads to a double fibration

$$\begin{array}{ccc} & q \quad Y \quad p & \\ & \swarrow \qquad \searrow & \\ Z & & X. \end{array}$$

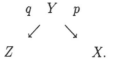

There are a number of possible technical difficulties that can arise (for example, if the distribution D drops in rank at some points in Y, Z will not be a manifold) and this definition of Z often only represents the starting point for investigation. The variety of difficulties and techniques available for circumventing them contribute to the diversity of methods for dealing with integrable systems. However, the general picture is roughly the same in most cases.

Thus, when the linear system can be prolonged to become equivalent to a distribution on a larger space, we know how to define a twistor correspondence. Conversely, if a system has a twistor correspondence of the standard type arising from a double fibration, there will be a linear system that can be prolonged to become a distribution on the space Y of the double fibration.

However, the linear systems for the full Einstein and Yang-Mills equations can be seen to be inequivalent to distributions on a larger (but finite-dimensional) space. One way to see this is to take the flat case of the linear system on the smallest space on which it is well defined (in this case this is X as there is no spectral parameter). Take the Fourier transform of its solutions. If it were possible to prolong the system to a distribution, the Fourier transform on the larger space would be that of functions constant along vector fields. When the vector field has constant coefficients, functions that are constant along it have Fourier transforms that are supported on linear subspaces of momentum space. However, after suitable gauge choices etc., the linear systems for the Einstein and Yang-Mills equations can be reduced to the wave equation. The Fourier transform of solutions of wave equations are supported on a cone in momentum space which cannot be the intersection with a linear subspace in a higher dimensional space. One can generalize this argument away from the flat case by looking at the propagation of singularities (which depend on the symbol of the operator) so that that the flatness assumption is irrelevant. (Alternatively, the functions that commute with an operator that extends to a vector field depend precisely on one fewer variables, whereas the only functions that commute with the Rarita-Schwinger operator are constants.)

These arguments also apply to the standard Lax pair of the KP equation. This is the most prominent example of an integrable system that has not yet been expressed as a reduction of the self-duality equations and this argument shows that either it cannot be a reduction of the self-duality equations, or that it has an alternate inequivalent Lax pair that can be prolonged to yield a distribution. This latter possibility seem very unlikely. For the trivial KP solution, the operators of the Lax pair are the heat operator $\partial_y - \partial_x^2$ (or the Schrodinger operator $i\partial_y - \partial_x^2$) together with scalar operators that each introduce a new variable corresponding to higher flows in the hierarchy. Solutions of these operators have Fourier transform supported on the parabola $p_y + p_x^2 = 0$, and so the Lax pair cannot be prolonged to become a distribution. (Alternatively, only the constant functions commute with these operators.) Thus, unless the KP equation has a completely different Lax pair that can be prolonged to yield a distribution—an unlikely possibility—it will be impossible to express it as a reduction of the self-duality equations. It will therefore not have a twistor correspondence of standard type.

It is also worth noting that the KP Lax pair differs from that of the self duality equations in not involving a spectral parameter (as in the second difference above). However, the KP Lax pair sides with the self-duality Lax pair as far as the third difference listed above is concerned.

3 Generalized twistor correspondences

The argument of the last section appears to eliminate the hope that the self-duality equations are the universal integrable systems and that all integrable systems (including the KP equations) should be obtainable as symmetry reductions thereof. However, much of the attraction of that philosophy has been that the rich structure and theory that surrounds integrable systems can be traced to the twistor correspondences that they must have as symmetry reductions of the self-duality equations. Indeed the focus on the self-duality equations could already be seen to be just a convenience; if one wants to think of the equations that arise from hyperkahler structures in $4n$-dimensions as integrable equations (the condition that a connection on a bundle be holomorphic with respect to all the complex structures associated to a hyper-Kahler manifold). These certainly fit very naturally into twistor theory and much of their study is based on these correspondences. It is also the case that one must work very hard to show that some of the more obscure systems in two-dimensions are reductions of the self-duality equations (especially systems associated to higher flows in hierarchies) whereas, again, they have a relatively straightforward twistor correspondence (see for example [5] or [2]).

Thus, the slogan that should underly the programme is

'Twistor correspondences underly the concept of integrability.'

The self-duality equations are prominent because there is sufficient freedom in them to yield, via symmetry reduction, perhaps all integrable systems in 1 and 2 dimensions and a large class in 3–dimensions.

Thus, rather than worry about whether we should be able to express the KP equation as a reduction of the self-duality equations, we should instead look for a twistor correspondence for it.

In the next subsection I review the standard correspondence between solutions of the Bogomoln'yi hierarchy and bundles on $\mathcal{O}(n)$ (twisted minitwistor space) in which the bundle will be described by its $\bar{\partial}$-operator. In the subsequent subsection I give a formulation of a construction for solutions of the KP equations in which the $\bar{\partial}$-operator of the standard twistor construction is generalized to a Dirac operator (or at least, a Dirac operator when restricted to a holomorphically emebedded sphere in minitwistor space). This formulation is derived from the so called d-bar approach to the KP equations developed by Fokas and Ablowitz (and also by Zakharov and Manakov), [1, 3, 12].

3.1 A $\bar{\partial}$ formulation of the minitwistor correspondence

We will be working on the complex line bundle $\mathcal{O}(n)$ of Chern class n on $\mathbb{CP}^1$ (a twisted analogue of the standard minitwistor space, $\mathcal{O}(2)$). We shall work with affine coordinates (μ, λ) with λ an affine coordinate on $\mathbb{CP}^1$ and μ a linear coordinate up the fibre of the line bundle chosen with respect to a section that vanishes to n'th order at $\lambda = \infty$.

Points of $\mathbb{C}^{n+1}$ correspond to holomorphically embedded $\mathbb{CP}^1$'s via the incidence relation

$$\mu = \sum_{i=0}^{n} t_i \lambda^i$$

where t_i are coordinates on $\mathbb{C}^{n+1}$.

Recall that a solution of the Bogomoln'yi hierarchy with gauge group $SL(r, \mathbb{C})$ on $\mathbb{C}^{n+1}$ is a collection of $2n$ Lie algebra valued fields $(A_i, B_{i+1}), i = 1, \ldots, n$ defined up to the gauge freedom

$$(A_i, B_{i+1}) \mapsto (g^{-1}A_i g - g^{-1}\partial_i g, g^{-1}B_{i+1}g - g^{-1}\partial_{i+1}g)$$

such that

$$[\partial_{i+1} - B_{i+1} + \lambda(\partial_i - A_i), \partial_{j+1} - B_{j+1} + \lambda(\partial_j - A_j)] = 0. \tag{3.1}$$

for all values of λ, [4].

The Ward correspondence provides a rank r holomorphic vector bundle E on $\mathcal{O}(n)$ from a solution of the $SL(r, \mathbb{C})$ Bogomoln'yi equations on $\mathbb{C}^{n+1}$. The bundle has to be topologically trivial from the Ward construction (it must be analytically trivial on at least one holomorphically embedded $\mathbb{CP}^1$ in $\mathcal{O}(n)$). Because the fibres of $\mathcal{O}(n)$ over $\mathbb{CP}^1$ are Stein, we can choose a global smooth trivialization that is holomorphic up the fibres. In such a trivialization the holomorphic structure is determined by a $\bar{\partial}$-operator of the form

$$\bar{\partial}_E = \bar{\partial}_0 + d\bar{\lambda} \otimes \alpha(\mu, \lambda, \bar{\lambda})$$

where $\bar{\partial}_0 = d\bar{\mu} \otimes \partial_{\bar{\mu}} + d\bar{\lambda} \otimes \partial_{\bar{\lambda}}$ is the $\bar{\partial}$-operator of the trivial bundle, and α is a smooth Lie algebra valued function on $\mathcal{O}(n)$ holomorphic in μ satisfying appropriate regularity conditions as $\lambda \to \infty$.

In order to reconstruct the solution of the Bogomoln'yi equation, we must first solve the auxiliary problem

$$(\bar{\partial}_E \psi)|_{\mu = \sum_i t_i \lambda^i} := \partial_{\bar{\lambda}} \psi + \alpha \psi = 0 \tag{3.2}$$

where $\psi := \psi(t_i, \lambda, \bar{\lambda})$ is an r-component vector valued function on the correspondence space $Y = \mathbb{C}^{n+1} \times \mathbb{CP}^1$. The solution ψ must be regular over the whole of the $\mathbb{CP}^1$ factor. This is the analogue of the Sparling equation appropriate to the Bogomoln'yi hierarchy.

Equation (3.2) generically has precisely r linearly independent solutions since, when restricted to a $\mathbb{CP}^1$, $\bar{\partial}_E$ has index r. These can be assembled into an $r \times r$ fundamental solution matrix Ψ satisfing equ. (3.2).

If we now differentiate equ. (3.2) along the vector fields $V_i = \partial_{i+1} - \lambda\partial_i, i = 1, \ldots, n$ we discover that $V_i\Psi$ are also solutions of equ. (3.2) since α depends on t_i only through μ which is annihilated by V_i (i.e. $\bar{\partial}_E$ commutes with the V_i). Thus we can put $V_i\Psi = \Psi\gamma_i$ and equ. (3.2) implies that γ_i are holomorphic in λ. Furthermore, since Ψ is smooth on $\mathbb{CP}^1$, the γ_i are smooth save for a simple pole at ∞ owing to that in V_i, and hence must have the form

$$\gamma_i = B_i + \lambda A_{i-1}$$

for some $A_0, \ldots, A_n, B_1, \ldots B_{n+1}$ independent of λ. Furthermore, these must satisfy the Bogomoln'yi equations (3.1) since the operators $\partial_{i+1} - B_{i+1} + \lambda(\partial_i - A_i)$ must be compatible as Ψ is annihilated by them.

3.2 A construction for solutions to the KP equation

Recall that the KP equation

$$4\partial_{t_1}\partial_{t_3}u - \partial_{t_1}^4 u + 12\partial_{t_1}(u\partial_{t_1}u) = 3\partial_{t_2}^2 u \tag{3.3}$$

for a function $u := u(t_1, t_2, t_3)$ arises as the consistency condition for the operators

$$L_2 = (\partial_{t_2} - \partial_{t_1}^2 + 2u) \text{ and } L_3 = \partial_{t_3} - \partial_{t_1}^3 + 3u\partial_{t_1} + 3\partial_{t_1}u - v$$

(the function v is determined in terms of u by the consistency conditions). More generally, the equations of the KP hierarchy arise when one imposes the consistency conditions $[L_i, L_j] = 0$ for the sequence of operators

$$L_i = \partial_{t_i} - \partial_{t_1}^i - \sum_{j=0}^{i-2} u_j\partial_{t_1}^j.$$

We now generalize equ. (3.2) by replacing the $\bar{\partial}$-operator by a Dirac operator. The resulting construction is nontrivial (in the sense of leading to solutions of nonlinear equations) even in the case of a scalar Dirac equation which will lead to solutions of the KP equation.

The analogue of the Sparling equation in this case will be

$$\not{D}_\alpha \underline{\psi} := \begin{pmatrix} \partial_{\bar{\lambda}} & \alpha \\ \bar{\alpha} & \partial_\lambda \end{pmatrix} \begin{pmatrix} \psi \\ \tilde{\psi} \end{pmatrix} = 0. \tag{3.4}$$

As before, $\underline{\psi}$ is to be thought of as a function on the correspondence space Y which we will now take to be $\mathbb{R}^{n+1} \times \mathbb{CP}^1$ (rather than $\mathbb{C}^{n+1} \times \mathbb{CP}^1$). In general, one could take α to be any function of $(\mu, \bar{\mu}, \lambda, \bar{\lambda})$ and equation (3.4) will be a parametrized Dirac equation on a Riemann sphere given by setting $\mu = \sum t_i \lambda^i$ (so that the t_i are the parameters). In order to make contact with the KP equation we set

$$\alpha = \exp(\mu - \bar{\mu}) f(\lambda, \bar{\lambda})$$

which, in effect, implies a symmetry along ∂_{t_0} in space-time. We also impose the reality condition $\overline{f(\lambda, \bar{\lambda})} = f(\bar{\lambda}, \lambda)$ so that the reflection of f in the real axis is its complex conjugate.

For simplicity we shall require that f vanishes in some neighbourhood of infinity. This implies that ψ is holomorphic in a neighbourhood of $\lambda = \infty$ (and $\tilde{\psi}$ will be anti-holomorphic). Thus ψ will have an expansion

$$\psi = \sum_{n \leq 0} \psi_n \lambda^n$$

in a neighbourhood of infinity.

As before, we know that equation (3.4) generically has precisely 2 linearly independent solutions as the operator has index 2. However, we can use two conjugations to reduce the solution space from $\mathbb{C}^2$ to $\mathbb{R}$. Since the equation is invariant under $(\psi, \tilde{\psi}) \to (\bar{\tilde{\psi}}, \overline{\psi})$, we can take $\tilde{\psi} = \bar{\psi}$ since, if the original solution we chose did not satisfy this condition, then $(\psi', \tilde{\psi}') = (\psi + \bar{\tilde{\psi}}, \tilde{\psi} + \overline{\psi})$ would be a solution that does. Similarly we can require that the solution should satisfy $\psi(\lambda, \bar{\lambda}) = \overline{\psi(\bar{\lambda}, \lambda)}$ since if $(\psi(\lambda, \bar{\lambda}), \tilde{\psi}(\lambda, \bar{\lambda}))$ is a solution then $(\overline{\psi(\bar{\lambda}, \lambda)}, \overline{\tilde{\psi}(\bar{\lambda}, \lambda)})$ is also. This reduces the solution space to $\mathbb{R}$. Finally we rescale the solution so that $\psi = 1$ at $\lambda = \infty$.

Thus we have a unique solution, $\underline{\psi}(t_i, \lambda)$ on the correspondence space satisfying $\not{D}_\alpha \underline{\psi} = 0$, the two reality conditions and normalized so that $\psi = 1$ at $\lambda = \infty$.

In the last section, in order to reconstruct the fields on space-time, we looked for operators on Y (the spin bundle with coordinates (t_i, λ)) that commuted with the operator $\partial_{\bar{\lambda}} + \alpha$ of the Sparling equation (3.2). The operators used there do not commute with $\not{D}_\alpha$ since $\not{D}$ contains derivatives with respect to λ as well as $\bar{\lambda}$. However, the following operators do commute with $\not{D}_\alpha$

$$W_i := \partial_{t_i} - \Lambda^i$$

where

$$\Lambda = \begin{pmatrix} \lambda & 0 \\ 0 & \bar{\lambda} \end{pmatrix}$$

so that $[W_i, \not{D}_\alpha] = 0$.

Thus, if $\underline{\psi}$ is a solution of equation (3.4), then $(W_1)^{r_1}(W_2)^{r_2}\dots(W_i)^{r_i}\underline{\psi}$ is also a solution for all $r_1,\dots,r_i \in \mathbb{Z}$. These expressions generally have poles of various degrees at $\lambda = \infty$. In the case of the Ward transform, multiplication by holomorphic functions commutes with the Sparling equation and so one could express $V_i\Psi$ as Ψ multiplied by some global holomorphic matrix function with a simple pole at ∞. Here, there is no simple relation between $W_i\underline{\psi}$ and $\underline{\psi}$ in the first instance.

However, one can find linear combinations of products of the W_i whose action on $\underline{\psi}$ yields a solution that is regular on $\mathbb{CP}^1$ and satisfies the reality conditions so that, by uniqueness of real solutions to equation (3.4), it must be proportional to the original solution by a function that is constant on the sphere. The first two such combinations are as follows

$$\begin{aligned} V_2 &:= W_2 + W_1^2 &&= \partial_2 + \partial_1^2 - 2\Lambda\partial_1 \\ V_3 &:= 2W_3 + 3W_1W_2 + W_1^3 &&= 2\partial_3 + 3\partial_1\partial_2 + \partial_1^3 - 3\Lambda(\partial_2 + \partial_1^2) \end{aligned}$$

and this continues to an infinite sequence of such operators V_n each involving all the time variables t_r for $r \le n$ and containing only Λ raised to the first power.

These operators are still singular at $\lambda = \infty$ owing to the presence of Λ. However Λ is multiplied by a nontrivial differential operator. Thus, since we have normalized ψ so that $\psi = 1 + \psi_{-1}\lambda^{-1} + \dots$, when V_2 acts on ψ, the only possible divergent term is $\lambda\partial_1\psi_0 = \sum_{n\le 0}\partial_1\psi_n\lambda^{1+n}$ and its conjugate, but these are regular since $\partial_1\psi_0 = 0$. Thus $V_2\underline{\psi}$ is a global regular solution of (3.4) on $\mathbb{CP}^1$ and also satisfies the two reality conditions. It must therefore be proportional to $\underline{\psi}$. Thus

$$V_2\underline{\psi} = 2u\underline{\psi}$$

and similarly

$$V_3\underline{\psi} = 2v\underline{\psi}$$

for some u and v that depend on t_i alone.

We have that $\exp(-\sum t_i\Lambda^i)W_i\exp(\sum t_i\Lambda^i) = \partial_i$. It follows that, since, $\underline{\psi}$ is annihilated by the operators $(W_2 + W_1^2 - 2u)$ and $(2W_3 + 3W_1W_2 + W_1^3 - 2v)$, then $\exp(-\sum t_i\Lambda^i)\underline{\psi}$ is annihilated by the operators $(\partial_2 + \partial_1^2 - 2u)$ and $(2\partial_3 + 3\partial_1\partial_2 + \partial_1^3 - 2v)$ which must therefore be consistent. By subtracting $3\partial_1$ composed with the first of these from the second, we see that these operators are equivalent to the operators of the KP Lax pair given above. Thus, u must satisfy the KP equations.

4 Further problems

The above is an adaptation of the 'd-bar' approach to the KP equations of Fokas and Ablowitz (see [1] and references therein and also [3, 12]). This approach is particularly natural for the KPII equation (the version of the KP equation given in (3.3)). The KPI equation differs from equation (3.3) by sending ∂_{t_2} to $i\partial_{t_2}$. Solutions of the KPI equation are more naturally obtained from a parametrized *non-local Riemann-Hilbert problem.* The problem is to find $\psi^+(t_i,\lambda)$ holomorphic on the upper half λ-plane and $\psi^-(t_i,\lambda)$ holomorphic on the lower half λ-plane such that

$$\psi^+(\lambda)=\psi^-(\lambda)+\int_{-\infty}^{\infty}K(\lambda,\lambda')\exp(\mu-\mu')\psi^-(\lambda')d\lambda'$$

where as before $\mu=\sum t_i\lambda^i$. A similar, but slightly simpler non-local Riemann-Hilbert problem is available by adapting the Segal and Wilson approach [9]. We require that ψ^∞ be holomorphic in a neighbourhood of $\lambda=\infty$, that ψ^0 be holomorphic on the complement of $\lambda=\infty$, and

$$\psi^\infty(\lambda)=\psi^0(\lambda)+\oint K(\lambda,\lambda')\exp(\mu)\psi^0(\lambda')d\lambda',$$

where the contour is a small circle about $\lambda=\infty$. The Kernel in both cases has to be chosen so that the problem is effectively a Fredholm equation with a unique solution. This works if K is 'small'. This connects with the Grassmannian of the Segal and Wilson approach by identifying K with the Kernel associated to a map from the Hilbert space of positive powers of λ to that of negative powers of λ. The graph of this map is the subspace W of $L^2(S^1)$ defining an element of the Segal Wilson Grassmannian. Thus $K(\lambda,\lambda')$ is defined modulo the addition of negative powers of λ multiplied by arbitrary functions of λ' and positive powers of λ' multiplied by arbitrary functions of λ.

There should be a large class of solutions for which the above procedures are all the same. For such solutions one might expect that the correspondence between the d-bar approach and the non-local Riemann-Hilbert approach is a natural generalization of the Čech-Dolbeault isomorphism.

An intriguing feature of these nonlocal Riemann-Hilbert problem approaches is that they are very similar to the 'copatching' relations obtained by Penrose [8]. These arose out of an attempt to dualize the standard twistor constructions, but foundered on the difficulty of making sense of patching relations in which the transition function was replaced by a pseudo-differential operator or alternatively an integral kernel with cohomological gauge freedom. However, the above constructions make perfectly good sense and an investigation of these ideas should yield new insights into those proposals.

It is clear that there are many ways to generalize the above, for example by dropping the reality conditions on f or the restrictive form of the ansatze on α, or coupling the Dirac operator to a vector bundle. It would be useful to find a definitive general statement of such constructions and see what equations arise. The most challenging problem, however, is to generalize these ideas to produce twistor-like correspondences for non-integrable equations that have a linear system.

Acknowledgements
I would like to thank Mark Ablowitz for explaining the d-bar approach to the KP equations to me and for hospitality in Boulder where these ideas were first put together. I would also like to thank Michael Singer and George Sparling for useful discussions.

References

[1] Ablowitz, M. & Clarkson, P.A. (1991) *Solitons, Nonlinear Evolution Equations and Inverse Scattering*, LMS lecture note series 149, C.U.P.

[2] Carey, A., Hannabus, K. Mason, L.J. and Singer, M.A. (1993) The Landau Lifschitz equations, elliptic curves and the Ward transform, Comm. Math. Phys., **154**, 25-47.

[3] Carroll, R. & Konopelchenko, B. (1992) $\overline{D}$ dressing and Sato theory, Preprint.

[4] Mason, L.J. & Sparling, G.A.J. (1992) Twistor correspondences for the soliton hierarchies, J. Geom. Phys. **8**, 243-271. See also: Nonlinear Schrodinger and Korteweg de Vries are reductions of self-dual Yang-Mills, Phys. Lett. **A 137** (1989) 29-33.

[5] Mason, L.J. & Singer, M.A. (1994) The twistor theory of equations of KdV type, to appear in Comm. Math. Phys.

[6] Mason, L.J. (1993) Twistor theory, self-duality and integrability, in Proceedings of NATO A.R.W., Exeter 1992, ed. P.A. Clarkson, Kluwer.

[7] Mason, L.J. & Woodhouse, N.M.J., *Twistor theory, self-duality and integrability*, in preparation. See also: Mason, L.J. & Woodhouse, N.M.J. (1993) Self-duality and the Painlevé transcendents, *Nonlinearity*, **6**, 569-81, and Maszczyck, R., Mason, L.J. & Woodhouse, N.M.J. (1994) Self-dual Bianchi metrics and the Painlevé transcendents, Class. Quant. Grav. **11**, 65-71.

[8] Penrose, R. (1981) 'A new angle on the googly graviton', and 'Concerning a Fourier contour integral', TN11 and Physical left-right symmetry and googlies, TN12.

[9] Segal, G.B. & Wilson, G. (1985) Loop Groups and equations of KdV type, Publ. Math. IHES, **65**, 5–65.

[10] Ward, R.S. (1977) On self-dual gauge fields, Phys. Lett. A 61, 81-82.

[11] Ward, R.S. (1985) Integrable and solvable systems and relations among them, Phil. Trans. Roy. Soc. Lond. **A315**, 451-457. See also: Multi-dimensional integrable systems, in *Field Theory, Quantum Gravity and Strings*, eds. H.J. de Vega and N. Sanchez, Lecture Notes in Physics, Vol. 246 (Springer, Berlin 1986) and 'Integrable systems in twistor theory' in *Twistors in Mathematics and Physics*, eds. T.N.Bailey and R.J.Baston L.M.S. Lecture Notes 156, C.U.P.

[12] Zakharov, V. & Manakov, S. (1985) Funkts. Anal. Appl. **19**, 89-101.

9
Relative Deformation Theory and Differential Geometry

Sergey A. Merkulov

School of Mathematics and Statistics, University of Plymouth

Plymouth, Devon PL4 8AA, United Kingdom

1 Introduction

Let X be a compact complex submanifold of a complex manifold Y with normal bundle N such that $H^1(X, N) = 0$. In 1962 Kodaira proved that such a submanifold X belongs to the complete analytic family $\{X_t : t \in M\}$ of complex submanifolds X_t of Y with the moduli space M being a $(\dim_{\mathbb{C}} H^0(X, N))$-dimensional complex manifold. Moreover, there is a canonical isomorphism $k_t : T_tM \longrightarrow H^0(X_t, N_t)$ which associates a global section of the normal bundle N_t of $X_t \hookrightarrow Y$ to any tangent vector at the corresponding point $t \in M$.

About 14 years later Penrose (1976) studied the Kodaira complete family associated to an embedding of the projective line $X = \mathbb{CP}^1$ into a 3-dimensional complex manifold Y with normal bundle $N \cong \mathbb{C}^2 \otimes \mathcal{O}(1)$. His striking conclusion was that the moduli space M comes equipped canonically with very interesting geometric data — the 4-fold M has an induced conformal structure with self-dual Weyl tensor. He also explained what additional structures on Y permit one to pick out a metric from the conformal class which automatically satisfies Einstein's equations. This Penrose construction revealed deep and unexpected interconnections between multidimensional complex analysis from one side and differential geometry and mathematical physics from the other (Penrose 1977, Wells 1979, Manin 1981). Later a similar phenomenon has been observed for a number of other relative deformation problems $X \hookrightarrow Y$. Hitchin (1982) studied Kodaira moduli spaces of rational curves $X = \mathbb{CP}^1$ embedded into complex manifolds with normal bundle $N = \mathcal{O}(m)$ for $m = 1, 2$. He showed that in the case $m = 2$ the associated 3-dimensional Kodaira moduli spaces have an induced Einstein-Weyl structure. Kodaira moduli spaces of rational curves with normal bundle $N = \mathbb{C}^2 \otimes \mathcal{O}(1)$ have much to do with 4-dimensional self-dual Einstein equations (see, e.g., Penrose 1976, Ward 1980, Gindikin 1982, Hitchin 1982, Eastwood 1990), while those associated with relative deformations of $\mathbb{CP}^1$ with normal bundle $N = \mathbb{C}^{2k} \otimes \mathcal{O}(1)$, $k \geq 2$, are shown to have a sort of quaternionic structure depending on the choice of some additional structure on the ambient manifolds Y (Salamon 1982, Hitchin et al 1987, LeBrun 1989, Pedersen and Poon 1989, Bailey and Eastwood 1991). The next two examples deal with complete families of Legendre compact submanifolds of complex contact manifolds. LeBrun (1983) showed that moduli spaces of n-quadrics Q embedded into a $(2n+1)$-dimensional complex contact manifold Y with $N = J^1(\mathcal{O}(1)|_{Q \hookrightarrow \mathbb{CP}^{n+1}})$ have an induced complex conformal structure, and Bryant (1991) proved that the moduli space of Legendre rational curves in a complex contact 4-fold Y with normal bundle $N = \mathbb{C}^2 \otimes \mathcal{O}(2)$ always comes equipped with a torsion-free affine connection with exotic holonomy group G_3. Supersymmetric extensions of some of these results are obtained by Merkulov (1991, 1992*a*-*c*,1993*a*). It is worth noting that the ways the authors of the above mentioned papers obtained their results are very sensitive to both the holomorphy type of the submanifold under deformation and the holomorphy type of its normal bundle. The key ingredients of their constructions are Kodaira's isomorphism between $H^0(X, N_t)$ and the tangent space T_tM, and Penrose's idea to consider subspaces of $H^0(X, N_t)$ consisting of global sections having zeros of prescribed order. In some very specific situations such a subspace turns out to be an algebraic submanifold in TM with clear geometric meaning.

This paper presents a general analysis of Kodaira and Legendre moduli spaces of compact submanifolds which shows that the phenomenon of inducing rich geometric structures on these creatures is a very general one. In twistor theory one encounters a law of nature rather than a fortunate concurrence of circumstances. As a by-product we show that all the above mentioned twistor constructions follow from a couple of general theorems which say that moduli spaces of compact complex submanifolds often come equipped with families of torsion-free connections satisfying some natural integrability conditions. A surprising feature of this theory is how easily one can estimate the holonomy group of induced connections (see Theorems 2.9, 2.10 and 3.10) or check non-triviality of the resulting geometry (see, e.g., the "flatness" criteria, Theorems 2.11 and 2.13). All these theorems are proved in the old-fashioned style of explicit local coordinate parameterizations etc., very much in the spirit of the original Kodaira and Spencer (1958) and Kodaira (1962) approach to the deformation theory of complex structures. Therefore, the proofs are lengthy and dull, and, perhaps fortunately for the reader, the scope of this paper does not allow us to include them. Sometimes, a mathematical proposition enjoys a life in which the ideas and methods underlying its proof find no room (the applications of Kodaira's relative deformation theorem in twistor theory is an example). However it is worth noting that the proofs of theorems discussed in this paper can be of interest on their own — applying the developed machinery to a relative deformation problem, one could finally be left not only with a sentence about the existence of a torsion-free connection on a moduli space but with the explicit expression for this connection itself. The proofs as well as examples of their usage can be found in four papers Merkulov (1992*d*, 1993*b*,

1994) and Merkulov and Pedersen (1994).

2 Geometry of Kodaira moduli spaces

2.1 Existence of Kodaira moduli spaces

Let Y and M be complex manifolds and let $\pi_1 : Y \times M \longrightarrow Y$ and $\pi_2 : Y \times M \longrightarrow M$ be the natural projections. An analytic family of compact submanifolds of the complex manifold Y with the moduli space M is a complex submanifold $F \hookrightarrow Y \times M$ such that the restriction of the projection π_2 on F is a proper regular map (regularity means that the rank of the differential of $\boldsymbol{\nu} \equiv \pi_2 \mid_F : F \longrightarrow M$ is equal at every point to $\dim M$). Thus the family F has the structure of a double fibration

$$Y \overset{\boldsymbol{\mu}}{\leftarrow} F \overset{\boldsymbol{\nu}}{\rightarrow} M$$

where $\boldsymbol{\mu} \equiv \pi_1 \mid_F$. For each $t \in M$ the compact complex submanifold $X_t = \boldsymbol{\mu} \circ \boldsymbol{\nu}^{-1}(t) \hookrightarrow Y$ is said to belong to the family M. If $F \hookrightarrow Y \times M$ is an analytic family of compact submanifolds, then, for any $t \in M$, there is a natural linear map

$$\boldsymbol{k}_t : T_t M \longrightarrow H^0(X_t, N_t),$$

from the tangent space at t to the vector space of global holomorphic sections of the normal bundle $N_t = TY|_{X_t} / TX_t$ to the submanifold $X_t \hookrightarrow Y$.

An analytic family $F \hookrightarrow Y \times M$ of compact submanifolds is called *complete* if the Kodaira map $\boldsymbol{k}_t$ is an isomorphism at each point t in the moduli space M. It is called *maximal* if for any other analytic family $\tilde{F} \hookrightarrow Y \times \tilde{M}$ such that $\boldsymbol{\nu}^{-1}(t) = \tilde{\boldsymbol{\nu}}^{-1}(\tilde{t})$ for some points $t \in M$ and $\tilde{t} \in \tilde{M}$ there is a neighbourhood $\tilde{U} \subset \tilde{M}$ of the point $\tilde{t}$ and a holomorphic map $f : \tilde{U} \longrightarrow M$ such that $\tilde{\boldsymbol{\nu}}^{-1}(\tilde{t}') = \boldsymbol{\nu}^{-1}\left(f(\tilde{t}')\right)$ for every $\tilde{t}' \in \tilde{U}$. Here the equality $\boldsymbol{\nu}^{-1}(t) = \tilde{\boldsymbol{\nu}}^{-1}(\tilde{t})$ means that $\boldsymbol{\mu} \circ \boldsymbol{\nu}^{-1}(t)$ and $\tilde{\boldsymbol{\mu}} \circ \tilde{\boldsymbol{\nu}}^{-1}(\tilde{t})$ are the same submanifolds of Y.

In 1962 Kodaira proved that *if $X \hookrightarrow Y$ is a compact complex submanifold with normal bundle N such that $H^1(X, N) = 0$, then X belongs to the complete analytic family $\{X_t : t \in M\}$ of compact submanifolds X_t of Y. The family is maximal and its moduli space is of complex dimension* $\dim_{\mathbb{C}} H^0(X, N)$.

2.2 Alpha subspaces of moduli spaces

Let $F \hookrightarrow Y \times M$ be a complete family of compact complex submanifolds. For any point $y \in Y' \equiv \cup_{t \in M} X_t$, there is an associated subset $\boldsymbol{\nu} \circ \boldsymbol{\mu}^{-1}(y)$ in M. It is not difficult to show that such a subset is always an analytic subspace of M. Moreover, if the natural evaluation map

$$H^0(X_t, N_t) \otimes \mathcal{O}_{X_t} \longrightarrow N_t$$

is an epimorphism for all $t \in \nu \circ \mu^{-1}(y)$, then the subspace $\nu \circ \mu^{-1}(y) \subset M$ has no singularities, i.e. it is a submanifold. In general, we denote the manifold content of $\nu \circ \mu^{-1}(y)$ by α_y and, following the terminology of the seminal paper by Penrose (1976), call it an *alpha subspace* of M.

2.3 Two natural sheaves on M

Let $F \hookrightarrow Y \times M$ be a complete analytic family of compact submanifolds. The Kodaira theorem asserts the existence of a sheaf isomorphism $k : TM \longrightarrow \nu_*^0(N_F)$, where N_F denotes the normal bundle of F in $Y \times M$. It is easy to see that k^{-1} induces canonically the following morphisms of $\mathcal{O}_M$-modules

$$\begin{aligned} \phi_1 : \quad \nu_*^0(N_F \otimes \odot^2 N_F^*) &\longrightarrow TM \otimes \odot^2 \Omega^1 M, \\ \phi_2 : \quad \nu_*^0(N_F \otimes N_F^*) &\longrightarrow TM \otimes \Omega^1 M, \end{aligned}$$

whose images we denote by $\Lambda_{H^0(X, N \otimes \odot^2 N^*)}$ and $\Lambda_{H^0(X, N \otimes N^*)}$ respectively. Thus, the locally free sheaves $TM \otimes \odot^2 \Omega^1 M$ and $TM \otimes \Omega^1 M$ on any Kodaira moduli space M come equipped canonically with $\mathcal{O}_M$-submodules $\Lambda_{H^0(X, N \otimes \odot^2 N^*)}$ and, respectively, $\Lambda_{H^0(X, N \otimes N^*)}$ which may fail to be locally free in general.

2.4 Vector bundles on Y and affine connections on M

Let $X \hookrightarrow Y$ be a pair of complex manifolds satisfying the conditions of the Kodaira theorem and let M be (a domain in) the associated complete moduli space. After the work by Ward (1977) on instanton solutions of Yang-Mills equations much attention has been paid to holomorphic vector bundles E on Y which are trivial on submanifolds X_t for all $t \in M$. Their geometric role is that they generate vector bundles on M together with linear connections which are integrable on alpha subspaces of M (see, e.g., Manin 1981,1988). In this subsection we find a geometric meaning of holomorphic vector bundles on Y which, when restricted to X_t, $t \in M$, are canonically isomorphic to the normal bundle N_t of the embedding $X_t \hookrightarrow Y$, i.e. $E|_{X_t} = N_t$. It is shown that such bundles often generate torsion-free affine connections on M such that alpha subspaces are totally geodesic. We shall start with the simplest statement in this class, and then discuss what happens when its cohomology restrictions are violated.

Proposition 2.1 *Let $F \hookrightarrow Y \times M$ be a complete family of compact submanifolds. Suppose there is a holomorphic vector bundle E on Y such that $E|_{X_t} = N_t$ for all t in some domain*

$M_0 \subseteq M$. If $H^0(X_t, N_t \otimes \odot^2 N_t^*) = H^1(X_t, N_t \otimes \odot^2 N_t^*) = 0$ for each $t \in M_0$, then M_0 comes equipped canonically with an induced torsion-free affine connection such that, for every $y \in Y'$, the associated alpha subspace $\alpha_y \cap M_0$ is totally geodesic.

For any given relative deformation problem $X \hookrightarrow Y$, it is not difficult to identify a large class of holomorphic vector bundles E on the ambient manifold Y which have the property required by Proposition 2.1. Indeed, take any holomorphic distribution $E \subset TY$ on Y which is transverse to X and has rank equal to $\operatorname{codim} X$. Then, for all t in a sufficiently small neighbourhood M_0 of the point $t_0 \in M$ corresponding to X, the submanifolds $X_t \hookrightarrow Y$ remain transverse to E, and one has a *canonical* isomorphism $E|_{X_t} = N_{X_t}$.

The situation considered in Proposition 2.1 is not something artificial and alien to the twistor theory. The first example is just the original Penrose (1976) non-linear graviton construction. Recall that Penrose established a one-to-one correspondence between "small" 4-dimensional complex Riemannian manifolds M_0 with self-dual Riemann tensor and 3-dimensional complex manifolds Y equipped with the following data:

(i) a submanifold $X \simeq \mathbb{C}P^1$ with normal bundle $N \simeq \mathbb{C}^2 \otimes \mathcal{O}_{\mathbb{C}P^1}(1)$;

(ii) a fibration $\pi : Y \longrightarrow X$;

(iii) a global non-vanishing section of the "twisted" determinant bundle $\det(V_\pi) \otimes \pi^*(\mathcal{O}(2))$, where V_π is the vector bundle of π-vertical vector fields.

Let us look at this data from the point of view of Proposition 2.1. The distribution V_π is clearly transversal to X, and one easily checks that $H^0(X_t, N_t \otimes \odot^2 N_t^*) = H^1(X_t, N_t \otimes \odot^2 N_t^*) = 0$. Therefore, according to Proposition 2.1, the data (i) and (ii) imply that there is a torsion-free affine connection induced on M_0 which satisfies natural integrability conditions. What is the role of datum (iii)? It will be shown in subsection 2.7 that the datum (iii) ensures that this induced connection is precisely the Levi-Civita connection of a holomorphic metric with self-dual Riemann tensor. Therefore, the non-linear graviton construction is one of the manifestations of the phenomenon envisaged by Proposition 2.1.

Another manifestation is provided by relative deformation problems studied in the context of quaternionic geometry. Let Y be a complex $(2n+1)$-dimensional manifold equipped with a holomorphic contact structure, i.e. with a maximally non-degenerate rank $2n$ holomorphic distribution $D \subset TY$. Let X be a rational curve embedded into Y transversely to D and with normal bundle $N = \mathbb{C}^{2n} \otimes \mathcal{O}(1)$. Then Proposition 2.1 says that the $4n$-dimensional Kodaira moduli space M_0 comes equipped with a unique torsion-free affine connection satisfying natural integrability conditions. This is in accordance with well-known results in twistor theory. The case $n = 1$ has been studied by Ward (1980) and Hitchin (1982) who showed that M_0 has an induced complex Riemannian metric satisfying self-dual Einstein equations with non-zero scalar curvature. The case $n \geq 2$ has been investigated by LeBrun (1989), Pedersen and Poon (1989) and Bailey and Eastwood (1991) who proved that the moduli space M_0 comes equipped canonically with a torsion-free connection compatible with the induced (complexified) quaternionic Kähler structure on M. In Merkulov (1994) the sphere S^4 was studied along the lines suggested by the proof of Proposition 2.1 with the conclusion that the induced torsion-free connection is exactly the Levi-Civita connection of the symmetric metric on S^4 thus demonstrating the equivalence of two different twistor methods.

2.5 Families of affine connections on moduli space

What happens when $H^0(X_t, N_t \otimes \odot^2 N_t^*) \neq 0$? In this case, as we noted in subsection 2.3, the locally free $\mathcal{O}_M$-module $TM \otimes \odot^2 \Omega^1 M$ comes equipped with a non-zero $\mathcal{O}_M$-submodule $\Lambda_{H^0(X,N\otimes\odot^2 N^*)}$.

If $\nabla_1 : TM \longrightarrow TM \otimes \Omega^1 M$ and $\nabla_1 : TM \longrightarrow TM \otimes \Omega^1 M$ are two torsion-free affine connections on M, then their difference, $\nabla_1 - \nabla_2$, is a global section of $TM \otimes \odot^2 \Omega^1 M$. We say that ∇_1 and ∇_2 are $\Lambda_{H^0(X,N\otimes\odot^2 N^*)}$*-equivalent*, if

$$\nabla_1 - \nabla_2 \in H^0(M, \Lambda_{H^0(X,N\otimes\odot^2 N^*)}).$$

We define a $\Lambda_{H^0(X,N\otimes\odot^2 N^*)}$*-connection* on M as a collection of ordinary torsion-free affine connections $\{\nabla_i \mid i \in I\}$ on an open covering $\{U_i \mid i \in I\}$ of M which, on overlaps $U_{ij} = U_i \cap U_j$, have their differences in $H^0(U_{ij}, \Lambda_{H^0(X,N\otimes\odot^2 N^*)})$. Locally, a $\Lambda_{H^0(X,N\otimes\odot^2 N^*)}$-connection is the same thing as a $\Lambda_{H^0(X,N\otimes\odot^2 N^*)}$-equivalence class of affine connections, but globally they are different — the obstruction for existence of a $\Lambda_{H^0(X,N\otimes\odot^2 N^*)}$-connection on M lies in $H^1(M, TM \otimes \odot^2 \Omega^1 M/\Lambda)$, while the obstruction for existence of a $\Lambda_{H^0(X,N\otimes\odot^2 N^*)}$-equivalence class of affine connections is an element of $H^1(M, TM \otimes \odot^2 \Omega^1 M)$. A submanifold of M is said to be totally geodesic relative to a $\Lambda_{H^0(X,N\otimes\odot^2 N^*)}$-connection if it is totally geodesic relative to *each* of its local representatives ∇_i.

Proposition 2.2 *Let $F \hookrightarrow Y \times M$ be a complete family of compact submanifolds. Suppose there is a holomorphic vector bundle E on Y such that $E|_{X_t} = N_t$ for all t in some domain $M_0 \subseteq M$. If $H^1(X_t, N_t \otimes \odot^2 N_t^*) = 0$ for each $t \in M_0$, then M_0 comes equipped canonically with an induced $\Lambda_{H^0(X,N\otimes\odot^2 N^*)}$-connection such that, for every $y \in Y'$, the associated alpha subspace $\alpha_y \cap M_0$ is totally geodesic.*

This proposition is still amenable to generalizations. It will be shown in the next subsection that there is no need to ask for the vanishing of the whole cohomology group $H^1(X_t, N_t \otimes \odot^2 N_t^*)$. The statement remains true if only one of its elements vanishes, the one which compares to second order two embeddings, $X_t \hookrightarrow Y$ and $X_t \hookrightarrow E|_{X_t}$, with isomorphic normal bundles. Meanwhile, we consider another equivalence class of affine connections which is often induced on Kodaira moduli spaces.

Let $\boldsymbol{sym}$ denote the natural projection .

$$\boldsymbol{sym} : TM \otimes \Omega^1 M \otimes \Omega^1 M \to TM \otimes \Omega^1 M \odot \Omega^1 M.$$

Associated with a subsheaf $\Lambda_{H^0(X,N\otimes N^*)} \subset TM \otimes \Omega^1 M$ there is a concept of a $\Lambda_{H^0(X,N\otimes N^*)}$-*connection* on M which is, by definition, a collection of ordinary torsion-free affine connections $\{\nabla_i \mid i \in I\}$ on an open covering $\{U_i \mid i \in I\}$ of M which, on overlaps $U_{ij} = U_i \cap U_j$, have their differences in $H^0\left(U_{ij}, \boldsymbol{sym}(\Lambda_{H^0(X,N\otimes N^*)} \otimes \Omega^1 M)\right)$. If $\Lambda_{H^0(X,N\otimes N^*)}$ happens to be the structure sheaf $\mathcal{O}_M$ embedded diagonally into $TM \otimes \Omega^1 M$, then a $\Lambda_{H^0(X,N\otimes N^*)}$-connection is nothing but a torsion-free projective connection on M.

Theorem 2.3 *Let $F \hookrightarrow Y \times M$ be a complete family of compact submanifolds. Suppose there is a holomorphic vector bundle E on Y such that $E|_{X_t} = N_t$ for all t in some domain $M_0 \subseteq M$. If $H^1(X_t, N_t \otimes N_t^*) = 0$ for each $t \in M_0$, then M_0 comes equipped canonically with an induced $\Lambda_{H^0(X,N\otimes N^*)}$-connection such that, for every $y \in Y'$, the associated alpha subspace $\alpha_y \cap M_0$ is totally geodesic.*

Note that the condition $H^1(X_t, N_t \otimes N_t^*) = 0$ only says that N_t is a rigid vector bundle on X_t (see Kodaira and Spencer 1958). In all known twistor constructions this condition is satisfied.

2.6 Climbing up the tower of infinitesimal neighbourhoods

Let Y be a complex manifold and X a complex submanifold of Y. Let $\mathcal{O}_Y$ and $\mathcal{O}_X$ denote the structure sheaves of manifolds Y and X respectively, and J denote the ideal of functions on Y which vanish on X,

$$0 \longrightarrow J \longrightarrow \mathcal{O}_Y \longrightarrow \mathcal{O}_X \longrightarrow 0.$$

According to Griffiths (1966), the mth-order infinitesimal neighborhood of X in Y is the ringed space $X^{(m)} = (X, \mathcal{O}_X^{(m)})$ with the structure sheaf $\mathcal{O}_X^{(m)}$ defined by the following exact sequence

$$0 \longrightarrow J^{m+1} \longrightarrow \mathcal{O}_Y \longrightarrow \mathcal{O}_X^{(m)} \longrightarrow 0.$$

With the $(m+1)$th-order infinitesimal neighborhood there is naturally associated an $\mathcal{O}_X^{(m)}$-module $\mathcal{O}_X^{(m)}(N^*)$ defined by

$$0 \longrightarrow J^{m+2} \longrightarrow J \longrightarrow \mathcal{O}_X^{(m)}(N^*) \longrightarrow 0.$$

It is clear that $\mathcal{O}_X^{(m)}(N^*)$ is an ideal subsheaf of $\mathcal{O}_X^{(m+1)}$,

$$0 \longrightarrow \mathcal{O}_X^{(m)}(N^*) \longrightarrow \mathcal{O}_X^{(m+1)} \longrightarrow \mathcal{O}_X \longrightarrow 0,$$

consisting of all nilpotent elements. Motivated by the fact that in the case $m = 0$ this sheaf coincides precisely with the conormal bundle N^* of the embedding $X \hookrightarrow Y$ we call the sheaf $\mathcal{O}_X^{(m)}(N^*)$ the *mth-order conormal sheaf* of the embedding $X \hookrightarrow Y$, cf. Eastwood and LeBrun (1992). Note that all these considerations still make sense if X is a point in Y. In this case we denote the dual of the $\mathcal{O}_X^{(m)}(N^*)$ by $T_X^{(m)}Y$ and call it the mth-order tangent space at the point $X \in Y$. When $m = 0$, this is the usual tangent space to the manifold Y.

The first order conormal sheaf fits into the exact sequence of $\mathcal{O}_X^{(1)}$-modules

$$0 \longrightarrow N^* \odot N^* \longrightarrow \mathcal{O}_X^{(1)}(N^*) \longrightarrow N^* \longrightarrow 0.$$

Let E be a holomorphic vector bundle on Y. Then its restriction $E_X^{(1)} = E|_{X^{(1)}}$ to the first order infinitesimal neighbourhood of X in Y is an $\mathcal{O}_X^{(1)}$-module which fits into the exact sequence (Manin 1981)

$$0 \longrightarrow N^* \otimes E|_X \stackrel{i}{\longrightarrow} E_X^{(1)} \longrightarrow E|_X \longrightarrow 0.$$

If in addition

$$E|_X = N^*,$$

one can define the quotient $\mathcal{O}_X^{(1)}$-module

$$\mathcal{O}_X^{(1)}(E) \equiv E_X^{(1)}/i(\wedge^2 N^*),$$

which fits into the exact sequence of $\mathcal{O}_X^{(1)}$-modules

$$0 \longrightarrow N^* \odot N^* \longrightarrow \mathcal{O}_X^{(1)}(E) \longrightarrow N^* \longrightarrow 0,$$

of the same type as $\mathcal{O}_X^{(1)}(N^*)$.

Lemma 2.4 *Let $X \hookrightarrow Y$ be a complex submanifold and E a holomorphic bundle on Y such that $E|_X = N^*$. Then the difference $\left[\mathcal{O}_X^{(1)}(E) - \mathcal{O}_X^{(1)}(N^*)\right] \in \mathrm{Ext}_{\mathcal{O}_X^{(1)}}(N^*, N^* \odot N^*)$ is a locally free $\mathcal{O}_X$-module.*

Corollary 2.5 $\Delta(X, E) \equiv \left[\mathcal{O}_X^{(1)}(E) - \mathcal{O}_X^{(1)}(N^*)\right] \in H^1(X, N \otimes N^* \odot N^*)$.

Now we can reformulate Proposition 2.2 in the form

Theorem 2.6 *Let $F \hookrightarrow Y \times M$ be a complete family of compact submanifolds. Suppose there is a holomorphic vector bundle E on Y such that $E|_{X_t} = N_t^*$ for all t in some domain $M_0 \subseteq M$. If $\Delta(X_t, E) = 0$ for each $t \in M_0$, then M_0 comes equipped canonically with an induced $\Lambda_{H^0(X,N\otimes\odot^2N^*)}$-connection such that, for every $y \in Y'$, the associated alpha subspace $\alpha_y \cap M_0$ is totally geodesic.*

Suppose that a Kodaira moduli space M has an induced $\Lambda_{H^0(X,N\otimes\odot^2N^*)}$-connection, which, as we know, is locally a family of affine connections. Is there some way of picking out a particular connection from this family using only the holomorphic embedding data $X_t \hookrightarrow Y$? By construction, the cohomology class $\Delta(X_t, E) \in H^0(X_t, N_t \otimes \odot^2 N_t^*)$ vanishes if and only if the extension

$$0 \longrightarrow N_t^* \odot N_t^* \longrightarrow \left[\mathcal{O}_{X_t}^{(1)}(E) - \mathcal{O}_{X_t}^{(1)}(N_t^*)\right] \longrightarrow N_t^* \longrightarrow 0$$

admits a global splitting. The set of all global splittings of this extension is a principal homogeneous space for the group $H^0(X_t, N_t \otimes \odot^2 N_t^*)$. It can be shown that to choose any particular splitting,

$$\gamma : \left[\mathcal{O}_{X_t}^{(1)}(E) - \mathcal{O}_{X_t}^{(1)}(N_t^*)\right] \longrightarrow N_t^* \odot N_t^* \quad \oplus \quad N_t^*,$$

is the same thing as to choose any particular connection at $t \in M_0$ from the canonically induced $\Lambda_{H^0(X,N\otimes\odot^2N^*)}$-equivalence class.

Theorem 2.7 *Let $F \hookrightarrow Y \times M$ be a complete family of compact submanifolds and $\pi_1 : Y \times M \longrightarrow Y$ a natural projection. Suppose that there is a holomorphic vector bundle E on Y such that $\pi_1^*(E)|_F = N_F^*$ and $\Delta(F, \pi_1^*(E)) = 0$. Then any splitting of the exact sequence of the locally free $\mathcal{O}_F$-modules*

$$0 \longrightarrow N_F^* \odot N_F^* \longrightarrow \left[\mathcal{O}_F^{(1)}(\pi_1^*(E)) - \mathcal{O}_F^{(1)}(N_F^*)\right] \longrightarrow N_F^* \longrightarrow 0$$

induces canonically a torsion-free connection on M such that, for every point $y \in Y'$, the associated alpha subspace $\alpha_y \subset M$ is totally geodesic.

2.7 Fibred complex manifolds

Suppose that the ambient manifold Y has the structure of a holomorphic fibration over its compact submanifold X, i.e. there is a submersive holomorphic map

$$\pi : Y \longrightarrow X.$$

If $H^1(X, N) = 0$, then, by Kodaira's theorem, X belongs to the complete family $F \hookrightarrow Y \times M$ of compact submanifolds. The submanifold X is transverse to the distribution

V_π of π-vertical vector fields on Y, and so is X_t for every t in some neighbourhood M_0 of the point $t_0 \in M$ which corresponds to X. Thus Y has the structure of a holomorphic fibration $Y \longrightarrow X_t$ for $t \in M_0$, and, by Theorems 2.3 and 2.7, one should expect that M_0 comes equipped with induced torsion-free connections. There is however a peculiarity of the present class of relative deformation problems which deserves a further investigation.

One can show that, provided Y has the structure of a holomorphic fibration over its submanifold X_t, one has the equality

$$\mathcal{O}^{(1)}_{X_t}(V^*_\pi) - \mathcal{O}^{(1)}_{X_t}(N^*_t) = \mathcal{O}^{(1)}_{X_t}(N^*_t)$$

in $\mathrm{Ext}_{\mathcal{O}^{(1)}_X}(N^*, N^* \odot N^*)$. By Lemma 2.4 this means that the first order conormal sheaf $\mathcal{O}^{(1)}_{X_t}(N^*)$ is locally free on X_t, and the obstruction to a global splitting of the following exact sequence of locally free $\mathcal{O}_{X_t}$-modules,

$$0 \longrightarrow N^*_t \odot N^*_t \longrightarrow \mathcal{O}^{(1)}_{X_t}(N^*_t) \longrightarrow N^*_t \longrightarrow 0,$$

is precisely the cohomology class $\left[\mathcal{O}^{(1)}_{X_t}(V^*_\pi) - \mathcal{O}^{(1)}_{X_t}(N^*_t)\right] \in H^0(X_t, N_t \otimes \odot^2 N^*_t)$ of Corollary 2.5.

It is worth noting that in the situation under consideration the mth-order conormal sheaf $\mathcal{O}^{(m)}_X(N^*_t)$ is locally free on X for any $m \in \mathbb{N}^+$, and there is a canonical linear map

$$k^{(m)} : T^{(m)}_t M \longrightarrow H^0\left(X_t, \mathcal{O}^{(m-1)}_X(N_t)\right)$$

from the mth-order tangent space at $t \in M_0$ to the vector space of global sections of the $(m-1)$th-order normal sheaf $\mathcal{O}^{(m)}_X(N_t) = \left(\mathcal{O}^{(m)}_X(N^*)\right)^*$. In general this map is neither injective nor surjective. The map $k^{(0)}$ is the original Kodaira correspondence.

Summarizing the above discussion, we see that, by Theorem 2.7, any splitting of the exact sequence of the locally free $\mathcal{O}_F$-modules

$$0 \longrightarrow N^*_F \odot N^*_F \longrightarrow \mathcal{O}^{(1)}_F(N^*_F) \longrightarrow N^*_F \longrightarrow 0$$

induces canonically a torsion-free connection on M_0 such that, for every point $y \in Y'$, the associated alpha subspace $\alpha_y \cap M_0$ is totally geodesic. Let us call this an *induced connection*. Holonomy groups of induced connections can be estimated with surprising ease.

Theorem 2.8 *Let ∇ be an induced connection on the moduli space M_0. Then the curvature tensor R^∇ is a global section of the sheaf $\Lambda_{H^0(X, N\otimes N^*)} \otimes \Omega^2 M \cap K$, where K is the kernel of the natural anti-symmetrisation map*

$$0 \longrightarrow K \longrightarrow TM \otimes \Omega^1 M \otimes \Omega^2 M \longrightarrow TM \otimes \Omega^3 M \longrightarrow 0.$$

This fact combined with the theorem by Ambrose and Singer (1953) can be used to prove the following theorems.

Theorem 2.9 *Let ∇ be an induced connection on the moduli space M_0. If the function*

$$f : t \longrightarrow \dim H^0(X_t, N_t \otimes N_t^*)$$

is constant on M_0, then the holonomy algebra of ∇ is a subalgebra of the finite dimensional Lie algebra $H^0(X, N \otimes N^*)$.

Theorem 2.10 *Let ∇ be an induced connection on the moduli space M_0. Suppose that the cohomology groups $H^0(X_t, N_t^*)$ vanish, the function*

$$f : t \longrightarrow \dim H^0(X_t, N_t \otimes N_t^*)$$

is constant on M_0, and there is a holomorphic line bundle L on X such that the bundle $\pi^*(L) \otimes \det V_\pi^*$ admits a nowhere vanishing holomorphic section. Then the holonomy algebra of ∇ is a subalgebra of $H^0(X, N \otimes_0 N^*)$, where $\otimes_0$ denotes trace-free tensor product.

Examples.

1. Let $X = \mathbb{CP}^1$ be the projective line embedded into a 3-dimensional complex manifold Y with normal bundle $N \simeq \mathbb{C}^2 \otimes \mathcal{O}(1)$. If Y has the structure of a holomorphic fibration over X, then Proposition 2.1 states that there is an induced connection ∇ on the moduli space M_0. By Theorem 2.9, the holonomy algebra of ∇ is contained in $H^0(X, N \otimes N^*) = M_{2,2}(\mathbb{C}) \subset co(4, \mathbb{C})$, where $M_{2,2}(\mathbb{C})$ is the space of 2×2 complex matrices and $co(4, \mathbb{C}) = sl(2, \mathbb{C}) + sl(2, \mathbb{C}) + \mathbb{C}$ the complexified conformal algebra. Since ∇ is torsion-free, this fact implies that ∇ is a complex Weyl connection on the 4-dimensional complex conformal manifold M_0 which has the anti-self-dual parts of the Weyl tensor and the antisymmetrized Ricci tensor vanishing and satisfies the Einstein-Weyl equations. In fact any such connection arises locally in this way (Penrose 1976, Boyer 1988, Pedersen and Swann 1993).

2. Let the pair $(X = \mathbb{CP}^1, Y)$ be the same as in Example 1 and assume that the bundle $\det V_\pi^* \otimes \pi^*(\mathcal{O}(2))$ admits a nowhere vanishing holomorphic section. By Theorem 2.10, the holonomy group of the induced connection ∇ on the 4-dimensional parameter space M_0 is a subgroup of $SL(2, \mathbb{C}) \subset SO(4, \mathbb{C}) = SL(2, \mathbb{C}) \times_{\mathbb{Z}_2} SL(2, \mathbb{C})$. This means that ∇ is the Levi-Civita connection of a complex Riemannian metric on M_0 which is Ricci-flat and has the anti-self-dual part of the Weyl tensor vanishing. Again any such connection arises locally in this way (Penrose 1976).

3. Let $X = \mathbb{CP}^1$ be the projective line embedded into a $(2k + 1)$-dimensional complex manifold Y with normal bundle $N \simeq \mathbb{C}^{2k} \otimes \mathcal{O}(1)$, $k \geq 2$. The Kodaira moduli space

M is then a $4k$-dimensional complex manifold possessing a complexified almost quaternionic structure. If Y has the structure of a holomorphic fibration over X, then, by Proposition 2.1 and Theorem 2.9, there is an induced connection ∇ on $M_0 \subset M$ with holonomy in $GL(2k, \mathbb{C})$ which implies that ∇ is a complexified Obata connection.

The above analysis of the first order normal bundle $\mathcal{O}^{(1)}_{X_t}(N_t)$ results in the conclusion that any splitting of the associated exact sequence

$$0 \longrightarrow N_t \longrightarrow \mathcal{O}^{(1)}_{X_t}(N_t) \xrightarrow{\alpha_t} \odot^2 N_t \longrightarrow 0,$$

i.e. a morphism of $\mathcal{O}_{X_t}$-modules

$$\gamma_t : \odot^2 N_t \longrightarrow \mathcal{O}^{(1)}_{X_t}(N_t)$$

such that $\alpha_t \circ \gamma_t = id$, induces a torsion-free affine connection ∇ at $t \in M_0$. Let us now take one step upstairs and consider the second order normal bundle $\mathcal{O}^{(2)}_{X_t}(N_t)$. It fits into an exact sequence

$$0 \longrightarrow \mathcal{O}^{(1)}_{X_t}(N_t) \xrightarrow{\beta_t} \mathcal{O}^{(2)}_{X_t}(N_t) \longrightarrow \odot^3 N_t \longrightarrow 0$$

which implies

$$0 \longrightarrow N_t \longrightarrow \mathcal{O}^{(2)}_{X_t}(N_t)/\beta_t \circ \gamma_t(\odot^2 N_t) \longrightarrow \odot^3 N_t \longrightarrow 0.$$

Thus with any splitting γ_t one can associate canonically a cohomology class $\chi_t \in H^1(X_t, N_t \otimes \odot^3 N_t^*)$ which is by definition the obstruction to a global splitting of the latter exact sequence. The following "almost flatness" criterion reveals its geometric meaning.

Theorem 2.11 *Let ∇ be an induced connection on the moduli space M_0. If the cohomology classes $\chi_t \in H^1(X_t, N_t \otimes \odot^3 N_t^*)$ vanish for all $t \in M_0$, then the curvature tensor R^∇ is a global section of the sheaf $\phi\left(\Lambda_{H^0(X, N \otimes \odot^2 N^*)} \otimes \Omega^1 M\right)$, where ϕ is the composition*

$$\phi : TM \otimes \odot^2 \Omega^1 M \otimes \Omega^1 M \longrightarrow TM \otimes \Omega^1 M \otimes \Omega^1 M \otimes \Omega^1 M \longrightarrow TM \otimes \Omega^1 M \otimes \Omega^2 M$$

of the natural embedding with the antisymmetrization over last two indices.

Therefore, to generate non-trivial differential-geometric structures on Kodaira moduli spaces one may use relative deformation problems $X \hookrightarrow Y$ with normal bundle such that $H^1(X, N \otimes \odot^2 N^*) = 0$, but $H^1(X, N \otimes \odot^3 N^*) \neq 0$.

2.8 Projective structures on moduli spaces of compact hypersurfaces

An affine connection on M determines a family of geodesic curves. Two connections ∇ and $\hat{\nabla}$ are called *projectively equivalent* if they have the same geodesics considered as unparameterized paths. In a local coordinate chart this condition reads (Hitchin 1982)

$$\hat{\Gamma}^{\alpha}_{(\beta\gamma)} = \Gamma^{\alpha}_{(\beta\gamma)} + b_{\beta}\,\delta^{\alpha}_{\gamma} + b_{\gamma}\,\delta^{\alpha}_{\beta}$$

for some 1-form $b = b_{\alpha}dt^{\alpha}$. The following result by Merkulov and Pedersen (1994) shows that Kodaira moduli spaces of compact hypersurfaces very often come equipped with such a projective structure.

Theorem 2.12 *Let $F \hookrightarrow Y \times M$ be a complete family of compact hypersurfaces such that $H^1(X_t, \mathcal{O}_{X_t}) = 0$ for all $t \in M$. Then the moduli space M comes equipped canonically with a torsion-free projective connection such that, for every point $y \in Y'$, the associated alpha subspace $\alpha_y \subset M$ is totally geodesic.*

The following result provides a clear criterion for local flatness of such an induced projective connection.

Theorem 2.13 *The projective Weyl tensor of an induced projective connection on M is represented by a global section of $\nu^1_*(N^*_F) \otimes \Omega^1 M$.*

Thus, to generate non-trivial torsion-free projective connections one can use codimension 1 relative deformation problems $X \hookrightarrow Y$ such that $H^1(X, N) = H^1(X, \mathcal{O}_X) = 0$, but $H^1(X, N^*) \neq 0$.

2.9 Almost-Grassmanian structures on Kodaira moduli spaces

Let $F \hookrightarrow Y \times M$ be a complete family of compact submanifolds. Suppose that there is a holomorphic vector bundle L_F on F such that $N_F \simeq \mathbb{C}^n \otimes L_F$, $n \in \mathbb{N}$ (where the symbol $\simeq$ stands for non-canonical isomorphism, i.e. N_F and $\mathbb{C}^n \otimes L_F$ are supposed only to represent the same comology class in $H^1(F, GL(n, \mathcal{O}_F))$). Suppose also that X is rigid. Then the family $F \hookrightarrow Y \times M$ is locally trivial, i.e. each point $t \in M$ has a neighbourhood U such that $\nu^{-1}(U) \simeq U \times X$. If in addition to the rigidity of X the cohomology group $H^1(X, \mathcal{O}_X)$ vanishes, then all restrictions of the holomorphic line bundle L_F to the submanifolds $X_t = \mu \circ \nu^{-1}(t)$, $t \in M$, are isomorphic to each other. Under these conditions the direct image functor ν_* applied to the normal bundle N_F factorises the tangent bundle TM into the tensor product,

$$TM = \nu^0_*(N_F) = \nu^0_*(N_F \otimes L^*_F) \otimes_{\mathcal{O}_M} \nu^0_*(L_F)) = S \otimes_{\mathcal{O}_M} \tilde{S},$$

of two holomorphic vector bundles $S = \nu^0_*(N_F \otimes L^*_F)$ and $\tilde{S} = \nu^0_*(L_F)$ with rank$S = n$ and rank$\tilde{S} = \dim H^0\left(\nu^{-1}(t_0), L_F|_{\nu^{-1}(t_0)}\right)$. Therefore the moduli space M comes equipped canonically with an almost-Grassmanian structure (Akivis 1980, Manin 1988, Bailey and Eastwood 1991). Note that the sheaf $\Lambda_{H^0(X,N\otimes N^*)}$ is isomorphic to End S.

Theorem 2.14 *Let $X \hookrightarrow Y$ be a compact complex rigid submanifold with normal bundle $N \simeq \mathbb{C}^n \otimes L$ for some holomorphic line bundle L on X and natural number $n \geq 2$. If*

$$H^1(X, L) = H^1(X, \mathcal{O}_X) = H^1(X, L^*) = H^0(X, L^*) = 0,$$

the Kodaira moduli space M of relative deformations of X comes equipped canonically with

(i) an almost-Grassmanian structure

$$TM = S \otimes \tilde{S}$$

with rank $S = n$ *and* rank $\tilde{S} = \dim H^0(X, L)$, *and*

(ii) a $\Lambda_{H^0(X,N\otimes N^*)}$-connection such that, for every point $y \in Y'$, the associated alpha subspace $\alpha_y \subset M$ is totally geodesic.

3 Existence and geometry of Legendre moduli spaces

3.1 Complex contact manifolds

Let Y be a complex $(2n+1)$-dimensional manifold. A complex contact structure on Y is a rank $2n$ holomorphic subbundle $D \subset TY$ of the holomorphic tangent bundle to Y such that the Frobenius form

$$\begin{aligned} \Phi : D \times D &\longrightarrow TY/D \\ (v, w) &\longrightarrow [v, w] \bmod D \end{aligned}$$

is non-degenerate. Define the line bundle L by the exact sequence

$$0 \longrightarrow D \longrightarrow TY \longrightarrow L \longrightarrow 0$$

and let $\theta : TY \longrightarrow L$ be a canonical projection. Then one can define equivalently a complex contact structure on Y as a "twisted" 1-form $\theta \in H^0(Y, L \otimes \Omega^1 M)$ satisfying the condition

$$\theta \wedge (d\theta)^n \neq 0.$$

A complex n-dimensional submanifold X of a complex contact manifold Y is called a Legendre submanifold if $TX \subset D$. LeBrun (1991) showed that normal bundle of *any* Legendre submanifold $X \hookrightarrow Y$ is given by

$$N = J^1 L_X,$$

where L_X is the restriction to X of the contact line bundle L. Therefore, the normal bundle of a Legendre submanifold always fits into the exact sequence

$$0 \longrightarrow \Omega^1 X \otimes L_X \longrightarrow N \xrightarrow{pr} L_X \longrightarrow 0.$$

3.2 Existence theorem

Let Y be a complex contact manifold. An analytic family $F \hookrightarrow Y \times M$ of compact submanifolds of Y is called an analytic family of Legendre submanifolds if, for any point $t \in M$, the corresponding subset $X_t = \mu \circ \nu^{-1}(t) \hookrightarrow Y$ is a Legendre submanifold. According to Kodaira (1962), there is a natural linear map

$$k_t : T_t M \longrightarrow H^0(X_t, N_t).$$

We say that the analytic family $F \hookrightarrow Y \times M$ of compact Legendre submanifolds is *complete at a point* $t \in M$ if the composition

$$s_t : T_t M \xrightarrow{k_t} H^0(X_t, N_t) \xrightarrow{pr} H^0(X_t, L_{X_t})$$

provides an isomorphism between the tangent space to M at the point t and the vector space of global sections of the contact line bundle over X_t. The analytic family $F \hookrightarrow Y \times M$ is called *complete* if it is complete at each point of the moduli space.

Lemma 3.1 *If an analytic family $F \hookrightarrow Y \times M$ of compact Legendre submanifolds is complete at a point $t_0 \in M$, then there is an open neighbourhood $U \subseteq M$ of the point t_0 such that the family $F \hookrightarrow Y \times M$ is complete at all points $t \in U$.*

We say that an analytic family $F \hookrightarrow Y \times M$ of compact Legendre submanifolds is *maximal at a point* $t_0 \in M$, if, for any other analytic family $\tilde{F} \hookrightarrow Y \times \tilde{M}$ of compact Legendre submanifolds such that $\nu^{-1}(t_0) = \tilde{\nu}^{-1}(\tilde{t}_0)$ for a point $\tilde{t}_0 \in \tilde{M}$, there exists a neighbourhood $\tilde{U} \subset \tilde{M}$ of $\tilde{t}$ and a holomorphic map $f : \tilde{U} \longrightarrow M$ such that $f(\tilde{t}_0) = t_0$ and $\tilde{\mu} \circ \tilde{\nu}^{-1}(\tilde{t}') = \mu \circ \nu^{-1}\left(f(\tilde{t}')\right)$ for each $\tilde{t}' \in \tilde{U}$. The family $F \hookrightarrow Y \times M$ is called *maximal* if it is maximal at each point t in the moduli space M.

Lemma 3.2 *If an analytic family of compact Legendre submanifolds* $F \hookrightarrow Y \times M$ *is complete at a point* $t_0 \in M$, *then it is maximal at the point* t_0.

The map $s_t : T_tM \longrightarrow H^0(X_t, L_{X_t})$ studied by the previous two Lemmas will also play a fundamental role in our study of the rich geometric structures induced canonically on moduli spaces of complete analytic families of compact Legendre submanifolds described by the following theorem.

Theorem 3.3 *Let* X *be a compact complex Legendre submanifold of a complex contact manifold* (Y, L). *If* $H^1(X, L_X) = 0$, *then there exists a complete analytic family* $F \hookrightarrow Y \times M$ *of compact Legendre submanifolds containing* X. *This family is maximal and* $\dim M = \dim_{\mathbb{C}} H^0(X, L_X)$.

This theorem is proved by working in local coordinates adapted to the contact structure and expanding the defining functions of nearby compact Legendre submanifolds in terms of local coordinates on the moduli space M. This is much in the spirit of the original proof of Kodaira's theorem of the existence and completeness of compact submanifolds of complex manifolds. The only (though very essential) novelty in our case is that the infinite sequence of obstructions to agreements on overlaps of formal power series is situated now in $H^1(X, L_X)$ rather than in $H^1(X, N)$.

Examples.

1. Let X be the projective line $\mathbb{CP}^1$ and $\mathcal{O}_X$ its structure sheaf. Then the total space Y of the jet bundle $J^1\mathcal{O}_X$ has a canonical complex contact structure (Arnold 1978). The line X identified with the zero section of $J^1\mathcal{O}_X$ is a compact Legendre submanifold. The corresponding normal bundle fits into the short exact sequence

$$0 \longrightarrow \mathcal{O}_X(-2) \longrightarrow N \longrightarrow \mathcal{O}_X \longrightarrow 0,$$

which implies the long exact sequence

$$0 \longrightarrow H^0(X, N) \xrightarrow{pr} H^0(X, \mathcal{O}_X) \longrightarrow H^1(X, \mathcal{O}_X(-2)) \longrightarrow H^1(X, N) \longrightarrow 0.$$

Since the monomorphism pr is onto, $H^1(X, N) = H^1(X, \mathcal{O}_X(-2)) \neq 0$ and the Kodaira theorem is not suitable here. However Theorem 3.3 does apply, and there is a 1-dimensional moduli space M of compact Legendre submanifolds of Y. It is clear that M is isomorphic to the complex line $\mathbb{C}$.

2. Let (Y, L) be a 3-dimensional complex contact manifold and $X = \mathbb{CP}^1$ its Legendre submanifold such that $L_X = \mathcal{O}_X(1)$. Then the normal bundle of $X \hookrightarrow Y$ fits into the exact sequence

$$0 \longrightarrow \mathcal{O}_X(-1) \longrightarrow N \longrightarrow \mathcal{O}_X(1) \longrightarrow 0.$$

Since the only non-trivial extension of $\mathcal{O}_X(1)$ by $\mathcal{O}_X(-1)$ is isomorphic to $\mathcal{O}_X \oplus \mathcal{O}_X$, the normal bundle N is globally trivial. Both Kodaira's theorem and theorem 3.3 predict one and the same 2-dimensional moduli space M.

3. Let (Y, L) be a 3-dimensional complex contact manifold and $X = \mathbb{CP}^1$ its Legendre submanifold such that $L_X = \mathcal{O}_X(2)$. Then

$$0 \longrightarrow \mathcal{O}_X \longrightarrow N \longrightarrow \mathcal{O}_X(2) \longrightarrow 0,$$

and $N \simeq \mathbb{C}^2 \otimes \mathcal{O}_X(1)$. The Kodaira theorem implies that the moduli space of all possible deformations of X in Y is a 4-dimensional complex manifold which, according to Penrose (1976), comes equipped canonically with a half-flat conformal structure. Theorem 3.3 says that there is a 3-dimensional moduli space M of deformations of X in the class of Legendre submanifolds. LeBrun (1983) showed that such a moduli space M always comes equipped with a conformal structure, and any 3-dimensional conformal structure can be described locally in this way.

3.3 Interconnections between Kodaira and Legendre moduli spaces

If $X \hookrightarrow Y$ is a complex submanifold, there is an exact sequence of vector bundles

$$0 \longrightarrow N^* \longrightarrow \Omega^1 Y\Big|_X \longrightarrow \Omega^1 X \longrightarrow 0,$$

which induces a natural embedding $\mathbb{P}(N^*) \hookrightarrow \mathbb{P}(\Omega^1 Y)$ of total spaces of the associated projectivised bundles. The manifold $\hat{Y} = \mathbb{P}(\Omega^1 Y)$ carries a natural contact structure such that the constructed embedding $\hat{X} = \mathbb{P}(N^*) \hookrightarrow \hat{Y}$ is a Legendre one (Arnold 1978)[1]. Indeed, the contact distribution $D \subset T\hat{Y}$ at each point $\hat{y} \in \hat{Y}$ consists of those tangent vectors $V_{\hat{y}} \in T_{\hat{y}}\hat{Y}$ which satisfy the equation $< \hat{y}, \tau_*(V_{\hat{y}}) >= 0$, where $\tau : \hat{Y} \longrightarrow Y$ is a natural projection and the angular brackets denote the pairing of 1-forms and vectors at $\tau(\hat{y}) \in Y$. Since the submanifold $\hat{X} \subset \hat{Y}$ consists precisely of those projective classes of 1-forms in $\Omega^1 Y|_X$ which vanish when restricted on TX, we conclude that $T\hat{X} \subset D|_{\hat{X}}$.

Proposition 3.4 *If* $\{X_t \hookrightarrow Y \mid t \in M\}$ *is a complete analytic family of compact submanifolds, then the associated family* $\left\{\hat{X}_t = \mathbb{P}(N_t^*) \hookrightarrow \mathbb{P}(\Omega^1 Y) \mid t \in M\right\}$ *of projectivized conormal bundles is a complete analytic family of compact Legendre submanifolds.*

[1] The author is grateful to Claude LeBrun for drawing his attention to Arnold's construction.

This proposition implies that the class of Kodaira moduli spaces is a proper subclass of the class of Legendre moduli spaces. Therefore, when we study Legendre moduli spaces, we are in fact in a more general situation that the one invented by Kodaira.

3.4 Alpha subspaces of Legendre moduli spaces

Let $F \hookrightarrow Y \times M$ be a complete family of compact Legendre submanifolds and $\pi_1 : Y \times M \longrightarrow Y$ the natural projection on the first factor. For any point $y \in Y' \equiv \cup_{t \in M} X_t$, there is an associated subset $\boldsymbol{\nu} \circ \boldsymbol{\mu}^{-1}(y)$ in M. It is not difficult to prove that such a subset is always an analytic subspace of M. Moreover, if the natural evaluation map (cf. subsection 2.2)

$$H^0(X_t, L_{X_t}) \otimes \mathcal{O}_{X_t} \longrightarrow L_{X_t}$$

is an epimorphism for all $t \in \boldsymbol{\nu} \circ \boldsymbol{\mu}^{-1}(y)$, then the subspace $\boldsymbol{\nu} \circ \boldsymbol{\mu}^{-1}(y) \subset M$ has no singularities, i.e. it is a submanifold. In general, we denote the manifold content of $\boldsymbol{\nu} \circ \boldsymbol{\mu}^{-1}(y)$ by α_y and call it an *alpha subspace* of M.

3.5 Induced connections on Legendre moduli spaces

Let $F \hookrightarrow Y \times M$ be a complete analytic family of compact Legendre submanifolds. If $L_F = \pi_1^*(L)|_F$, then, for any point $t \in M$, we have an isomorphism

$$L_F|_{\boldsymbol{\nu}^{-1}(t)} \simeq L_{X_t},$$

and, by Theorem 3.3, there is an isomorphism of sheaves

$$\boldsymbol{s} : TM \xrightarrow{\simeq} \boldsymbol{\nu}_*^0(L_F)$$

on M. If N_F is the normal bundle of $F \hookrightarrow Y \times M$, then, for any point $t \in M$, we have

$$N_F|_{\boldsymbol{\nu}^{-1}(t)} \simeq N_t,$$

where N_t is the normal bundle of the submanifold X_t. The Kodaira map

$$\boldsymbol{k} : TM \longrightarrow \boldsymbol{\nu}_*^0(N_F)$$

is not isomorphic in general.

There exists a natural morphism of sheaves of $\mathcal{O}_M$-modules

$$\phi : \boldsymbol{\nu}_*^0\left(L_F \otimes \odot^2 N_F^*\right) \longrightarrow TM \otimes \odot^2 \Omega^1 M.$$

Indeed, if χ is a germ of $\nu_*^0(L_F \otimes \odot^2 N_F^*)$ at a point $t \in M$, then the action

$$\phi(\chi) : \odot^2 TM \longrightarrow TM$$

of the corresponding germ $\phi(\chi) \in TM \otimes \odot^2 \Omega^1 M$ on $\odot^2 TM$ may be described as follows. First we note that there is a natural morphism

$$\begin{aligned} \lambda : \nu_*^0 N_F \odot_{\nu^{-1}(\mathcal{O}_M)} \nu_*^0 N_F &\longrightarrow \nu_*^0 (N_F \odot_{\mathcal{O}_F} N_F), \\ \sigma_1 \odot_{\nu^{-1}(\mathcal{O}_M)} \sigma_2 &\longrightarrow \sigma_1 \odot_{\mathcal{O}_F} \sigma_2, \end{aligned}$$

defined by the pointwise symmetric tensor product of germs of global sections of N_F over the germ of the submanifold $\nu^{-1}(t) \subset F$, $t \in M$. Combining this map with the Kodaira map $\boldsymbol{k}$, we obtain a natural composition

$$TM \odot TM \xrightarrow{\boldsymbol{k} \odot \boldsymbol{k}} \nu_*^0 N_F \odot_{\nu^{-1}(\mathcal{O}_M)} \nu_*^0 N_F \xrightarrow{\lambda} \nu_*^0 (N_F \odot N_F) \xrightarrow{\chi} \nu_*^0 (L_F) \xrightarrow{s^{-1}} TM$$

which explains the action of $\phi(\chi)$ on $TM \odot TM$. The image $\phi(\nu_*^0(L_F \otimes N_F^* \odot N_F^*)) \subseteq TM \otimes \Omega^1 M \odot \Omega^1 M$ is a subsheaf of $\mathcal{O}_M$-modules which we denote by $\Lambda_{H^0(X, L_X \otimes \odot^2 N^*)}$. By analogy with subsection 2.5, one defines a $\Lambda_{H^0(X, L_X \otimes \odot^2 N^*)}$-connection on M as a collection of ordinary torsion-free affine connections $\{\nabla_{\mathrm{i}} \mid \mathrm{i} \in I\}$ on an open covering $\{U_{\mathrm{i}} \mid \mathrm{i} \in I\}$ of M which, on overlaps $U_{\mathrm{ij}} = U_{\mathrm{i}} \cap U_{\mathrm{j}}$, have their differences in $H^0(U_{\mathrm{ij}}, \Lambda_{H^0(X, L_X \otimes \odot^2 N^*)})$.

Proposition 3.5 *Let $F \hookrightarrow Y \times M$ be a complete family of compact Legendre submanifolds. If $H^1(X_t, L_{X_t} \otimes \odot^2 N_t^*) = 0$ for each $t \in M$, then M comes equipped canonically with an induced $\Lambda_{H^0(X, L_X \otimes \odot^2 N^*)}$-connection such that, for every $y \in Y'$, the associated alpha subspace α_y is totally geodesic.*

Example 1. Let (Y, L) be a $(2n+1)$-dimensional complex contact manifold and $X \hookrightarrow Y$ a Legendre submanifold which is isomorphic to a non-degenerate quadratic hypersurface (quadric) in $\mathbb{CP}^{n+1}$. Suppose also that $L_X \simeq \mathcal{O}_{\mathbb{CP}^{n+1}}(1)|_{i(X)}$, where $i : X \hookrightarrow \mathbb{CP}^{n+1}$ is the standard embedding. According to LeBrun (1983), the $(n+2)$-dimensional moduli space M of the associated complete family of Legendre submanifolds comes equipped canonically with a holomorphic conformal structure $[g_{\alpha\beta}]$. On the other hand, $H^1(X_t, L_{X_t} \otimes \odot^2 N_t^*) = 0$, and so, by Proposition 3.5, the moduli space M has a distinguished torsion-free $\Lambda_{H^0(X, L_X \otimes \odot^2 N^*)}$-connection. How are these canonical structures related to each other?

The conormal bundle of $X \hookrightarrow Y$ fits into the exact sequence (LeBrun 1983, Manin 1988)

$$0 \longrightarrow N^* \longrightarrow \Omega^1 M \otimes_{\mathbb{C}} \mathcal{O}_X \longrightarrow L_X[-2] \longrightarrow 0,$$

where numbers in square brackets are conformal weights and $\Omega^1 M$ denotes the cotangent vector space to the point in the moduli space M corresponding to the submanifold X.

Then

$$0 \longrightarrow L_X \otimes \odot^2 N^* \longrightarrow L_X \otimes_{\mathbb{C}} \Omega^1 M \odot_{\mathbb{C}} \Omega^1 M \longrightarrow \Omega^1 M \otimes_{\mathbb{C}} L_X^{\otimes 2}[-2] \longrightarrow 0,$$

and hence

$$\begin{array}{rcl} H^0(X, L_X \otimes \odot^2 N^*) = \ker : \Omega^1 M \odot \Omega^1 M \otimes TM & \longrightarrow & \Omega^1 M \otimes \Omega^1 M \odot_0 \Omega^1 M[2] \\ f^\gamma_{\alpha\beta} & \longrightarrow & f^\delta_{\alpha\beta}\, g_{\gamma\delta} + f^\delta_{\alpha\gamma}\, g_{\beta\delta} - \dfrac{2\, g_{\beta\gamma}}{n+2} f^\delta_{\alpha\delta} \end{array}$$

where $\odot_0$ denotes the trace-free symmetric tensor product. Thus two local affine connections, $\Gamma^\alpha_{\beta\gamma}$ and $\tilde{\Gamma}^\alpha_{\beta\gamma}$, on M belong to one and the same $\Lambda_{H^0(X, L_X \otimes \odot^2 N^*)}$-equivalence class if and only if

$$\tilde{\Gamma}^\gamma_{\alpha\beta} - \Gamma^\gamma_{\alpha\beta} = p_\alpha\, \delta^\gamma_\beta + p_\beta\, \delta^\gamma_\alpha - g_{\alpha\beta}\, p^\gamma$$

for some 1-form $p = p_\alpha\, dt^\alpha$. Since null geodesics of $[g_{\alpha\beta}]$ are totally geodesic relative to the canonical $\Lambda_{H^0(X, L_X \otimes \odot^2 N^*)}$-connection, we conclude that this $\Lambda_{H^0(X, L_X \otimes \odot^2 N^*)}$-connection is just what one could expect — locally it is exactly the set of all torsion-free affine connections on M which preserve the conformal structure (such connections are called Weyl connections).

Theorem 3.5 implies

Corollary 3.6 *Let $F \hookrightarrow Y \times M$ be a complete family of compact Legendre submanifolds. If*

$$H^0(X_t, L_{X_t} \otimes \odot^2 N_t^*) = H^1(X_t, L_{X_t} \otimes \odot^2 N_t^*) = 0$$

for each $t \in M$, then M comes equipped canonically with an induced torsion-free connection such that, for every $y \in Y'$, the associated alpha subspace α_y is totally geodesic.

Example 2. Let (Y, L) be a 3-dimensional complex contact manifold and $X = \mathbb{CP}^1$ its Legendre submanifold with $L_X = \mathcal{O}(k)$, $k \in \mathbb{N}^+$. Then the normal bundle N is isomorphic to $\mathbb{C}^2 \otimes \mathcal{O}_X(k-1)$. It is easy to check that both cohomology groups $H^0(X, L_X \otimes \odot^2 N^*)$ and $H^1(X, L_X \otimes \odot^2 N^*)$ vanish if and only if $k = 3$. By Theorem 3.3, the Legendre moduli space generated by $X \hookrightarrow Y$ is a 4-dimensional complex manifold M. By Corollary 3.6, M comes equipped with a distinguished torsion-free affine connection. This connection was introduced and studied by Bryant (1991).

3.6 Infinitesimal neighbourhoods of Legendre submanifolds

Suppose now that (Y, L) is a complex contact manifold and $X \hookrightarrow Y$ a complex Legendre submanifold. We are interested in this paper in a certain $\mathcal{O}_X^{(1)}$-submodule of the first order conormal sheaf $\mathcal{O}_X^{(1)}(N^*)$ which encodes some information about the Legendre nature of the embedding $X \hookrightarrow Y$. As we know, $\mathcal{O}_X^{(1)}(N^*)$ fits into the exact sequence of $\mathcal{O}_X^{(1)}$-modules

$$0 \longrightarrow N^* \odot N^* \longrightarrow \mathcal{O}_X^{(1)}(N^*) \xrightarrow{g_1} N^* \longrightarrow 0.$$

Since

$$0 \longrightarrow L_X^* \longrightarrow N^* \xrightarrow{g_2} TX \otimes L_X^* \longrightarrow 0,$$

the kernel, $\mathcal{L}_X^{(1)}(N^*)$, of the composition $g_2 \circ g_1$ fits into the exact sequence of $\mathcal{O}_X^{(1)}$-modules

$$0 \longrightarrow \mathcal{L}_X^{(1)}(N^*) \longrightarrow \mathcal{O}_X^{(1)}(N^*) \xrightarrow{g_2 \circ g_1} TX \otimes L_X^* \longrightarrow 0,$$

and has the extension structure

$$0 \longrightarrow N^* \odot N^* \longrightarrow \mathcal{L}_X^{(1)}(N^*) \longrightarrow L_X^* \longrightarrow 0$$

we are interested in.

If L is the contact line bundle on Y, then the restriction of its dual to the first order infinitesimal neighbourhood of X in Y,

$$L_X^{*(1)} \equiv L^*|_{X^{(1)}},$$

fits into the exact sequence of $\mathcal{O}_X^{(1)}$-modules

$$0 \longrightarrow N^* \otimes L_X^* \xrightarrow{i_1} L_X^{*(1)} \longrightarrow L_X^* \longrightarrow 0.$$

Let **Id** be the canonical global section of $N \otimes N^* = \mathrm{End}(N)$ whose value at each point of X is the multiplicative identity in the corresponding stalk of endomorphism groups. Then the global section

$$\mathbf{1} \equiv pr \otimes id(\mathbf{Id}) \in H^0(X, L_X \otimes N^*),$$

where $pr : N \longrightarrow L_X$ is the canonical projection, defines a canonical monomorphism of $\mathcal{O}_X^{(1)}$-modules

$$\begin{array}{rccccc} i_2 : & N^* \otimes L_X^* & \longrightarrow & (N^* \otimes L_X^*) \otimes (L_X \otimes N^*) & \cong & N^* \otimes N^* \\ & f & \longrightarrow & f \otimes \mathbf{1} & & \end{array}$$

Consider the composition

$$i_3 : N^* \otimes L_X^* \xrightarrow{j} N^* \otimes L_X^* \oplus N^* \otimes L_X^* \xrightarrow{i_1 \oplus i_2} L_X^{*(1)} \oplus N^* \otimes N^*,$$

where

$$\begin{array}{rcl} j : N^* \otimes L_X^* & \longrightarrow & N^* \otimes L_X^* \;\oplus\; N^* \otimes L_X^* \\ f & \longrightarrow & f \oplus (-f). \end{array}$$

The quotient sheaf of $\mathcal{O}_X^{(1)}$-modules,

$$\tilde{L}_X^{*(1)} \equiv \left(L_X^{*(1)} \;\oplus\; N^* \otimes N^* \right) / i_3 (N^* \otimes L_X^*)$$

fits into the commutative diagram

$$\begin{array}{ccccccccc} & & 0 & & 0 & & & & \\ & & \downarrow & & \downarrow & & & & \\ 0 & \longrightarrow & N^* \otimes L_X^* & \xrightarrow{i_1} & L_X^{*(1)} & \longrightarrow & L_X^* & \longrightarrow & 0 \\ & & \downarrow{\scriptstyle i_2} & & \downarrow & & \| & & \\ 0 & \longrightarrow & N^* \otimes N^* & \xrightarrow{i} & \tilde{L}_X^{*(1)} & \longrightarrow & L_X^* & \longrightarrow & 0 \end{array}$$

with rows and columns exact. Then the sheaf of $\mathcal{O}_X^{(1)}$-modules,

$$\mathcal{L}_X^{(1)}(L^*) \equiv \tilde{L}_X^{*(1)} / i \left(\wedge^2 N^* \right),$$

fits into the exact sequence

$$0 \longrightarrow N^* \odot N^* \longrightarrow \mathcal{L}_X^{(1)}(L^*) \longrightarrow L_X^* \longrightarrow 0.$$

Lemma 3.7 *Let $X \hookrightarrow Y$ be a complex Legendre submanifold of a complex contact manifold (Y, L). Then the difference*

$$\left[\mathcal{L}_X^{(1)}(N^*) - \mathcal{L}_X^{(1)}(L^*) \right] \in \mathrm{Ext}^1_{\mathcal{O}_X^{(1)}}(L_X^*, N^* \odot N^*)$$

is a locally free sheaf of $\mathcal{O}_X$-modules.

Corollary 3.8 $\left[\mathcal{L}_X^{(1)}(N^*) - \mathcal{L}_X^{(1)}(L^*) \right] \in H^1(X, L_X \otimes \odot^2 N^*)$.

This conclusion about a cohomology class in $H^1(X, L_X \otimes \odot^2 N^*)$ canonically associated to a Legendre embedding $X \hookrightarrow Y$ is not a surprise — it is in accord with LeBrun's (1991) abstract machinery of infinitesimal fattenings of Legendre submanifolds. What is really achieved by the above discussion is that this canonical cohomology class has a clear interpretation as the obstruction to a global splitting of an exact sequence of some natural locally free sheaves on X. This interpretation is crucial for the next result.

All we used in the construction of extensions $\mathcal{L}_X^{(1)}(N^*)$ and $\mathcal{L}_X^{(1)}(L^*)$ is that the normal bundle of $X \hookrightarrow Y$ has a canonical projection $N \longrightarrow L|_X$, where L is a line bundle on Y. If F is a complete family of compact Legendre submanifolds of a complex contact manifold (Y, L), then the normal bundle of the submanifold $F \hookrightarrow Y \times M$ also has the canonical projection $N_F \longrightarrow L_F \equiv \pi_1^*(L)|_F$ and we can construct analogously sheaves of $\mathcal{O}_F^{(1)}$-modules $\mathcal{L}_F^{(1)}(N_F^*)$ and $\mathcal{L}_F^{(1)}(L_F^*)$ and prove that $\left[\mathcal{L}_F^{(1)}(N_F^*) - \mathcal{L}_F^{(1)}(L_F^*)\right] \in H^1(F, L_F \otimes N_F^* \odot N_F^*)$. For each $t \in M$,

$$\left[\mathcal{L}_F^{(1)}(N_F^*) - \mathcal{L}_F^{(1)}(L_F^*)\right]\Big|_{\nu^{-1}(t)} \cong \left[\mathcal{L}_{X_t}^{(1)}(N_t^*) - \mathcal{L}_{X_t}^{(1)}(L^*)\right].$$

Theorem 3.9 *Let $F \hookrightarrow Y \times M$ be a complete family of Legendre moduli spaces. Then a splitting (if any) of the exact sequence of the locally free sheaves*

$$0 \longrightarrow N_F^* \odot N_F^* \longrightarrow \left[\mathcal{L}_F^{(1)}(N_F^*) - \mathcal{L}_F^{(1)}(L_F^*)\right] \longrightarrow L_F \longrightarrow 0$$

induces a torsion-free connection on M such that, for every $y \in Y'$, the associated alpha subspace α_y is totally geodesic.

This theorem is a generalization of Proposition 3.5. Torsion-free affine connections generated on Legendre moduli spaces by Theorem 3.9 are called *induced connections.*

3.7 Holonomy of induced connections

One can use the fact that the normal bundle N of a Legendre submanifold $X \hookrightarrow Y$ is always of the form $J^1 L_X$ to show that the vector space $H^0(X, L_X \otimes N^*)$ has a canonical Lie algebra structure. There is a natural representation of this finite-dimensional Lie algebra on the vector space $H^0(X, L_X)$: any vector $s \in H^0(X, L_X \otimes N^*)$ defines a map

$$\begin{array}{rccc} \psi_s: & H^0(X, L_X) & \longrightarrow & H^0(X, L_X) \\ & \alpha & \longrightarrow & < s, j^1(\alpha) >, \end{array}$$

where the angular brackets stand for the natural pointwise L_X-valued pairing of a global section of $L_X \otimes N^*$ with a global section of N. We have, for any $s, t \in H^0(X, L_X \otimes N^*)$,

$$\psi_s \circ \psi_t - \psi_t \circ \psi_s = \psi_{[s,t]}.$$

Theorem 3.10 *Let ∇ be an induced connection on a Legendre moduli space M. If the function*

$$f : t \longrightarrow \dim H^0(X_t, L_{X_t} \otimes N_t^*)$$

is constant on M, then the holonomy algebra of ∇ is a subalgebra of the Lie algebra $H^0(X, L_X \otimes N^*)$.

It is very easy to check using this Theorem that in the case of LeBrun's family of quadrics (see example 1 in subsection 3.4) the Lie algebra $H^0(X, L_X \otimes N^*)$ is precisely the conformal algebra, thus confirming once again that induced torsion-free connections are Weyl connections. In addition, the holonomy group of the canonical connection on the space of Legendre rational curves, also discussed in subsection 3.4, is easily found to be Bryant's exotic group $G_3 \subset GL(4, \mathbb{C})$ (exotic in the sense that it is missing in the Berger (1955) list of admissible holonomies of torsion-free affine connections).

One of the main conclusions of this section is that Legendre moduli spaces often come equipped with induced differential-geometric structures which, due to Theorem 3.10, can be easily classified and explicitly described.

Acknowledgements. It is a pleasure to thank Stephen Huggett, Henrik Pedersen and Paul Tod for many valuable discussions and helpful remarks. Thanks are also due to the Department of Mathematics and Computer Science of Odense University for hospitality and financial support.

References

Akivis, M. A. (1980). Webs and almost-Grassmanian structures, *Dokl. Akad. Nauk USSR, 252*: 267-270.

Ambrose, W., and Singer, I. M. (1953). A theorem on holonomy, *Trans. Amer. Math. Soc., 79*: 428-443.

Arnold, V.I. (1978). Mathematical Methods of Classical Mechanics, Springer, Berlin Heidelberg New York [Russian: Nauka, Moscow, 1974].

Bailey, T. N., and Eastwood, M. G. (1991). Complex paraconformal manifolds – their differential geometry and twistor theory, *Forum Math., 3*: 61-103.

Berger, M. (1955). Sur les groupes d'holonomie des variétés á connexion affine et des variétés Riemanniennes, *Bull. Soc. Math. France, 83*: 279-330.

Boyer, C. P. (1988). A note on hyperhermitian four-manifolds, *Proc. Amer. Math. Soc., 102*: 157-164.

Bryant, R. (1991). Two exotic holonomies in dimension four, path geometries, and twistor theory, *Proc. Symposia in Pure Mathematics, 83*: 33-88.

Eastwood, M. G. (1990). The Penrose transform, *Twistors in mathematics and physics* (T. N. Bailey and R. J. Baston, eds.), Cambridge University Press, Cambridge, p. 87-103.

Eastwood, M. G., and LeBrun, C. R. (1992). Fattening complex manifolds: curvature and Kodaira-Spencer maps, *J. Geom. Phys.*, *8*: 123-146.

Griffiths, P. A. (1966). The extension problem in complex analysis II: embeddings with positive normal bundle, *Amer. J. Math.*, *88*: 366-446.

Gindikin, S. G. (1982). Integral geometry and twistors, *Lect. Notes Math.*, *970*: 2-42.

Hitchin, N. (1982). Complex manifolds and Einstein's equations, *Lect. Notes Math.*, *970*: 73-99.

Hitchin, N., Karlhede, A., Lindström, U., and Roček, M. (1987). HyperKähler metrics and supersymmetry, *Commun. Math. Phys.*, *108*: 535-589.

Kodaira, K. (1962). A theorem of completeness of characteristic systems for analytic families of compact submanifolds of complex manifolds, *Ann. Math.*, *75*: 146-162.

Kodaira, K., and Spencer, D. C. (1958). On deformations of complex analytic structures I-II, *Ann. Math.*, *67*: 328-465 .

LeBrun, C. R. (1983). Spaces of complex null geodesics in complex-Riemannian geometry, *Trans. Amer. Math. Soc.*, *284*: 209-321.

LeBrun, C. R. (1989). Quaternionic-Kähler manifolds and conformal geometry, *Math. Ann.*, *284*: 353-376.

LeBrun, C. R. (1991). Thickenings and conformal gravity, *Commun. Math. Phys.*, *139*: 1-43.

Manin, Yu. I. (1981). Gauge fields and holomorphic geometry, *Itogi Nauki i Tekhniki, Seriya Sovremennye Problemy Matematiki, 17*: 3-55 [In Russian]

Manin, Yu. I. (1988). Gauge Field theory and Complex Geometry, Springer, Berlin Heidelberg New York [Russian: Nauka, Moscow, 1984].

Merkulov, S. A. (1991). Twistor transform of Einstein-Weyl superspaces, *Class. Quantum Grav.*, *8*: 2149-2162.

Merkulov. S. A. (1992a). Supersymmetric nonlinear graviton, *Funct. Anal. and Its Applic.*, *26*: 89-90.

Merkulov, S. A. (1992b). Superconformal geometry in three dimensions, *J. Math. Phys.*, *33*: 735-757.

Merkulov, S. A. (1992c). Quaternionic, quaternionic Kähler, and hyper-Kähler supermanifolds, *Lett. Math. Phys.*, *25*: 7-16.

Merkulov, S. A. (1992d). Deformation theory of compact submanifolds of complex fibred manifolds and affine connections, *Odense preprint.*

Merkulov, S. A. (1993a). Self-dual Einstein supermanifolds and supertwistor theory, *Spinors, Twistors, Clifford Algebras and Quantum Deformations* (Z. Oziewicz et al, eds.), Kluwer Academic Publishers, p. 141-146.

Merkulov, S. A. (1993c). Existence and geometry of Legendre moduli spaces. *Odense preprint.*

Merkulov, S. A. (1994). Geometry of relative deformations I. Article in this volume.

Merkulov, S. A., and Pedersen, H. (1994). Geometry of relative deformations II. Article in this volume.

Pedersen, H., and Poon, Y. S. (1989), Twistorial construction of quaternionic manifolds, *Proc. VIth Int. Coll. on Diff. Geom. , Cursos y Congresos, Univ. Santiago de Compostela, 61*: 207-218.

Pedersen, H., and Swann, A. (1993). Riemannian submersions, four-manifolds and Einstein-Weyl geometry, *Proc. London Math. Soc., 66*: 381-399.

Penrose, R. (1976). Non-linear gravitons and curved twistor theory, *Gen. Rel. Grav., 7*: 31-52.

Penrose, R. (1977). The twistor programme, *Rep. Math. Phys., 12*: 65-76.

Salamon, S. M. (1982). Quaternionic Kähler manifolds, *Invent. Math., 67*: 143-171.

Ward, R. S. (1977). On self-dual gauge fields, *Phys. Lett., 61A*: 81-82.

Ward, R. S. (1980). Self-dual space-times with cosmological constant, *Commun. Math. Phys., 78*: 1-17.

Wells, R.O., Jr. (1979). Complex manifolds and mathematical physics, *Bull. A.M.S., 1* 296-336.

10
Self-Duality and Connected Sums of Complex Projective Planes

HENRIK PEDERSEN
Department of Mathematics and Computer Science,
Odense University, DK-5230 Odense M, Denmark

1 INTRODUCTION

Since the discovery by Poon (1986) that $2\mathbb{CP}^2 = \mathbb{CP}^2 \# \mathbb{CP}^2$ carries a self-dual structure, an enormous number of results on self-duality and $n\mathbb{CP}^2$ has appeared through the work of Donaldson and Friedman (1989), Floer (1991), LeBrun (1991a, 1991b, 1991c, 1992, 1993), LeBrun and Poon (1992), Taubes (1992), Joyce (1993), Kim (1993) and Kim and Pontecorvo (1993).

I want to report on joint work with Poon on issues very much related to the results of the gentlemen mentioned above. First, I will discuss a theorem (Pedersen and Poon, 1993a) giving a sufficient condition for a self-dual 4-manifold to be diffeomorphic to $n\mathbb{CP}^2$. Second, I shall briefly mention some results on self-dual structures with symmetries on $n\mathbb{CP}^2$ (Pedersen and Poon, 1993c).

2 SELF-DUALITY AND SMOOTH STRUCTURES ON $n\mathbb{CP}^2$

An oriented conformal 4-manifold $(M, [g])$ is said to be *self-dual* if its Weyl conformal curvature satisfies $W = *W$, where $*$ is the Hodge star operator. If $(M, [g])$ satisfies this conformally invariant curvature condition, its *twistor space* is a complex 3-manifold Z, whose underlying smooth 6-manifold is the total space of the bundle $S(\Lambda^2_-) \xrightarrow{\pi} M$ (Atiyah, Hitchin and Singer, 1978). Here $S(\Lambda^2_-)$ is the sphere bundle of the bundle of anti-self-dual 2-forms. Equivalently, the smooth bundle $Z \xrightarrow{\pi} M$ is isomorphic to the bundle $P(V^-) \xrightarrow{\pi} M$ of projectivised half spinors.

The anticanonical bundle $K^{-1} = \Lambda^{3,0}TZ$ of Z has a square root $K^{-1/2}$ which, when restricted to a *twistor line*, $\pi^{-1}(x) = \mathbb{CP}^1_x$, coincides with the degree 2 holomorphic bundle $\mathcal{O}(2)$.

The antipodal map on the twistor lines induces a *real structure* on Z and allows us to think of the space of holomorphic sections $H^0(Z, \mathcal{O}(K^{-1/2}))$ as a real vector space. The zero set in Z of a real holomorphic section of $K^{-1/2}$ is called a real *degree* 2 *divisor*. On

$Z = \mathbb{CP}^3$, the twistor space of the canonical conformal structure on the 4-sphere, the degree 2 divisors are just quadric surfaces.

Recall that we may divide conformal structures into positive, negative and zero types according to the sign of the scalar curvature s_g of the Yamabe metric g in the conformal class (Schoen, 1984). The theorem I want to discuss may now be stated as follows (Pedersen and Poon, 1993a).

THEOREM 2.1 *Let $(M,[g])$ be a compact, simply connected, oriented and self-dual conformal 4-manifold of positive type. Then, if the twistor space admits a real degree 2 divisor, M is diffeomorphic to $n\mathbb{CP}^2$, where n is the signature of M.*

Remarks 1) The condition on the scalar curvature is almost free. Due to a theorem of Gauduchon (1991) the existence of the divisor implies that the conformal structure is of non-negative type. If $s_g = 0$, Pontecorvo (1991) has proved that the opposite oriented space $\overline{M}$ is a scalar-flat Kähler surface — and this situation is well known.

2) A few words of motivation may be in order. From the Weitzenböck decomposition (Hitchin, 1980)

$$D^*D = \nabla^*\nabla + \frac{1}{3}s_g$$

of the operator $D = d|_{\Lambda^2_-}$ we get kern $D = 0$ so $b_- = 0$ and the intersection form is positive definite. It then follows from the celebrated theorem of Donaldson (1983) that the intersection form is diag $(1,\ldots,1)$. Thus, M is *homeomorphic* to $n\mathbb{CP}^2$ (Freedman, 1982). It is therefore natural to ask whether M is also *diffeomorphic* to $n\mathbb{CP}^2$. This kind of speculation about diffeomorphism type has also appeared in connection with work on the algebraic dimension of twistor spaces: if the twistor space Z is Moishezon (i.e. the algebraic dimension is equal to 3) then M is simply connected (Campana, 1991) and of positive type (Poon, 1988). The argument above then gives that M is homeomorphic to $n\mathbb{CP}^2$. In (Pedersen and Poon, 1993b) we proved the existence of deformations of known twistor spaces containing degree 2 divisors but no degree 1 divisors. However, if furthermore the deformed twistor spaces are assumed to be Moishezon we proved there had to be degree 1 divisors. This may suggest that the Moishezon property forces the existence of low degree divisors. We are far from being able to prove that Moishezon twistor spaces are *diffeomorphic* to $n\mathbb{CP}^2$, but after Theorem 2.1 such a result would follow if it could be established that Moishezon twistor spaces always carry either degree 1 or degree 2 divisors.

3) It is natural to estimate the number of degree 2 divisors, i.e. to estimate the dimension h^0 of the vector space $H^0(Z,\mathcal{O}(K^{-1/2}))$ (Poon, 1986; Hitchin, 1981). By the Riemann-Roch-Hirzebruch formula, we may express the Euler characteristic $\chi(K^{-1/2}) = \sum_{i=0}^{3}(-1)^i h^i(Z,\mathcal{O}(K^{-1/2}))$ in terms of the Chern numbers of Z. As Z is a 2-sphere bundle over M, the Leray-Hirsch Theorem allows $\chi(K^{-1/2})$ to be expressed in terms of the signature $\tau(M)$ and the Euler characteristic $\chi(M)$ of M. As M is homeomorphic to $n\mathbb{CP}^2$ we have $\chi(M) = 2 + \tau(M)$ and get

$$\chi(K^{-1/2}) = 2(5 - \tau(M)).$$

Using Serre duality we have

$$H^2(Z, \mathcal{O}(K^{-1/2})) \cong H^1(Z, \mathcal{O}(K^{3/2})).$$

The bundle $K^{3/2}$ has degree -6 on each twistor line, so $H^1(Z, \mathcal{O}(K^{3/2}))$ corresponds, via the Penrose transform, to solutions of a zero rest-mass field equation (Hitchin, 1980). However, $s_g > 0$ implies that there are no such solutions. Thus we have $h^2(Z, \mathcal{O}(K^{-1/2})) = 0$. Also, since $\mathcal{O}(-6)$ has no sections, we get

$$h^3(Z, \mathcal{O}(K^{-1/2})) = h^0(Z, \mathcal{O}(K^{3/2})) = 0.$$

From these vanishing results we obtain the estimate

$$h^0 \geq 2(5 - \tau).$$

COROLLARY 2.1 *If a compact, simply connected and self-dual conformal* 4*-manifold* $(M, [g])$ *of positive type has signature* $\tau \leq 4$ *then* M *is diffeomorphic to* $\tau\mathbb{CP}^2$.

3 A SKETCH OF THE PROOF OF THEOREM 2.1

Assume we have a real degree 2 divisor $S = [s]$, $s \in H^0(Z, \mathcal{O}(K^{-1/2}))$.

1) If S is reducible then $S = D + \overline{D}$ where D is a degree 1 divisor and $\overline{D}$ is the conjugate divisor. Moreover, D contains exactly one real twistor line $\mathbb{CP}^1_\infty$ (LeBrun and Poon, 1992; Poon, 1992a), and as the self-intersection number of this line is equal to 1, D is the blow-up of $\mathbb{CP}^2$ at n points (Barth, Peters and Van de Ven, 1984).

The twistor projection (see Figure 3.1) $\pi|_D : D \to M$ collapses $\mathbb{CP}^1_\infty$ to a point and gives a diffeomorphism $D \cong M$. Now, recall that blowing up a point amounts to cutting out a disk around the point and attaching the disk bundle of $\mathcal{O}(-1) \to \mathbb{CP}^1$. On the other hand, the bundle given by projecting $\mathbb{CP}^2 \backslash \{p\}$ onto a line not containing p is diffeomorphic to $\mathcal{O}(-1) \to \mathbb{CP}^1$. So blowing up a point of a manifold amounts to taking a connected sum with $\mathbb{CP}^2$. Therefore, M is diffeomorphic to $n\mathbb{CP}^2$.

2) Suppose S is irreducible. Such an example is that of a quadric surface

$$S = Q : z_0^2 + z_1^2 + z_2^2 + z_3^2 = 0$$

in $\mathbb{CP}^3$, the twistor space of S^4. It is easily seen that Q contains an S^1-family of lines. If these lines are removed from Q we obtain a double cover $M^+ \cup M^-$ of $S^4 \backslash S^1$ and the twistor projection

$$\pi : M^+ \cup S^1 \to S^4$$

is a smooth compactification of M^+.

We shall indicate how an irreducible degree 2 divisor S embeds a quadric in our picture. First, we claim that S must be non-singular. If $z \in S = [s]$ is a singular point then s vanishes to second order at z. Consider the twistor line $\mathbb{CP}^1_x$ through z, where $\pi(z) = x$.

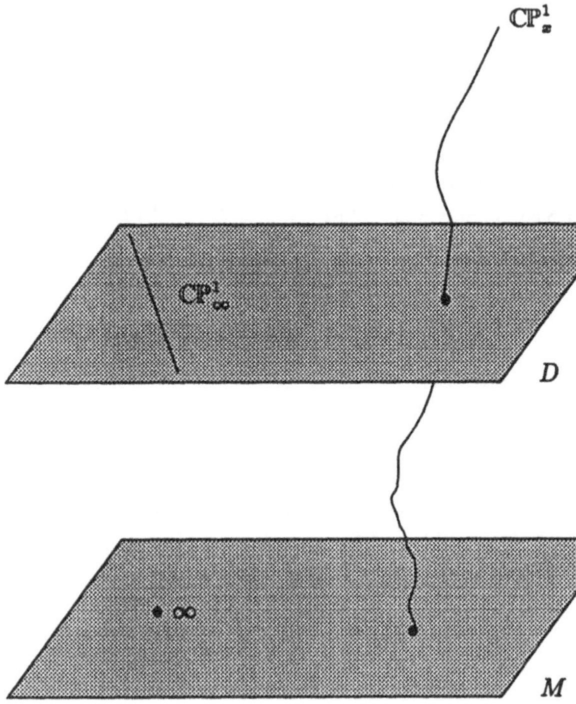

Figure 3.1: The twistor projection from a degree 1 divisor D onto M.

Then the restriction of s onto $\mathbb{CP}^1_x$ is a second order polynomial in an affine coordinate with a double root. However, s being a real section, the roots must be antipodal points. Therefore, s vanishes identically on $\mathbb{CP}^1_x$ so $\mathbb{CP}^1_x \subseteq S$. In fact, the surface must be singular on the whole line as illustrated in Figure 3.2: it is easily seen that the restriction of the 1-jet j^1s of s is a section of the normal bundle of $\mathbb{CP}^1_x$. Therefore, by the correspondence of Kodaira (1962), j^1s represents an element of the tangent space T_xM. Since j^1s is assumed to vanish at x it corresponds to a real null vector (Penrose, 1976). Thus, j^1s vanishes identically on the line.

Now blow up Z along $\mathbb{CP}^1_x$. The proper transform $\hat{S}$ of S intersects the exceptional quadric Q in a curve C and its conjugate $\overline{C}$ (Figure 3.3). In $\hat{S}$ we compute the self-intersections $C^2 = 1 = \overline{C}^2$. However, this contradicts a corollary of the Hodge-Riemann bilinear relations commonly called the index theorem (Griffiths and Harris, 1978). As S can only be singular along finitely many twistor lines this proves that S is non-singular.

The smooth surface S of degree 2 must contain a twistor line $\mathbb{CP}^1_x$ because otherwise S would be a double cover of the simply connected space M. Moreover, we may easily see that the self-intersection number is equal to zero. This means essentially that S is a ruled surface. However, some of the curves C in the ruling may be reducible $C = \sum_i a_i C_i$. But it follows from the strong Castelnuovo-Enriques Criterion that S is the blow-up of a Hirzebruch surface Σ_d (Griffith and Harris, 1978).

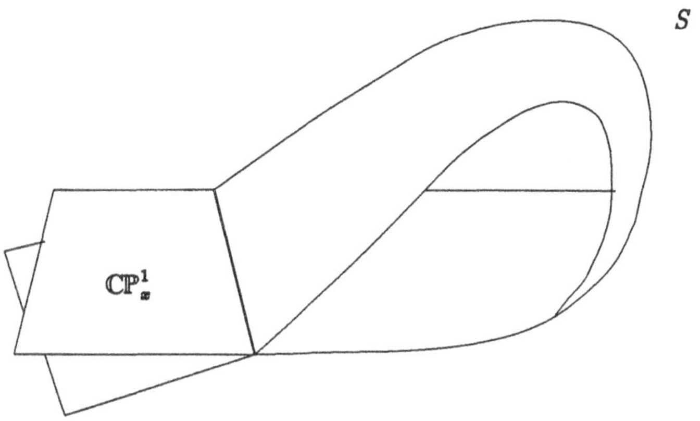

Figure 3.2: If S is singular at a point it is singular along the twistor line through that point.

We shall see that because of the real structure, Σ_d is in fact a quadric surface. If a reducible curve $\sum a_i C_i$ is real, each C_i cannot be real as $C_i \cdot (-K_S) = 1$ and K_S is real. By Zariski's Lemma (Barth, Peters and Van de Ven, 1984) $C_i \cdot \overline{C}_i = 0$, so the pair of curves $C_i, \overline{C}_i$ may be blown down simultaneously, and the resulting ruled surface inherits a real structure without real points. By construction there is a unique line E_∞ with negative intersection. On the other hand, E_∞ intersects a generic real fiber in only one point so E_∞ is not a real curve, i.e. $E_\infty \neq \overline{E}_\infty$. This forces Σ_d to be a quadric.

Now it is easily seen that the restriction of the twistor fibration to S makes M diffeomorphic to $S^4 \# r\mathbb{CP}^2 \cong r\mathbb{CP}^2$ and the theorem is proved.

4 SYMMETRIC SELF-DUAL STRUCTURES ON $n\mathbb{CP}^2$

The construction of general examples of self-duality is no longer an issue due to the work mentioned in the introduction. However, we may look for examples supporting a symmetry group G. If $\dim G \geq 3$, Poon (1992b) has classified such spaces. They are either some well-known conformally flat spaces or $\mathbb{CP}^2$.

It remains to find more examples with $T^2 = S^1 \times S^1$-symmetry and structures with non-semi-free S^1-actions (the semi-free case is treated by the hyperbolic Ansatz of LeBrun, 1993). In Pedersen and Poon (1993c) we have adapted the Donaldson-Friedman construction to do such a job.

When X_1 and X_2 are two compact self-dual manifolds with points $x_i \in X_i$, $i = 1, 2$ and when there is an orientation reversing isometry $\psi : T_{x_1}X_1 \to T_{x_2}X_2$ we get induced isometries

$$\psi_\pm : \Lambda^2_{\pm,x_1} \longrightarrow \Lambda^2_{\mp,x_2}$$

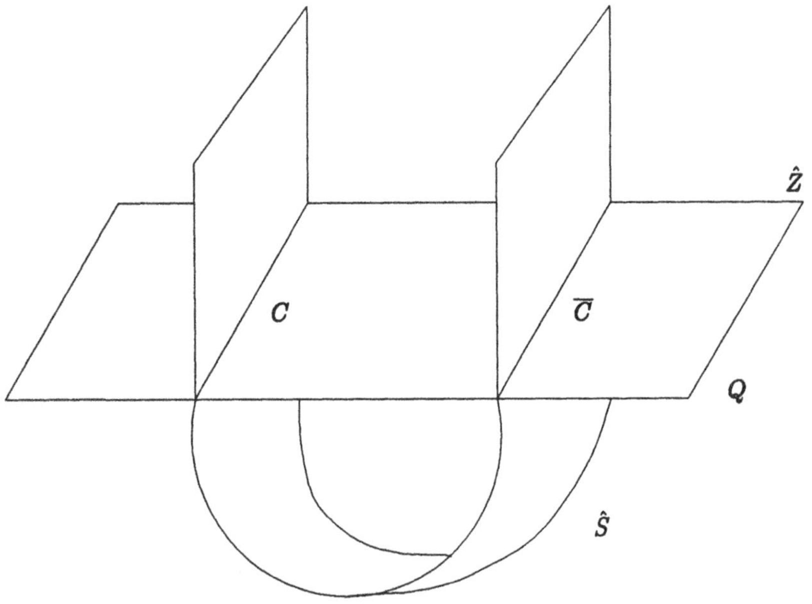

Figure 3.3: The blow-up picture.

and an induced holomorphic isomorphism

$$f = \psi_+ \times \psi_- : P(V_{x_1}^+) \times P(V_{x_1}^-) \to P(V_{x_2}^-) \times P(V_{x_2}^+).$$

Consider the blow-up $b_i : \tilde{Z}_i \to Z_i$, $i = 1, 2$ of the twistor space Z_i of X_i along the twistor line over x_i, $i = 1, 2$. Then the exceptional divisor of the blow-up is $Q_i = P(V_{x_i}^+) \times P(V_{x_i}^-)$, $i = 1, 2$. Using the identification f we define the space

$$Z = \tilde{Z}_1 \cup_f \tilde{Z}_2$$

which is singular along $Q = Q_1 = f^{-1}(Q_2)$. If $H^2(Z_i, \Theta_i) = 0$, where Θ_i is the sheaf of holomorphic tangents to Z_i, then Z can be deformed into a smooth twistor space (Donaldson and Friedman, 1989).

Assume now that a group G acts as a symmetry group for X_i fixing x_i, $i = 1, 2$. Then we have the isotropy representations

$$i : G \to SO(4)$$

with respect to any metric in the conformal class for which G is an isometry group (Obata, 1971). Then if the isotropy representations are intertwined by the orientation-reversing gluing the unobstructed smoothing carries the symmetry group G. We may also estimate the number of smoothing parameters in terms of orbit-data.

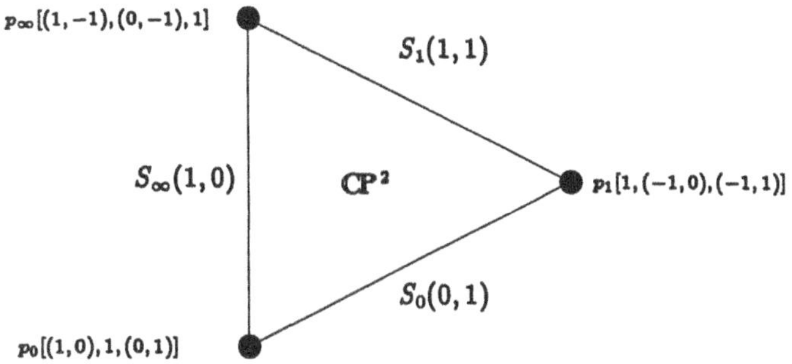

Figure 4.1: Orbit-data on $\mathbb{CP}^2$ with T^2-action.

Consider for example $\mathbb{CP}^2$ with the torus action

$$\begin{array}{rcl} T^2 \times \mathbb{CP}^2 & \rightarrow & \mathbb{CP}^2 \\ ((\theta,\psi),[z_1,z_2,z_3]) & \mapsto & [e^{i\theta}z_1, z_2, e^{i\psi}z_3]. \end{array}$$

Then we may represent the relevant orbit-data as in Figure 4.1.

Here $p_\infty[(1,-1),(0,-1),1]$ is shorthand notation for the action

$$\begin{array}{rcl} T^2 \times \mathbb{CP}^2 & \rightarrow & \mathbb{CP}^2 \\ ((\theta,\psi),[z_1,z_2,z_3]) & \mapsto & [e^{i(\theta-\psi)}z_1, e^{-i\psi}z_2, z_3] \end{array}$$

It also describes the isotropy representation at the fixed point $p_\infty = [0,0,1]$ as

$$\begin{array}{rcl} i: T^2 & \rightarrow & SO(4) \\ (\theta,\psi) & \mapsto & \left(\begin{array}{cc|cc} \cos(\theta-\psi) & -\sin(\theta-\psi) & 0 & 0 \\ \sin(\theta-\psi) & \cos(\theta-\psi) & 0 & 0 \\ \hline 0 & 0 & \cos\psi & \sin\psi \\ 0 & 0 & -\sin\psi & \cos\psi \end{array}\right). \end{array}$$

Also $S_1(1,1)$ labels the 2-sphere joining the fixed points p_1 and p_∞. On this sphere the torus acts with S^1-stabilizer described as a subgroup of T^2 by the integers $(1,1)$. Now construct $2\mathbb{CP}^2$ by attaching T^2-equivariantly a $\mathbb{CP}^2$ to the point p_∞. Then continue as indicated in Figure 4.2.

The conformal structure obtained in this way can be seen to coincide with a solution by LeBrun (1991a) corresponding to a symmetric monopole in hyperbolic 3-space. However, we may continue to put these equivariant Lego bricks together in different ways to obtain new T^2-symmetric examples and also examples with non-semi-free S^1-action (Pedersen and Poon, 1993c).

It remains to compare our solutions with the work of Joyce (1993), who uses an Ansatz based on Ashtekar's formulation of the self-dual Einstein equations (Ashtekar, Jacobsen and Smolin, 1988).

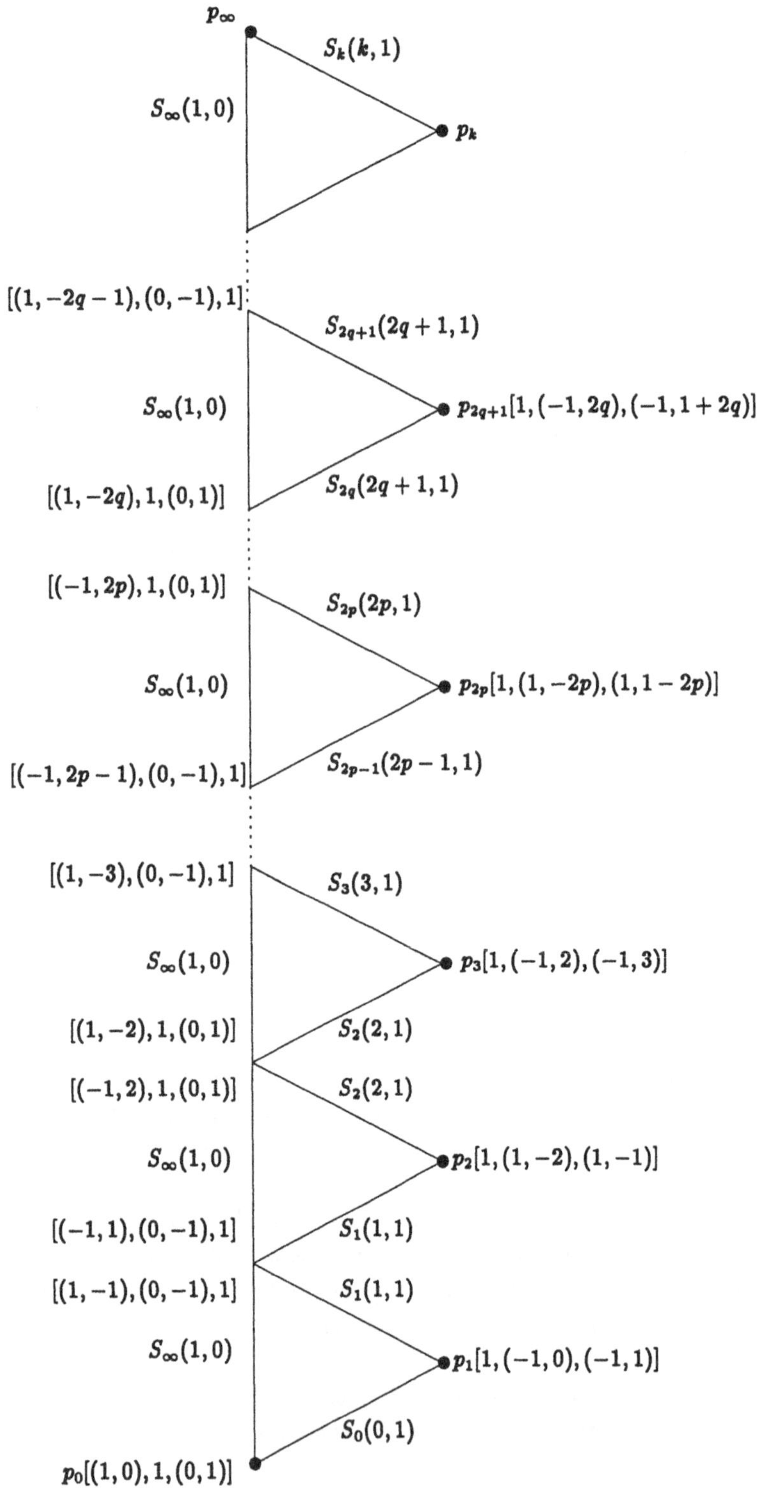

Figure 4.2: Orbit-data on $n\mathbb{CP}^2$ with T^2-action.

Combining the work of Orlik and Raymond (1974) on manifolds with smooth T^2-actions, the work of Poon (1993) on anti-self-dual Hermitian surfaces and LeBrun's observations on the topology of self-dual 4-manifolds (LeBrun, 1986) we are able to prove the following observation: if a non-negative type self-dual 4-manifold has torus symmetry with fixed points then the manifold is homeomorphic to $n\mathbb{CP}^2$. However, it is still an open question if the T^2-symmetry forces the conformal structure to be of non-negative type.

References

Ashtekar, A., Jacobsen, T. and Smolin, L. (1988). A New Characterization of Half-Flat Solutions to Einstein's Equations, *Commun. Math. Phys.*, **115**: 631.

Atiyah, M. F., Hitchin, N. J. and Singer, I. M. (1978). Self-Duality in Four-Dimensional Riemannian Geometry, *Proc. Roy. Soc. Lond.*, **A 362**: 425.

Barth, W., Peters, C and Van de Ven, A. (1984). Compact Complex Surfaces, *Ergebnisse der Mathematik und ihrer Grenzgebiete, 3. Folge, Band 4* (Springer) Berlin-Heidelberg.

Campana, F. (1991). On Twistor Spaces of the Class C, *J. Differential Geometry*, **33**: 541.

Donaldson, S. K. (1983). An Application of Gauge Theory to Four Dimensional Topology, *J. Differential Geometry*, **17**: 279.

Donaldson, S. and Friedman, R. (1989). Connected Sums of Self-Dual Manifolds and Deformations of Singular Spaces, *Nonlinearity*, **2**: 197.

Floer, A. (1991). Self-Dual Conformal Structures on $\ell\mathbb{CP}^2$, *J. Differential Geometry*, **33**: 551.

Freedman, M. H. (1982). The topology of Four-Dimensional Manifolds, *J. Differential Geometry*, **17**: 357.

Gauduchon, P. (1991). Structures de Weyl et Théorems d'Annulation sur une Variété Conforme Autoduale, *Ann. Sc. Norm. Sup. Pisa* **18**: 563.

Griffiths, P. and Harris, J. (1978). Principles of Algebraic Geometry, (John Wiley & Sons), New York.

Hitchin, N. J. (1980). Linear Field Equations on Self-Dual Spaces, *Proc. R. Lond. Soc. A* **370**: 173.

Hitchin, N. J. (1981). Kählerian Twistor Spaces, *Proc. Lond. Math. Soc.* **43**: 133.

Joyce, D. (1993). Explicit Construction of Self-Dual 4-Manifolds, preprint.

Kim, J. (1993). On the Scalar Curvature of Self-Dual Manifolds, *Math. Ann.* **297**: 235.

Kim, J. and Pontecorvo, M. (1993). Relative Singular Deformations with Applications to Twistor Theory, Preprint.

Kodaira, K. (1962). A Theorem of Completeness of Characteristic Systems for Analytic Families of Compact Submanifolds of Complex Manifolds, *Ann. Math.*, **75**: 146.

LeBrun, C. (1986). On the Topology of Self-Dual 4-Manifolds, *Proc. Amer. Math. Soc.*, **98**: 637.

LeBrun, C. (1991a). Explicit Self-Dual Metrics on $\mathbb{CP}^2\#\cdots\#\mathbb{CP}^2$, *J. Differential Geometry*, **34**: 223.

LeBrun, C. (1991b). Anti-Self-Dual Hermitian Metrics on Blown-Up Hopf Surfaces, *Math. Ann.*, **289**: 383.

LeBrun, C. (1991c). Scalar-Flat Kähler Metrics on Blown-Up Ruled Surfaces, *J. reine u. angew. Math.*, **420**: 161.

LeBrun, C. (1992). Twistors, Kähler Manifolds and Bimeromorphic Geometry I, *J. Amer. Math. Soc.*, **5**: 289.

LeBrun, C. (1993). Self-Dual Manifolds and Hyperbolic Geometry. In Einstein Metrics and Yang-Mills Connections (Sanda 1990), 99–131, *Lecture Notes in Pure and Appl. Math.*, **145**, Dekker, New York.

LeBrun, C. and Poon, Y. S. (1992). Twistors, Kähler Manifolds, and Bimeromorphic Geometry II, *J. Amer. Math. Soc.*, **5**: 317.

Obata, M. (1971). The Conjectures on Conformal Transformations of Riemannian Manifolds, *J. Differential Geometry*, **6**: 247.

Orlik, P. and Raymond, F. (1974). Actions of the Torus on 4-Manifolds II, *Topology*, **13**: 89.

Pedersen, H. and Poon, Y. S. (1993a). Self-Duality and Differentiable Structures on the Connected Sum of Complex Projective Planes, *Proc. Amer. Math. Soc.*, to appear.

Pedersen, H. and Poon, Y.S. (1993b). A Relative Deformation of Moishezon Twistor Spaces. Preprint.

Pedersen, H. and Poon, Y. S. (1993c). Equivariant Connected Sums of Compact Self-Dual Manifolds, Preprint.

Penrose, R. (1976). Non-linear Gravitons and Curved Twistor Theory, *Gen. Relativ. Grav.*, **7**: 31.

Pontecorvo, M. (1992). Algebraic Dimension of Twistor Spaces and Scalar Curvature of Anti-Self-Dual Metrics, *Math. Ann.*, **291**: 113.

Poon, Y. S. (1986). Compact Self-Dual Manifolds with Positive Scalar Curvature, *J. Differential Geometry*, **24**: 97.

Poon, Y. S. (1988). Algebraic Dimension of Twistor Spaces, *Math. Ann.*, **282**:621.

Poon, Y. S. (1992a). On The Algebraic Structure of Twistor Spaces, *J. Differential Geometry*, **36**:451.

Poon, Y. S. (1992b). Conformal Transformations of Compact Self-Dual Manifolds, *International J. Math*, to appear.

Poon, Y. S. (1993). Anti-Self-Dual Hermitian Surfaces. Preprint.

Schoen, R. (1984). Conformal Deformation of a Riemannian Metric to Constant Scalar Curvature, *J. Differential Geometry*, **20**: 478.

Taubes, C. (1992). The Existence of Anti-Self-Dual Conformal Structures, *J. Differential Geometry*, **36**: 163.

11
Twistors and the Einstein Equations

Roger Penrose
Mathematical Institute
Oxford, OX1 3LB, U.K.

Abstract It has been proposed that the appropriate global definition of a twistor, applicable to general curved vacuum space-times, would be as a charge for a massless field of helicity $^3/_2$. In flat space-time, using the Dirac form of these potentials, these twistor charges arise as the "gauge freedom of the second kind" in a long exact sequence involving the first and second potentials for the field.

A construction due to Ward is recalled, in which potentials for massless fields can act as partial connections on non-linear bundles, integrable on β-planes. This is generalized, in the case of helicity $^3/_2$, to provide a full connection on a vector bundle of rank 3, leading to an expression whereby the usual Rarita-Schwinger potential is supplemented by a second potential.

1. Twistors in curved space-time

From the very beginning, twistor theory has been motivated by the desire for a holomorphic underpinning to space-time structure (cf. Penrose 1986). When (null projective) twistors are identified with light rays in space-time, this twistor holomorphicity shows up in the conformal structure of the celestial sphere and in the shear-free property of families of light rays. Yet general relativity has a basic requirement that (vacuum) space-times be conformally curved, in general. This leads to an essential incompatibility with the foregoing manifestations of twistor holomorphicity, because the shear-free condition is not generally preserved along light rays in the presence of conformal curvature and, accordingly, the conformal structures of the local and asymptotic celestial spheres disagree with one another. It might thus *appear* to be the case that there is a deep-seated incompatibility between twistor-holomorphic requirements and the principles of general relativity.

If this were really so, it would, of course, be "bad news" for twistor theory's aspirations to occupy significant place in the basic formulation of Nature's laws — especially since Einstein's vacuum equations are now observationally confirmed to a degree greater than that of any other physical theory. But despite these initially unpromising-looking conflicts, there have emerged, from time to time, some suggestive — though as yet somewhat limited — relationships between twistor holomorphicity and the specific form of the Einstein vacuum equations. In the first place, the Hamiltonian that governs the transformation between the flat twistor spaces on the two sides of a plane or spherically fronted gravitational impulsive wave (as is determined by the way in which the wave deflects light rays) is specified in terms of a holomorphic function (Penrose and MacCallum 1972), this being a characteristic feature of the Einstein equations holding along the wave. More strikingly, the $\mathcal{H}$-space and non-linear graviton constructions (Newman 1976, Penrose 1976, 1992a) provide direct links between twistor holomorphicity and the Einstein vacuum equations, albeit only in the case of (anti-) self-dual Weyl curvature. These are perhaps the only unambiguous manifestations of twistor holomorphicity in the non-linear Einstein vacuum equations, although in the weak field linear limit of Einstein's equations, twistor holomorphicity provides a complete description through the standard contour integral and cohomology expressions (cf. Penrose 1969, Penrose and Rindler 1986, Eastwood, Penrose, and Wells 1981). As is well known, the non-linear graviton provides a successful "non-linearization" of the helicity -2, or anti-self-dual case (twistor function homogeneity +2); moreover the "googly" construction provides a very tentative "non-linearization" of the helicity +2, or self-dual case (twistor function homogeneity -6). The ambitwistor approach to the full Einstein vacuum equations has had some success within its own terms (LeBrun 1990). Hypersurface twistors provide a means that can in principle lead to a formulation of the full vacuum equations in twistor terms (Mason 1985, Mason and Penrose 1989, cf. also Penrose 1990a). The Ward-Woodhouse-Mason construction for axisymmetric stationary vacuums (Ward 1983, Woodhouse and Mason 1988) provides a non-standard route to a twistor-holomorphic description of an interesting family of solutions, but it is not clear how this might be generalized to handle general vacuum fields.

Indeed, it is not clear from *any* of this how one might envisage a twistor concept that applies to Einstein vacuum equations in complete generality. It is for this kind of reason that we may feel encouraged by the existence of a another approach to the twistorial description of the Einstein equations.

2. Twistors as charges for helicity $^3/_2$ massless fields

In Penrose (1990b, 1991, 1992b, and 1993) the suggestion was explored that one should combine the following two facts concerning massless fields of helicity $^3/_2$:

(i) the appropriate field equations governing massless fields of helicity $^3/_2$ are consistent if and only if the space-time is Ricci-flat;

(ii) in Minkowski space-time $\mathbb{M}$, the space of conserved charges for massless fields of helicity $^3/_2$ is its twistor space $\mathbb{T}$.

In accordance with (i) and (ii), it would seem that all we need do, in order to construct the elusive twistor space $\mathcal{T}$ for a general Ricci-flat space-time $\mathcal{M}$, is to find the space of charges for massless fields of helicity $^3/_2$ on $\mathcal{M}$. The hope would be that, as was the case for the non-linear graviton construction, the complex space $\mathcal{T}$ would be completely equivalent to the space-time $\mathcal{M}$, although related to it only in a non-local way. One might also hope that the general such $\mathcal{T}$ could be constructed by the use of free functions, and that twistor theory would thereby establish itself as providing a general procedure for analysing the Einstein vacuum equations. If these aims could be realized, then there would be a powerful case for considering that the physically "correct" twistor concept had at last been found.

However, it turns out that there are several profound difficulties in implementing this idea (see, in particular, Penrose 1992b). An over-riding obstacle to finding a "*T*", according to the above ideas is that it ought *not* to be a vector space (as is the case with the non-linear graviton), whereas in any of the normal definitions of a "charge" for a linear field would indeed provide us with a vector-space structure. The most promising way of circumventing this problem would seem to be to find way in which helicity $^3/_2$ fields can play some kind of active role as a *connection* (cf. Penrose 1991), in addition to its passive role in which the twistor charge acts as its source. Combining such active and passive roles might give rise to the desired non-linearity and thus break the unwanted vector-space structure.

In the present note, I shall consider both the passive and active aspects of fields of helicity $^3/_2$. However, much of what I shall have to say will be concerned with such fields in Minkowski space-time $\mathbb{M}$.

3. The Dirac and Rarita-Schwinger formulations

In $\mathbb{M}$, a field of helicity $^3/_2$ is described (using the 2-component spinor notation, cf. Penrose and Rindler 1984), as a solution of the equation

$$\nabla^{AA'}\varphi_{A'B'C'} = 0,$$

where

$$\varphi_{A'B'C'} = \varphi_{(A'B'C')},$$

with round brackets denoting symmetrization in the usual way. In a general space-time, these equations are inconsistent because of Buchdahl conditions relating $\varphi_{A'B'C'}$ to the (self-dual) conformal curvature $\Psi_{A'B'C'D'}$:

$$\Psi_{A'B'C'D'}\,\varphi^{A'B'C'} = 0.$$

For this reason, a potential $\sigma^{A}{}_{B'C'}$ (modulo a gauge freedom) is employed, instead of a gauge-invariant field.

In *flat* space-time $\mathbb{M}$, two such descriptions have been given, namely that of Dirac (1936) and of Rarita and Schwinger (1941). According to the Dirac formulation, we have

$$\nabla_{AA'}\sigma^{A}{}_{B'C'} = \varphi_{A'B'C'} \quad \text{and} \quad \nabla^{BB'}\sigma^{A}{}_{B'C'} = 0,$$

where $\sigma^{A}{}_{B'C'}$ is symmetric in and B′ and C′. The gauge freedom for $\sigma^{A}{}_{B'C'}$ is

$$\sigma^{A}{}_{B'C'} \mapsto \sigma^{A}{}_{B'C'} + \nabla^{A}{}_{B'}\nu_{C'}$$

for which $\nu_{A'}$ satisfies the Weyl (anti-)neutrino equation

$$\nabla^{AB'}\nu_{B'} = 0.$$

In the Rarita-Schwinger form (in 2-component spinor notation), no symmetry condition is imposed on $\sigma^{A}{}_{B'C'}$, and we have the pair of field equations

$$\nabla^{B'(C}\sigma^{A)}{}_{B'C'} = 0, \quad \nabla_{C(C'}\sigma^{C}{}_{A')}{}^{C'} = 0,$$

with the same gauge freedom as above, except that $\nu_{A'}$ is not now restricted to satisfy any equation. (The Dirac description is equivalent to that of Rarita-Schwinger, in a special choice of gauge.) The second of the above equations tells us that $\nabla_{C(C'}\sigma^{C}{}_{A')B'}$ is totally symmetric in its primed indices, and we can now define the gauge-invariant field $\varphi_{A'B'C'}$ by

$$\nabla_{C(C'}\sigma^{C}{}_{A')B'} = \varphi_{A'B'C'}.$$

In both the Dirac and Rarita-Schwinger formulations, the equations for the potential and the gauge freedom are still consistent in *curved* space-time $\mathcal{M}$, if (and only if) $\mathcal{M}$ is Ricci-flat. However, in each case, the "field" $\varphi_{A'B'C'}$, as defined above, is not gauge invariant. Thus, one must be content with a "potential-modulo-gauge" description. The fact that the Ricci-flat condition is necessary and sufficient for the consistency of the gauge freedom follows immediately from the spinor form of the Ricci identities (as given in Penrose and Rindler 1984). The consistency of the field equations on $\sigma^{A}{}_{B'C'}$ also follows from the Ricci-flat condition (see, for example, Buchdahl 1958, Deser and Zumino 1976, Julia 1982).

4. Twistors as gauge freedom of the second kind

In flat space-time $\mathbb{M}$, one can see the role of twistors as "charges" for massless fields of helicity $^3/_2$ very succinctly (in the Dirac form of the field equations) when a "second potential" is introduced, and then twistors appear as providing the "gauge freedom of the second kind". The Dirac-type second potential would be a quantity $\rho^{AB}{}_{C'}$, symmetric in A and B, where

$$\nabla_{BB'}\rho^{AB}{}_{C'} = \sigma^{A}{}_{B'C'} \quad \text{and} \quad \nabla^{CC'}\rho^{AB}{}_{C'} = 0.$$

The extra gauge freedom in $\rho^{AB}{}_{C'}$, over and above that in $\sigma^{A}{}_{B'C'}$ is given by

$$\rho^{AB}{}_{C'} \mapsto \rho^{AB}{}_{C'} + \nabla^{B}{}_{C'}\mu^{A},$$

where μ^A satisfies the Weyl neutrino equation

$$\nabla_{AC'}\mu^{A} = 0.$$

Combining the two gauge freedoms together, we find that

$$\rho^{AB}{}_{C'} \mapsto \rho^{AB}{}_{C'} + \varepsilon^{AB}\nu_{A'} + \nabla^{B}{}_{C'}\mu^{A},$$

and instead of satisfying the free Weyl neutrino equation, μ^A must have the gauge field $\nu_{B'}$ (for $\sigma^{A}{}_{B'C'}$) as a "source":

$$\nabla_{AC'}\mu^{A} = 2\nu_{C'}$$

(from the symmetry of $\rho^{AB}{}_{C'}$).

The *gauge freedom of the second kind* is (as with standard electromagnetism) that freedom in the gauge fields which does not affect the potentials. Thus, this freedom, in the present situation, is given when $\nu_{C'}$ is a constant and is related to μ^A by

$$\varepsilon^{AB}\nu_{A'} + \nabla^{B}{}_{C'}\mu^{A} = 0,$$

that is to say, the pair $(\omega^A, \pi_{A'})$ is a *twistor*, where

$$\omega^A = \mu^A, \quad \pi_{A'} = i\,\nu_{A'}.$$

The various relationships between these gauge quantities fit into an exact sequence:

$$0 \to \{\text{twistors}\} \to \{\text{gauge fields}\} \to \{\text{potentials}\} \to \{\text{fields}\} \to 0$$

where "{twistors}" stands for the space of pairs $(\omega^A,\pi_{A'})$, related by $\nabla_{AA'}\omega^B=-i\varepsilon_A{}^B\pi_{A'}$; "{gauge fields}" for the space of pairs $(\mu^A,\nu_{C'})$, related by $\nabla_{AC'}\mu^A=2\nu_{C'}$, $\nabla^{AC'}\nu_{C'}=0$; "{potentials}" for the space of $(\rho^{AB}{}_{C'},\sigma^A{}_{B'C'})$, each symmetric and related by $\nabla_{BB'}\rho^{AB}{}_{C'}=\sigma^A{}_{B'C'}$, $\nabla^{CC'}\rho^{AB}{}_{C'}=0$, $\nabla^{BB'}\sigma^A{}_{B'C'}=0$; and "{fields}" by the space of symmetric $\varphi_{A'B'C'}$, subject to $\nabla^{AA'}\varphi_{A'B'C'}=0$. The role of the twistors as "charges" for the fields $\varphi_{A'B'C'}$ can then be understood in terms of the cohomology of this sequence (cf. Penrose 1992b).

5. The Ward non-linear connection

In an early Twistor Newsletter (see Ward 1979) Richard Ward showed how to encode the information of a helicity s $(>{}^1/_2)$ massless field into a non-linear partial connection on the (unprimed) spin-bundle over $\mathbb{CM}$ (complexified Minkowski space) which is integrable over β-planes. Let λ_A be the "fibre coordinate", and define a partial connection by the operator

$$W_{A'} = \lambda^A \nabla_{AA'} - \rho_{A'}\Upsilon,$$

where Υ is the Euler operator along the fibres:

$$\Upsilon = \lambda^A\,\partial/\partial\lambda^A.$$

Note that the operator $W_{A'}$ is defined only in the tangent directions to the β-plane defined by λ_A at the point in question. We require that $W_{A'}$ be integrable on each β-plane (as defined by λ_A), and this condition amounts to

$$\lambda^A \nabla_{AA'}\rho^{A'} = \rho_{A'}\Upsilon\rho^{A'},$$

where $\tau_{A'}$ is defined on the bundle — so it depends on λ_A as well as on the space-time point. Let us take $\rho_{A'}$ to have homogeneity degree m in λ_A and to have the polynomial form

$$\rho_{A'} = \lambda_B \ldots \lambda_D \, \rho_{A'}{}^{B \ldots D},$$

where $\rho_{A'}{}^{B \ldots D}$ (symmetric in the m indices B,..., D) is now a field depending only on the space-time point. Then the condition of integrability on β-planes becomes

$$\nabla^{A'(A} \rho_{A'}{}^{B \ldots D)} = 0.$$

This last relation is the equation for a (conformally invariant) penultimate potential for the massless field $\varphi_{A'B' \ldots D'}$ of helicity $s=(m+1)/2$, the field $\varphi_{A'B' \ldots D'}$ (symmetric in its $m+1$ indices) being defined by

$$\varphi_{A'B' \ldots D'} = \nabla^{B}{}_{(B'} \ldots \nabla^{D}{}_{D'} \rho_{A')B \ldots D},$$

and consequently satisfying

$$\nabla^{AA'} \varphi_{A'B' \ldots D'} = 0.$$

Apart from the final remarks concerning the field $\varphi_{A'B' \ldots D'}$, all these considerations apply in a conformally self-dual complex space-time as well as in flat space-time. A version of the expression for $\varphi_{A'B' \ldots D'}$ holds also if the space-time is conformally flat.

6. The case of helicity $^3/_2$

The case relevant to our present situation is given when $m=2$, and we have

$$W_{A'} = \lambda^A \nabla_{AA'} - \lambda_B \lambda_C \, \rho_{A'}{}^{BC} \lambda^A \, \partial / \partial \lambda^A$$

If we consider this as a differential operator acting on spinor fields λ_D defined on the space-time, we can write this

$$W_{A'}\,\lambda_D = \lambda^A\,\nabla_{AA'}\,\lambda_D - \lambda_B\lambda_C\,\rho_{A'}{}^{BC}\,\lambda_D.$$

Setting

$$W_{A'} = \lambda^A\,\Omega_{AA'},$$

where the operator $\Omega_{AA'}$ acts on spinor fields λ_D, we can write

$$\lambda^A\,\Omega_{AA'} = \lambda^A\,\nabla_{AA'} - \lambda_B\lambda_C\,\rho_{A'}{}^{BC}.$$

Integrability on β-planes yields

$$\nabla^{A'(A}\rho_{A'}{}^{BC)} = 0,$$

and in flat space-time, we have the *field*

$$\varphi_{A'B'C'} = \nabla^B{}_{(B'}\nabla^C{}_{C'}\rho_{A')BC}.$$

However, the operator $\Omega_{AA'}$ only acts on λ^A if it differentiates in a direction within the self-dual 2-plane corresponding to λ^A. If we wish $\Omega_{AA'}$ to provide a genuine connection on the spin-bundle, we require its definition to be generalized so that it can act in *any* direction. We can take

$$\Omega_{AA'} = \nabla_{AA'} - \lambda^C\,\rho_{A'AC}.$$

Note that in the above definition of $\Omega_{AA'}$ it is necessary that the part of $\rho_{A'AC}$ that is *skew* in AC be known, whereas for $W_{A'}$ it was not. This comes about because we are now concerned with obtaining a genuine connection on the spin-bundle (albeit still a non-linear one) for which differentiation in directions other than just those compatible with λ^A are required. This suggests that the "$\rho_{A'AB}$" that is being used here should *not* be thought of as the direct analogue of the second potential arising in the *Dirac* formulation of helicity $3/2$, but of some kind of analogous second potential for the *Rarita-Schwinger* formulation.

We cannot simply ignore this skew part of $\rho_{A'AC}$ because the *gauge* transformations

$$\lambda_A \mapsto \tilde{\lambda}_A = \chi\,\lambda_A$$

are accompanied by

$$\rho_{A'AB} \mapsto \tilde{\rho}_{A'AB} = \rho_{A'AB} + \nabla_{AA'}\xi_B$$

where ξ_A depends only on the space-time point and not on the fibre coordinate, but where χ depends on both. Consistency requires

$$\chi = (1 + \lambda_A \xi^A)^{-1},$$

which gives our spin-bundle a somewhat "crazy" non-linear structure.

7. Description in terms of a vector bundle of rank 3

Rather than having non-linear fibres, we might prefer to pass to a description in terms of a genuine vector bundle of rank 3, as follows. In place of a fibre coordinate λ_A (rank 2) we can use

$$(\eta_A, \zeta) \quad \text{where} \quad \lambda_A = \zeta^{-1} \eta_A.$$

Then

$$\Omega_{AA'}(\eta_B, \zeta) = (\nabla_{AA'}\eta_B, \nabla_{AA'}\zeta_{'} - \eta^C \rho_{A'AC}).$$

The gauge transformations are

$$(\eta_B, \zeta) \mapsto (\eta_B, \zeta) = (\eta_B, \zeta + \eta_A \xi^A).$$

Note that the gauge freedom on $\rho_{A'AB}$, namely

$$\rho_{A'AB} \mapsto \rho_{A'AB} + \nabla_{AA'}\xi_B,$$

gives

$$\rho_{A'(AB)} \mapsto \rho_{A'(AB)} + \nabla_{A(A'}\xi_{B)}$$

(which preserves the "field" $\varphi_{A'B'C'} = \nabla^B{}_{(B'}\nabla^C{}_{C'}\rho_{A')BC}$, in flat space-time) and

$$\rho_{A'C}{}^C \mapsto \rho_{A'C}{}^C + \nabla_{CA'}\xi^C.$$

The condition for $\Omega_{AA'}$ to be integrable on β-planes is

$$\nabla^{A'(A}\rho_{A'}{}^{B)C} = 0.$$

Note that, on the symmetric and skew parts of $\rho_{A'}{}^{BC}$ this is

$$\nabla^{A'(A}\rho_{A'}{}^{BC)} = 0$$

and

$$\nabla^{A'A}\rho_{A'(AB)} = {}^3/_2\ \nabla^{A'}{}_B\,\rho_{A'C}{}^C.$$

Note that $\rho_{A'C}{}^C$ constitutes an additional "gauge", which does not contribute to the field of helicity $^3/_2$.

This appears to be much more within the framework of the Rarita-Schwinger, rather than the Dirac formulation, and we may ask what should be the corresponding Rarita-Schwinger *first* potential $\sigma^A{}_{B'C'}$ corresponding to this $\rho^{BC}{}_{A'}$. In fact we find, rather surprisingly, that if $\sigma^A{}_{B'C'}$ is defined according to

$$\sigma^A{}_{B'C'} = \nabla_{CC'}\,\rho_{B'}{}^{AC}$$

(note ordering of indices!), then the Rarita-Schwinger equations

$$\nabla^{B'(C}\sigma^{A)}{}_{B'C'} = 0 \text{ and } \nabla_{C(C'}\sigma^C{}_{A')}{}^{C'} = 0$$

are satisfied in flat (or at least self-dual) space-time.

8. Further comments

Since the lecture on which this article is based was given, certain advances have been made. A long exact sequence has been found for the first and second Rarita-Schwinger-type potentials $\sigma^{A}{}_{B'C'}$ and $\rho_{A'}{}^{BC}$ given here. This has also been generalized, using a local-twistor formulation due to Lionel Mason, to give a more general form for $\rho_{A'}{}^{BC}$ in which its gauge freedom is unrestricted. (See Mason and Penrose 1994 for further details.) The active role of this extended Rarita-Schwinger-type second potential, and some of its interrelations with general relativity, are being explored.

References

Buchdahl, H.A. (1958) On the compatibility of relativistic wave equations for particles of higher spin in the presence of a gravitational field *Nuovo Cim.* **10**, 96-103.

Deser, S. and Zumino, B. (1976) Consistent supergravity *Phys. Lett.* **62B**, 335-7.

Dirac, P.A.M. (1936) Relativistic wave equations *Proc. Roy. Soc. (Lond.)* **A155**, 447-59.

Eastwood M.G., Penrose, R., and Wells, R.O., Jr. (1981) Cohomology and massless fields, *Comm. Math. Phys.* **78**, 305-51.

Julia, B. (1982) Système linéare associé aux équations d'Einstein Comptes. Rendus. Acad. Sci. Paris, Sér. II, 113-6.

LeBrun, C.R. (1990) Twistors, ambitwistors, and conformal gravity, in *Twistors in Mathematical Physics,* LMS Lec. Notes **156**, eds T.N. Bailey and R.J. Baston (Cambridge Univ. Press, Cambridge).

Mason, L.J. (1985) The structure and evolution of hypersurface twistor spaces. *Twistor Newsletter* **20**, 75-79.

Mason, L.J. and Penrose, R. (1989) A twistorial approach to the full vacuum equations. *Twistor Newsletter* **29**, 1-5.

Mason, L.J. and Penrose, R. (1994) Spin 3/2 fields and local twistors. Twistor Newsletter **37**, 1-6.

Newman, E.T. (1976) Heaven and its properties *Gen. Rel. Grav.* **7**, 107-11.

Penrose, R. (1969) Solutions of the zero rest-mass equations, J. *Math. Phys.* **10**, 38-9.

Penrose, R. (1976) Non-Linear gravitons and curved twistor theory *Gen. Rel. Grav.* **7**, 31-52.

Penrose, R. (1986) On the origins of twistor theory, in *Gravitation and Geometry,* (I. Robinson festschrift volume), eds. W. Rindler and A. Trautman (Bibliopolis, Naples).

Penrose, R. (1990a) Twistor theory after 25 years — its physical status and prospects, in *Twistors in Mathematical Physics,* LMS Lec. note ser. 156, eds T.N. Bailey and R.J. Baston (Cambridge Univ. Press, Cambridge).

Penrose, R (1990b) Twistors theory for vacuum space-times: a new approach. *Twistor Newsletter* **31**, 6-8.

Penrose, R (1991) Twistors as spin 3/2 charges continued: SL(3,C) bundles. *Twistor Newsletter* **33**, 1-6.

Penrose, R.(1992a) $\mathcal{H}$-space and Twistors, in *Recent Advances in General Relativity,* (Einstein Studies, Vol. 4) eds. Allen I. Janis and John R. Porter (Birkhäuser, Boston) 6-25.

Penrose, R. (1992b) Twistors as spin 3/2 charges, in *Gravitation and Modern Cosmology* (P.G. Bergmann's 75th birthday vol.) eds. A. Zichichi, N. de Sabbata, and N.Sánchez (Plenum Press, New York).

Penrose, R. (1993) Quantum Non-Locality and Complex Reality, in *The Renaissance of General Relativity* (in honour of D.W. Sciama) eds. G. Ellis, A. Lanza, and J. Miller (Cambridge Univ. Press, Cambridge).

Penrose, R. and MacCallum, M.A.H. (1972) Twistor theory: an approach to the quantization of fields and space-time, *Phys. Repts.* **6C**, 241-315.

Penrose, R. and Rindler, W. (1984) *Spinors and Space-Time,* Vol. 1: *Two-Spinor Calculus and Relativistic Fields* (Cambridge University Press, Cambridge).

Penrose, R. and Rindler, W. (1986) *Spinors and Space-Time,* Vol. 2: *Spinor and Twistor Methods in Space-Time Geometry* (Cambridge University Press, Cambridge.

Rarita, W and Schwinger, J. (1941) On the theory of particles with half-integer spin *Phys. Rev.* **60**, 61- .

Ward, R.S. (1979) The twisted photon: massless fields as bundles, in *Advances in Twistor Theory,* eds. L.P. Hughston and R.S. Ward (Pitman, London).

Ward, R.S. (1983) Stationary and axi-symmetric spacetimes, *Gen. Rel. Grav.* **15**, 105-9.

Woodhouse, N.M.J. and Mason, L.J. (1988) The Geroch group and non-Hausdorff twistor spaces *Nonlinearity* **1**, 73-114.

12
Remarks on the Period Mapping for 4-Dimensional Conformal Structures

Michael Singer Department of Mathematics, University of Edinburgh, King's Buildings, Mayfield Road, Edinburgh EH9 3JZ, Scotland

1 INTRODUCTION

A development of twistor theory, 4-dimensional conformal field theory, has led to an interest among twistor theorists in conformally flat manifolds, since their twistor spaces play the role in 4-dimensional CFT that Riemann surfaces take on in 2-dimensional CFT. Conformally flat manifolds are also the subject of a non-trivial branch of differential geometry. My purpose in this note is to describe a period mapping Φ, analogous to the period mapping for Riemann surfaces, which may prove to be important in the study of conformally flat 4-manifolds. This work is incomplete: we give the definition of Φ, but have not yet computed it in very many examples. This is partly because the definition is complicated. We introduce the condition of 'strong regularity' for conformal structures on a 4-manifold and show that Φ is much simpler for such structures. The second part of this paper shows the existence of interesting strongly regular conformally flat manifolds; this is the intended point of departure for a more detailed study of Φ.

Most of the work has been carried out in collaboration with Michael Eastwood; the reader is referred to Eastwood and Singer (1993), henceforth to be nicknamed FTS, for the proofs that have been omitted from this article. I also thank Claude LeBrun for some helpful suggestions.

Motivation: the case of dimension 2

In 2 dimensions, a conformally flat structure is the same thing as a conformal structure, and this is in turn the same thing as a complex structure. One way in which distinct conformal structures on a compact oriented surface S are distinguished is by the *period map*. To define this, note that $H^1(S, C)$ is a complex symplectic vector space (the cup product ω is skew and non-degenerate on this space), and hence of even dimension

$2g$, say. A conformal structure on S defines the complex subspace $H^{1,0}(S)$, the space of holomorphic 1-forms on S. This subspace has complex dimension g and is Lagrangian—that is, ω vanishes identically on it. Allowing the conformal structure to vary, we obtain the period map

$$\Pi : \{\text{conformal structures on } S\} \to L(H^1(S)) \tag{1}$$

where we have written $L(V)$ for the Grassmannian of Lagrangian subspaces of the symplectic vector space V. The Torelli theorem states that this map yields an imbedding of the moduli space of curves of a given topological type in $L(H^1)$ provided that the genus g is at least 2.

2 A PERIOD MAP FOR 4-DIMENSIONAL CONFORMAL MANIFOLDS

Remarkably, there is an analogous construction in (real) dimension 4. The analogy is clearest if one works with the twistor spaces.

Let M be a compact, oriented, anti-self-dual (ASD) manifold, and let Z be its twistor space. Thus Z is a manifold of real dimension 6 and once again, the middle-dimensional cohomology, $H^3(Z,C)$ is a complex symplectic vector space of dimension $2k$, say. (In fact, $k = b_1(M)$ because Z is an S^2-bundle over M, and so $H^3(Z)$ can be identified with $H^3(M) \oplus H^1(M)$.)

Since an ASD structure on M determines a complex structure on Z, it is natural to try to use the Dolbeault cohomology groups $H^{p,3-p}(Z)$ to define structures on $H^3(Z)$ which will distinguish inequivalent complex structures on Z (and hence inequivalent ASD structures on M). Of these groups, $H^{3,0} = H^{0,3} = 0$ on a twistor space, so the only available Dolbeault cohomology groups are $H^{1,2}$ and $H^{2,1}$. Unfortunately, there is no reason to expect a direct-sum decomposition

$$H^3(Z) = H^{1,2} \oplus H^{2,1}. \tag{2}$$

This would hold if Z were of Kähler type, but it is a result of Hitchin (1981) that the only (compact) Kählerian twistor spaces correspond to the 4-manifolds S^4 and CP_2.

In place of (2) one has the Fröhlicher spectral sequence; analysis of this gives a short exact sequence

$$0 \to K \to H^3(Z) \to H \to 0$$

where

$$K = \operatorname{coker}(\partial : H^{1,1} \to H^{2,1})$$

and

$$H = \ker(\partial : H^{1,2} \to H^{2,2}).$$

Wedge product of forms descends to define a perfect pairing of K and H. For reasons of bidegree, K and H are also Lagrangian subspaces of $H^3(Z)$.

Interpretation in terms of M

The Penrose transform gives a way to interpret K and H in terms of the conformal geometry of M. Set

$$D : \Lambda^1 \to \Lambda^1,\ D = dd^* - S$$

where

$$S = 2r - (2/3)sg.$$

(Here r is the Ricci tensor and s the scalar curvature of g.) Thus D is a self-adjoint non-elliptic differential operator of second order. Then d^*D yields a map $\Lambda^1 \to \Lambda^0$: use this to define

$$\Lambda^1 \xrightarrow{d^*D \oplus d_+} \Lambda^0 \oplus \Lambda_+ \tag{3}$$

where Λ_+ is the space of smooth self-dual two-forms on M and d_+ is the composition of d with projection onto this space.

Lemma 1 *The kernel, K^1 say, of (3) on a compact manifold is finite-dimensional.*

Proof Observe that on a compact manifold, if $\omega \in \Lambda^1$ then

$$d_+\omega = 0 \text{ iff } d\omega = 0 \text{ iff } \Delta d\omega = 0$$

where $\Delta = dd^* + d^*d$ is the Laplacian on 2-forms. Thus K^1 is also the kernel of

$$\Lambda^1 \xrightarrow{d^*D \oplus \Delta d_+} \Lambda^0 \oplus \Lambda_+ \tag{4}$$

But the third-order operator here is elliptic: its symbol is just $|\xi|^2$ times the symbol of the standard elliptic operator

$$\Lambda^1 \xrightarrow{d \oplus d_+} \Lambda^0 \oplus \Lambda_+.$$

QED

In general, K^1 is not equal to K; one has to take the quotient by the subspace

$$E = \{\omega \in \Lambda^1 : \omega = df, *Ddf = d\rho, \text{ for some } \rho \in \Lambda^2\}. \tag{5}$$

There is a similar definition of H which we shall not give here. The definition of H and K was motivated from twistor theory and hence appears at first to apply only to anti-self-dual conformal structures. However, we have

Theorem 2 *Let M be a compact oriented manifold with $b_1(M) = k$. Then for any (Riemannian) conformal structure on M, there exists a Lagrangian subspace $K \subset H^1(M) \oplus H^3(M)$. In particular, $\dim K = k$.*

Proof The reader is referred to Proposition 4.1 of FTS for details; we note only that the embedding of K is given by the map

$$\omega \longmapsto (\omega, *D\omega). \tag{6}$$

In the light of this, it is natural to make the following:

Definition The period map for conformal 4-manifolds is the map

$$\Phi : \{\text{(Riemannian) conformal structures on } M\} \to L(H^1(M) \oplus H^3(M))$$

which assigns to any conformal structure, the corresponding subspace K via (6).

Question/Remark Is there any reason to believe that Φ will be able to distinguish inequivalent conformal structures in some interesting class? Since the definition of the period map depends upon $H^1(M)$, it will clearly be of limited use in the study of ASD manifolds (for there are many compact, simply connected examples of these). However, there is only one compact, simply connected conformally flat manifold—S^4. This theorem (Kuiper 1949) suggests that the period map at least stands a chance of giving interesting information for conformally flat manifolds.

Example Conformally flat structures on $S^1 \times S^3$. We shall calculate Φ for conformally flat structures on $S^1 \times S^3$ that arise in the following way. Think of $S^1 \times S^3$ as the quotient space $M = U/\Gamma$ where $U = R^4 - \{0\}$ and Γ is the infinite cyclic group generated by the conformal motion

$$P : x \mapsto e^l Ax$$

where $l > 0$ and $A \in SO(4)$. P is not an isometry of the standard flat metric on U, so this metric does not descend to the quotient. However, as P is conformal, the flat metric does descend to define a flat conformal structure on M. A fundamental domain U_0 for the action of Γ can be taken to be the spherical shell which lies between the spheres in U with centre 0 and radii 1 and e^l; P identifies the two components of ∂U_0 with a twist given by the isometry A.

Because the definition of K is conformally invariant, we may calculate on U with the flat metric, provided that the quantities we write down are Γ-invariant. Consider

$$\omega = d \log r, \quad r = |x|.$$

Then ω is closed and d^*D-closed. Now $*D\omega$ is just dA, the 3-form on U which gives α, the area of the unit 3-sphere when integrated over any representative of the generator of $H_3(U, Z)$. On the other hand, when paired with the generator $c \in H_1(S^1 \times S^3, Z)$, ω gives the answer l. (This is because c is represented in U_0 by any curve which joins a point on ∂U_0 to its image by P.) So Φ detects the logarithm l of the scale factor, but not the twist given by A: it is sensitive only to 1 of the 3 moduli which parameterize conformally flat structures on $S^1 \times S^3$.

This approach, using universal covers and calculating with the flat metric does not seem easy to apply in more general situations. When the fundamental group grows larger, the construction of invariant representatives on the universal cover becomes difficult.

Example Flat tori. On a flat torus, the d^*D-closed representatives of H^1 are harmonic and hence the 1-forms with constant coefficients. They are thus annihilated by $*D$ and so in this case, $K = H^1(M)$.

Strongly regular manifolds

The definition of K in the previous section is complicated. On the other hand, the definition of K^1 as the kernel of the differential operator in (3) is relatively simple. We know, moreover, that K and K^1 are both finite-dimensional, so it is not unreasonable to begin with the study of conformal structures with $K = K^1$. This holds when the 'error term' E of (5) vanishes, i.e., when the equation

$$*Ddf = d\rho$$

has no solutions with $df \neq 0$. One circumstance under which this is the case is when the equation $Lf = d{*}Ddf = 0$ has no non-constant solutions. Here L is the standard conformally invariant modification of the square of the Laplacian (Eastwood & Singer 1985). Motivated by this remark, we make the following

Definition A 4-manifold M with Riemannian conformal structure $[g]$ is called *strongly regular* if the equation $Lf = 0$ has no non-constant solutions.

Remark The reason for the term 'strongly regular' is that a weaker notion called 'regularity' was introduced in FTS.

Proposition 3 *Let M be compact and strongly regular. Then*

$$K = \ker(d) \cap \ker(d^*D) \subset \Lambda^1.$$

If M is strongly regular, then, every class in $H^1(M)$ has a representative ω which satisfies the 'gauge-fixing condition' $d^*D\omega = 0$. This condition may be viewed as a conformally invariant replacement of the usual harmonic (or Hodge) condition $d^*\omega = 0$.

Definition of the period matrix for a strongly regular manifold

The 'period matrix' will just be the matrix of the map $A : H^1 \to H^3$ whose graph is K. We begin by choosing cycles $a_1, \ldots, a_k$ which give a basis for H_1, and cycles $b_1, \ldots, b_k$ giving a dual basis for H_3 ($a_i \cdot b_j = \delta_{ij}$). Now use Poincaré duality to represent each b_i as a closed differential 1-form, ω_i, which lies in K. Then

$$\int_{a_i} \omega_j = \delta_{ij}$$

and the period matrix is defined to be the $k \times k$ matrix

$$A_{ij} = \int_{b_i} *D\omega_j.$$

Notice that as the graph of A is Lagrangian, the matrix entries will be symmetric: $A_{ij} = A_{ji}$.

3 THE CONDITION OF STRONG REGULARITY

Strong regularity is not very well understood. We give here two results: one guarantees strong regularity under the hypothesis that a certain curvature quantity is positive semi-definite; the other shows that the set of strongly regular conformal structures is open.

Theorem 4 *Let M be a compact oriented 4-manifold with Riemannian metric g. If the tensor*

$$T_\lambda = sg - \lambda r$$

is positive semi-definite on M for some $\lambda \in (1, 3]$, then the conformal structure g is strongly regular. The conclusion remains true if T_1 is positive semi-definite provided that, in addition,

M supports no conformal Killing vectors;

or *T_1 is positive-definite at at least one point of M.*

Proof Apart from the statement with $\lambda = 1$, this is Theorem 5.5 of FTS. To get the case $\lambda = 1$, let

$$P = \mathrm{Hess}_0^* \mathrm{Hess}_0$$

where Hess_0 is the operator which assigns to a function the trace-free part of its Hessian. Then one compares P with L as in the proof of Theorem 5.5 of FTS:

$$P = (3/4)L - (1/2)d^* \cdot T_1 \cdot d.$$

If T_1 is positive semi-definite, then this shows that L is a positive operator. It also shows that L is strictly positive if either of the two additional conditions are true: for if f is in $\ker(P)$ then $\mathrm{grad}(f)$ is a conformal Killing vector.

Proposition 5 *If g_0 is a strongly regular metric on the compact manifold M, then any metric g that is sufficiently close to g_0 in, say, the C^∞ topology, is also strongly regular.*

Proof This is a consequence of standard elliptic theory: the map

$$[g] \mapsto \dim \ker(L_g)$$

is upper semi-continuous.

It is anticipated that strong regularity will turn out to be a generic condition, but this Proposition falls well short of such a result. In the next two sections, we shall consider strong regularity of conformally flat Kähler surfaces, and of connected sums of $S^1 \times S^3$.

4 STRONG REGULARITY ON KÄHLER MANIFOLDS

If M is a complex Kähler surface, with complex structure given by the operator $J : TM \to TM$, then M is ASD iff it has zero scalar curvature. It is now a remarkable fact that L becomes Lichnerowicz's fourth-order operator (Besse 1987 pp. 91–92). This enables one to conclude that if $Lf = 0$, then $X = \mathrm{grad}(f)$ is a holomorphic vector field ($\mathcal{L}_X J = 0$ ($\mathcal{L}$ = Lie derivative)). Thus M is strongly regular iff it supports no non-parallel holomorphic vector fields.

Now let Γ be the fundamental group of a compact Riemann surface of genus ≥ 2 and let us consider the Riemannian product $U = S^2 \times H$, where H is the hyperbolic plane equipped with the standard metric of curvature -1. Then U is conformally flat and Kähler, and by allowing Γ to act by isometries, one can obtain compact quotients. Now if Γ acts irreducibly in the first factor, the quotient M is the projectivization of a stable holomorphic vector bundle (Narasimhan & Seshadri 1965), and it follows that M supports no holomorphic vector fields. If, on the other hand, Γ acts trivially on the first factor, the quotient M is just the product $S^2 \times \Sigma$ (the projectivization, if you will, of the trivial C^2-bundle over Σ). This product has a 3-dimensional space of non-parallel holomorphic vector fields corresponding to the holomorphic automorphisms of the first factor. One can perturb the trivial representation to an irreducible representation by hand, as follows. If the genus of Σ is k, one has to find elements (a_i, b_i) of $SU(2)$ such that

$$a_1 b_1 a_1^{-1} b_1^{-1} \ldots a_k b_k a_k^{-1} b_k^{-1} = 1. \tag{7}$$

Let t be real and $X_1, \ldots, X_k$ be elements of the Lie algebra of $SU(2)$. Then (7) is satisfied by taking

$$a_i = \exp(tX_i), b_i = 1$$

and this representation will be irreducible for any value of $t \neq 0$, if the X_i are generically chosen. On the other hand, as $t \to 0$, this representation becomes trivial. To summarize:

Theorem 6 *There exist conformally flat structures on manifolds diffeomorphic to $S^2 \times \Sigma$ (where Σ is a compact Riemann surface of genus k) that are* not *strongly regular, but the generic perturbation of any such structure will be strongly regular.*

5 CONNECTED SUMS OF $S^1 \times S^3$

In this section we shall consider the class of conformally flat manifolds diffeomorphic to $k(S^1 \times S^3)$ that arise as quotients of an open subset U of R^4 by a discrete subgroup Γ of Conf_4 generated by k elements. We shall make some remarks about the moduli space of such structures and show that there exist strongly regular structures among them.

U/Γ versus connected sum

Let us begin by comparing the quotient construction just mentioned with the differential-geometric notion of connected sum. Select $2k$ distinct points, and partition them into two equal sets, $P_1, \ldots, P_k$ and $Q_1, \ldots, Q_k$. Near each of these points, make a conformal rescaling of the metric to the standard metric on the cylinder of some small radius r. Now for each i, identify the cylindrical ends corresponding to P_i and Q_i by means of a an isometry of S^3. In each case, a $15k - 15$-dimensional family of flat conformal structures is obtained, provided $k > 1$.

Indeed, the freedom in choosing k generators of Γ comprises $15k$ real numbers, but one gets equivalent conformal structures whenever Γ is conjugated by any fixed element of $SO(5,1)$. In the second description, the count is: $2k$ points of S^4, k isometries of S^3 (elements of $SO(4)$), and k lengths (how far you go along the cylinders before you identify); a total of $15k$ real numbers, as before. Once again, 15 has to be subtracted since equivalent conformal structures arise if the data is transformed by a conformal motion of S^4.

For a more detailed comparison of these constructions, fix attention on P_1 and Q_1. Identify $S^4 - Q_1$ with R^4 by stereographic projection (from Q_1) and let the image of P_1 be a. In the quotient construction, the identification corresponding to the pair (P_1, Q_1) is given by a conformal motion which fixes this pair. The most general such motion can be written down as follows.

Identify Conf_4 with the realization of $SO(5,1)$ that preserves the quadratic form with matrix

$$\begin{bmatrix} 0 & 0 & 1 \\ 0 & 1_4 & 0 \\ 1 & 0 & 0 \end{bmatrix}$$

and identify R^4 with the null vectors of the form

$$\begin{bmatrix} -|x|^2/2 \\ x \\ 1 \end{bmatrix}.$$

Then ∞ corresponds to the ray $[1,0,0]^t$ and any element that fixes ∞ has the form:

$$\begin{bmatrix} \lambda & r^t & \mu \\ 0 & A & s \\ 0 & 0 & \lambda^{-1} \end{bmatrix} \quad \text{where } s = -Ar/\lambda, \ \mu = -|r|^2/2\lambda. \tag{8}$$

Now for X in the Lie algebra of $SO(4)$ and real t, define $\gamma_{X,a}(t)$ by putting

$$\lambda = e^t, r(t) = (e^t - e^{-tX})a, s(t) = (e^{-t} - e^{tX})a, \mu(t) = a^t(\cosh(tX) - \cosh t)a$$

in (8). This is the most general 1-parameter subgroup of Conf_4 that fixes a and ∞. For large positive values of t, this element maps any small sphere with centre a to a large sphere.

To move from here to the connected sum, use $\gamma_{0,a}(t)$ to identify $R^4 - a$ with the cylinder $R \times S^3$. Thus $(t,y) \in R \times S^3$ is mapped to $x = a + e^t y \in R^4$. Here we think of y as a unit vector in R^4, $|y| = 1$. Then in the (t,y) coordinates, $\gamma_{X,a}(T)$ acts as

$$(t,y) \mapsto (t+T, Ay) \ (T \in R, \ A \in SO(4)).$$

This completes the comparison between U/Γ and the connected sum since it shows how $\gamma_{X,a}(T)$ gives rise to the 'gluing data' needed for the connected sum.

Our next task is to rescale the round metric on S^4 to the cylinder metric for $|t| \gg 0$. We shall show how to do this in the special case $a = 0$, which corresponds to P_1 and Q_1 being antipodal, so that T_2 is everywhere positive semi-definite. Then Theorem 2 will give strong regularity of the connected sum.

We begin by transferring the round metric to the cylinder. After stereographic projection to R^4, the round metric takes the form

$$g = \frac{4|dx|^2}{(1+|x|^2)^2}.$$

Pulling back to the cylinder with $\gamma_{0,0}(t)$, we obtain

$$(\cosh t)^{-2}[dt^2 + d\Omega_y^2] \tag{9}$$

where $|y| = 1$ and $d\Omega_y^2$ denotes the round metric on the unit y-sphere.

Note that for a cylinder, $T = 6dt^2 + 2(3-\lambda)|dy|^2$, while for (9), $T = 3(4-\lambda)g$; so T_2 is positive semi-definite on both these spaces. On the other hand, if the metric is rescaled, $\hat{g} = e^{2f}g$, then T changes as follows:

$$\hat{T} = T + [(6-\lambda)\Delta f \cdot g + 2\lambda \mathrm{Hess}(f)] + [(2\lambda - 6)|df|^2 \cdot g - 2\lambda df \otimes df]$$

Here $\mathrm{Hess}(f) = \nabla df$ and $\Delta f = -\mathrm{tr}(\mathrm{Hess} f)$.

To construct a strongly regular metric, take g to be the cylinder metric, f a function of t only and $\lambda = 2$. Then

$$\hat{T} = 6(1 - f'^2)dt^2 + 2(1 - 2f'' - f'^2)d\Omega^2 \tag{10}$$

and the problem is to find a smooth f such that:

$\hat{T}$ is positive semi-definite for all $t \geq 0$;

for $t \leq \alpha$, say, $\hat{g}$ is the metric on the sphere, i.e. $e^{-f} = \cosh t$;

for $t \geq \beta$, say, f is constant (giving the cylinder metric).

The first of these conditions yields two differential inequalities on f. The inequality which comes from the coefficient of dt^2 in (10) is easy to understand, while the more complicated coefficient of $d\Omega^2$ can be linearized:

Lemma 7 *If a is any real number and f any smooth real function of t, then*

$$a^{-1}f'' + f'^2 \leq 1 \text{ iff } \phi'' \leq a^2\phi$$

where

$$f = a^{-1}\log\phi$$

Applying this with $a = 1/2$, we transform our problem to that of finding a smooth positive function ϕ on $t \geq 0$ such that

$$-\phi/2 \leq \phi' \leq \phi/2 \text{ and } \phi'' \leq \phi/4 \tag{11}$$

and satisfying the boundary conditions

$$\phi = \begin{cases} (\cosh t)^{-1/2}, & t \leq \alpha \\ c > 0, & t \geq \beta \end{cases}$$

The next result says that in the presence of the boundary condition at β, the second inequality of (11) is stronger than the first.

Proposition 8 *Suppose that $\phi'' \leq \phi/4$ and $\phi = c$ for $t \geq \beta$. Then*

$$\phi(t) \leq c \cdot \cosh\frac{t-\beta}{2} \text{ for } t \leq \beta. \tag{12}$$

Furthermore, $\phi'' \geq -\phi/2$; so if ϕ is monotonically decreasing, then both parts of (11) are satisfied.

Proof We obtain (12) by setting $\phi(t) = e^{-t/2}\psi(t)$. Then we have $\psi'' \leq \psi'$, and integrating from t to β a couple of times gives the result. For the second part, consider $\chi = \phi' + \phi/2$. Then (11) gives

$$\frac{d}{dt}(\chi e^{-t/2}) \leq 0$$

so $\chi e^{-t/2}$ is decreasing. Since $\chi(\beta) > 0$, it follows that $\chi > 0$ for all $t \leq \beta$.

We use this to get 'weak solutions' to our problem of the form

$$\phi = \begin{cases} (\cosh t)^{-1/2}, & t \leq \alpha; \\ c \cdot \cosh\frac{t-\beta}{2}, & \alpha \leq t \leq \beta; \\ c, & t \geq \beta. \end{cases} \tag{13}$$

where α, β and c are related by

$$c \cdot \cosh\frac{\beta-\alpha}{2} = (\cosh\alpha)^{-1/2} \tag{14}$$

and

$$c \leq (\cosh \alpha)^{-3/2}. \tag{15}$$

The first of these gives the continuity of ϕ at $t = \alpha$: the second guarantees that the inequality remains satisfied (weakly) at α. Note that c can be made as small as one likes by the following procedure. Given c, choose $\alpha > 0$ so that (15) holds. Then choose $\beta > 0$ to satisfy (14). Note also that (13) defines a C^1 function if we take $c^2 = (\cosh \alpha)^{-3}$ and $\beta = 3\alpha$.

We now use regularization to go from our weak solution (13) to a C^∞ solution.

Lemma 9 *Suppose that ϕ_0 is continuous on $t \geq 0$ and L is a constant-coefficient linear ordinary differential operator. Suppose also that*

$$L\phi_0 \geq 0 \text{ for } t \geq 0 \text{ with strict inequality on } [0, a]$$

and that ϕ_0 is C^∞ on $[0, a]$. Then given $\delta > 0$, there exists a smooth function ϕ on $t \geq 0$ such that $L\phi \geq 0$ and $\phi = \phi_0$ on $[0, a - \delta]$.

Proof Use a smooth cut-off function to write $\phi_0 = \phi_1 + \phi_2$ where the support of ϕ_1 is contained in $[0, a]$ and that of ϕ_2 is contained in $t \geq a - \delta$. Thus $\phi_1 = \phi_0$ on $[0, a - \delta]$ and is smooth everywhere. We shall smooth ϕ_2 by convolution with a bump function ρ.

To be more precise, assume that ρ is smooth, non-negative, has total integral 1 and support contained in $|t| < \varepsilon$. Denote by f_ε the convolution $f * \rho$. Then

f_ε is smooth;

$f \geq 0$ implies $f_\varepsilon \geq 0$;

$(f_\varepsilon)' = (f')_\varepsilon$;

as $\varepsilon \to 0$, $f_\varepsilon \to f$ (and the same is true for all derivatives);

if f is constant on some interval then $f_\varepsilon =$ the same constant on a slightly smaller interval.

Now put $\phi = \phi_1 + (\phi_2)_\varepsilon$. Then ϕ is smooth and

$$L\phi = L\phi_1 + (L\phi_2)_\varepsilon$$

It is clear from this that $L\phi \geq 0$ on $[0, a - \delta]$ and $[a, \infty)$. But on $[a - \delta, a]$,

$$L\phi - L\phi_0 = (L\phi_2)_\varepsilon - L\phi_2.$$

Now if $L\phi_0 > \eta$ on $[0, a]$, and ε is chosen so small that

$$|(L\phi_2)_\varepsilon - L\phi_2| < \eta$$

then we have $L\phi > 0$ on $[a - \delta, a]$ as required.

This completes the proof that there exists a family of strongly regular conformally flat structures on $k(S^3 \times S^1)$ for any positive k. As we have pointed out, these structures correspond to taking P_i to be antipodal to Q_i, so they make up a family of dimension $11k - 15$. Proposition 5, however, guarantees that there exists a non-empty *open* set of strongly regular conformally flat structures on $k(S^1 \times S^3)$. Intuitively, these correspond to perturbing the arrangement of the P_i and Q_i so that they no longer come in antipodal pairs. It is possible to do this by hand, but the calculations are complicated; it seems much simpler to invoke Proposition 5.

REFERENCES

Besse, A. (1987) Einstein Manifolds, Springer-Verlag, Berlin etc.

Eastwood, M. G. & M. A. Singer (1985) A conformally invariant Maxwell gauge, *Phys. Lett.* **107A**: 73–74.

Eastwood, M. G. & M. A. Singer (1993) The Fröhlicher spectral sequence on a twistor space, *Jour. Differential Geometry* **38**: 653–669.

Hitchin, N. J. (1981) Kählerian twistor spaces, *Proc. London Math. Soc.* **43**: 133–150.

Kuiper, N. H. (1949) On conformally flat spaces in the large, *Ann. of Math.* **50**: 916–924.

Narasimhan, M. S. & Seshadri, C. S. (1965) Stable and unitary bundles on a compact Riemann surface, *Ann. of Math.* (2) **82**: 540–567.

13
Cohomogeneity-One Metrics with Self-Dual Weyl Tensor

K. P. Tod Mathematical Institute and St. John's College,
Oxford, United Kingdom

ABSTRACT

I review some of what is known about cohomogeneity-one (or 'spatially homogeneous') 4-dimensional metrics with self-dual Weyl tensor and give explicit expressions for a family of Bianchi-type IX Einstein metrics.

1 INTRODUCTION

My main aim in this article is to give explicit expressions for a family of Bianchi-type IX Einstein metrics with self-dual Weyl tensor. The existence of these metrics has been known, or at least strongly suspected, for some years. A description of their twistor spaces, and an explicit form for the conformal metric was given by Nigel Hitchin in a seminar in Oxford in October 1991 (see Hitchin 1993). What is new in this article is the correct choice of conformal scale to make the metrics Einstein. With the aim of making the subject more inviting, I shall recall the motivation for studying self-dual cohomogeneity-one metrics and review the Bianchi classification.

Before that, there are some questions of terminology to address. In general relativity, a space-time is 'spatially-homogeneous' if it has an isometry group which is transitive on space-like 3-surfaces; in Riemannian geometry, a space has 'cohomogeneity-one' if it has an isometry group transitive on codimension-one surfaces. Given the origins of twistor theory it has been common practice to retain the usage of general relativity even when speaking of Riemannian or complex 'space-times' and say 'spatially-homogeneous' even when the distinction between space and time is lost.

Next, the classification of 3-dimensional Lie algebras due to Bianchi uses the term 'Bianchi-type IX' for the Lie algebra of $SU(2)$, so relativists will say 'Bianchi-type IX' where mathematicians would say 'an action of $SU(2)$ transitive on 3-surfaces'.

Finally there is the problem of what to call a 'space-time' whose Weyl tensor is self-dual but where no restrictions are implied on the Ricci tensor. One finds 'half-conformally-flat' (Besse 1987), 'conformally-half-flat', 'right-(or left-)conformally-flat' (Penrose and Rindler 1986) or simply 'self-dual' (LeBrun 1991b) as well as circumlocutions like 'with self-dual Weyl tensor/conformal structure' such as I have used in my title. None is ideal!

There are three reasons for considering these metrics, whatever they are to be called.

Firstly, a 'spatially-homogeneous' metric is a function only of 'time' so that any field equations imposed on the metric reduce to a system of ODEs rather than a system of PDEs. On the one hand this makes them easier to solve, on the other hand it provides a test of the 'integrability metatheorem' (or 'dogma'). By this I mean the idea that any system of ODEs obtained from a system of PDEs corresponding to a geometrical problem solved by a twistor construction should itself be solvable (e.g. should satisfy the Painlevé test and have no movable singularities except for poles). Thus if one imposes field equations which have a twistor solution one anticipates obtaining a system of ODEs which can be solved explicitly while conversely if there is no twistor solution for the field equations then the system of ODEs should not be solvable, or should fail the Painlevé test, or could possibly have chaotic solutions.

Next there is the 'filling-in' problem. The standard round metric on the 3-sphere $\mathbf{S}^3$ can be 'filled-in' by the 4-dimensional hyperbolic metric $\mathbf{H}^4$, so that it is the metric at infinity and can be attached by conformally rescaling $\mathbf{H}^4$. One can ask, given a different 3-metric on $\mathbf{S}^3$ at infinity, can one fill it in with a canonical nice metric? According to Graham and Lee (1991), a smooth metric close to the standard one will fill in to give an Einstein metric on the ball. LeBrun (1982) had earlier shown that an analytic 3-metric will fill in part of the way with an Einstein metric which also has self-dual Weyl tensor, in that the conformal 4-metric is defined on a collar neighbourhood of the complexified 3-sphere at infinity. LeBrun's proof gives a construction for the twistor space $\mathcal{T}$ of the filled-in space; a twistor thought of as a totally-null 2-surface in the interior meets the (totally-umbilic) 3-sphere at infinity in a curve which is necessarily a null-geodesic of the 3-dimensional conformal structure; one therefore constructs $\mathcal{T}$ as the space of null geodesics of this conformal structure. Given $\mathcal{T}$ one finds the 4-metric by standard twistor methods (Ward 1980; LeBrun 1982) though this is still a non-trivial exercise.

It is not clear from the LeBrun construction which conformal 3-metrics on $\mathbf{S}^3$ will fill in to give complete self-dual Einstein metrics on the ball in $\mathbf{R}^4$ and which will not. For those that will, he proposed the term 'positive-frequency' (LeBrun 1991b). Pedersen (1986) studied the LeBrun construction for the left-invariant metric h on $\mathbf{S}^3$ with an extra symmetry (an action of $U(2)$ rather than $SU(2)$) which is given in terms of a basis (σ_i; $i = 1, 2, 3$) of invariant 1-forms by

$$h = {\sigma_1}^2 + {\sigma_2}^2 + \lambda{\sigma_3}^2, \quad \text{constant } \lambda, \tag{1.1}$$

($\mathbf{S}^3$ with this metric is sometimes called the Berger sphere). He found the twistor space for this case and the corresponding Einstein metric and showed that it filled in to give a complete metric (of cohomogeneity-one) on the ball. Thus the metric (1.1) is positive frequency, and it fills in with a family of nested Berger spheres, functions only of the radial coordinate.

It is now natural to ask how the general left-invariant metric on $\mathbf{S}^3$ fills in, and whether it has positive frequency. Since the Berger sphere fills in with a cohomogeneity-one 4-metric, this is the first thing to try for a 4-metric to fill in the general left-invariant 3-metric.

The final reason for looking at cohomogeneity-one metrics is to construct a framework for pigeon-holing all the many known such metrics, associated for example with the names of Taub-NUT, Mixmaster, Eguchi-Hanson, Belinski-Gibbons-Page-Pope, Atiyah-Hitchin, Gibbons-Pope, Pedersen-Poon etc. etc. All of these are Bianchi-type IX, but with different conditions on the curvature, or equivalently they satisfy different field equations.

This article is a more detailed version of the talk given at Seale Hayne. I am very grateful to Stephen Huggett and his colleagues for the invitation to participate in such an excellent and well-organised conference in such stimulating and original surroundings.

2 THE BIANCHI CLASSIFICATION

The Bianchi classification is for 3-dimensional, real Lie algebras, (Bianchi 1897) which, for us, will be the Lie algebras of isometry groups of cohomogeneity-one metrics. A metric may have an isometry group transitive on 3-surfaces without having a 3-dimensional subgroup which is transitive (think of e.g. $S^1 \times S^2$ with the product metric) but we won't consider these cases, which are usually called Kantowski-Sachs cosmologies rather than Bianchi cosmologies in the literature of relativity (see e.g. Kramer et al 1980 for more on this subject).

Suppose then that X_i, $i = 1, 2, 3$, form a basis for the Lie algebra. The structure constants $c_{ij}{}^k$ are defined by the commutation relations

$$[X_i, X_j] = c_{ij}{}^k X_k \tag{2.1}$$

and are subject to the relation

$$c_{[ij}{}^m c_{k]m}{}^n = 0 \tag{2.2}$$

which follows from the Jacobi identity

$$[[X_i, X_j], X_k] + \quad \text{cyclic sum} \quad = 0. \tag{2.3}$$

The problem is to classify the possible sets of structure constants and give canonical forms for them, given the freedom in the choice of basis:

$$X_i \rightarrow \hat{X}_i = L_i{}^j X_j, \quad L_i{}^j \in GL(3, \mathbf{R}) \tag{2.4}$$

This problem is solved, following Estabrook et al. (1968) and Ellis and MacCallum (1969), by first choosing a skew tensor ϵ^{ijk} and dualising the structure constants:

$$\frac{1}{2}\epsilon^{ijk} c_{ij}{}^m \equiv t^{km} = n^{km} + \epsilon^{kmn} a_n \tag{2.5}$$

Here t^{km} is split into its symmetric part n^{km} and its antisymmetric part, which is then dualised and written in terms of a vector a_i. The great virtue of this parametrisation is that (2.2) reduces to the simple equation

$$n^{ij} a_j = 0. \tag{2.6}$$

Under (2.3), n^{ij} transforms as a quadratic form, which can therefore be reduced to a diagonal matrix with entries 0, 1 or -1, and a_i transforms as a vector. The classification splits into two classes, Class A in which the vector a_i is zero and Class B in which a_i may be assumed to be non-zero only for $i = 1$, say $a_1 = a$. In each class, there are then various possibilities for the diagonal matrix $n^{ij} = \mathrm{diag}(n_1, n_2, n_3)$ where, in Class B, n_1 is zero by virtue of (2.6). Also in Class B the most general cases have a continous invariant:

$$h = \frac{a^2}{n_2 n_3} \tag{2.7}$$

which is not changed by (2.3). (This invariant is not a scalar invariant but is a ratio of tensors. If the vector a_i is non-zero then h may be obtained from the equation

$$c_{mi}{}^{p} c_{pj}{}^{m} = \frac{(h-1)}{2h} c_{mi}{}^{m} c_{pj}{}^{p}.)$$

The different types, in the nomenclature of Bianchi (1897) which is still used, are tabulated as follows:

Class A

type	n_1	n_2	n_3
I	0	0	0
II	1	0	0
VI_0	0	1	-1
VII_0	0	1	1
VIII	1	1	-1
IX	1	1	1

Class B

type	a	n_2	n_3
V	1	0	0
IV	1	0	1
III	1	1	-1
VI_h	$\sqrt{-h}$	1	-1
VII_h	$\sqrt{h}$	1	1

A great deal can be said about these tables, the groups which correspond to each type, and canonical 3-geometries which have corresponding isometry groups. For example, type IX is the Lie algebra of $SU(2)$ as remarked previously; type VIII corresponds to $SU(1,1)$; type II to Nil and type VI_0 to Sol in the terminology of Thurston. The connection with Thurston's 3-geometries (Thurston 1982) can also be deduced from Scott (1983) and was discussed recently by Fujiwara et al. (1993). However for our purposes it is the canonical forms of the structure constants which are important.

We seek metrics on some suitable 3-manifold Σ with the $\{X_i\}$ as a set of Killing vectors. Our considerations will always be local, so further questions about the 3-manifold Σ, and the globality of the Killing vectors are ignored. For more on these see e.g. Fujiwara et al. (1993) or Ashtekar and Samuel (1991). We next choose a basis $\{\sigma^i\}$ of invariant 1-forms on Σ:

$$\pounds_{X_i} \sigma^j = 0 \text{ for all } i, j.$$

A standard argument shows that there is enough freedom in the choice of the σ^i to impose the equation:

$$d\sigma^i = \frac{1}{2} c_{jk}{}^{i} \sigma^j \wedge \sigma^k \tag{2.8}$$

with the structure constants $c_{ij}{}^{k}$ as in (2.1). Now one may use the results of the tables to write out (2.8):

$$\text{for Class A: } d\sigma^1 = n_1 \sigma^2 \wedge \sigma^3, \quad d\sigma^2 = n_2 \sigma^3 \wedge \sigma^1, \quad d\sigma^3 = n_3 \sigma^1 \wedge \sigma^2 \tag{2.9A}$$

for Class B: $d\sigma^1 = 0, \quad d\sigma^2 = -a\sigma^1 \wedge \sigma^2 + n_2\sigma^3 \wedge \sigma^1, \quad d\sigma^3 = a\sigma^3 \wedge \sigma^1 + n_3\sigma^1 \wedge \sigma^2$ (2.9B)

For each type, one may introduce coordinates and solve (2.9) for the 1-forms. As a simple example of the method we consider type II: here σ^2 and σ^3 are exact so we may write $\sigma^2 = dy$, $\sigma^3 = dz$ for coordinate functions y and z; then, from (2.9A), $d\sigma^1 = \sigma^2 \wedge \sigma^3 = dy \wedge dz$, so that $\sigma^1 = dx + ydz$ introducing the third coordinate x.

Standard expressions for the 1-forms for each type are tabulated in e.g. Kramer et al. (1980). Given the 1-forms for a particular type, one writes down the 3-metric

$$h = h_{ij}\sigma^i\sigma^j \tag{2.10}$$

and this is the desired homogeneous 3-metric. The spatially-homogeneous or cohomogeneity-one 4-metric is obtained by stacking up the 3-metrics in a 'time'-dependent way:

$$g = dT^2 + h_{ij}(T)\sigma^i\sigma^j \tag{2.11}$$

In the next section, we shall study the results of imposing various field equations on the metric (2.11). Before that we briefly consider the issue of diagonalisability: clearly the 3-metric h can be diagonalised at one time T; will it then stay diagonal? Naturally this depends on the field equations. It is an important question as, for a diagonal metric, one is typically faced with only three non-linear second-order ODEs rather than six for a general metric.

The extrinsic curvature of a surface of constant T for the metric (2.11) can be defined as

$$K_i{}^j = h^{jk}\frac{dh_{ik}}{dT} \tag{2.12}$$

The momentum constraint of general relativity, also called the Codazzi equation, relates components of the 4-dimensional Einstein tensor $G_{\alpha\beta}$ to derivatives of $K_i{}^j$:

$$G_{0i} = D_jK_i{}^j - D_iK_j{}^j \tag{2.13}$$

where D_i denotes the intrinsic 3-dimensional covariant derivative. For a spatially-homogeneous vacuum or Einstein solution, the left-hand-side of (2.13) vanishes and the right-hand-side can be written in terms of the structure constants parametrised according to (2.4):

$$\frac{1}{3}\epsilon_{ijk}K_m{}^jn^{km} + (K_i{}^j - \frac{1}{3}K_m{}^m\delta_i{}^j)a_j = 0 \tag{2.14}$$

For types VIII and IX, this constraint implies that the extrinsic curvature is diagonal; thus if the 3-metric is diagonalised at one time, it will stay diagonal and all type VIII or IX Einstein metrics may be assumed to be diagonal, without loss of generality. For other types, and notably for Class B, the situation is different and more delicate. One may still impose as an ansatz that the metric is diagonal, but one may then be missing other possibilities.

3 ALTERNATIVE FIELD EQUATIONS

There are three kinds of restriction which have been imposed, in various combinations, on spatially-homogeneous metrics so we can conveniently display them as a Venn diagram (figure 3.1).

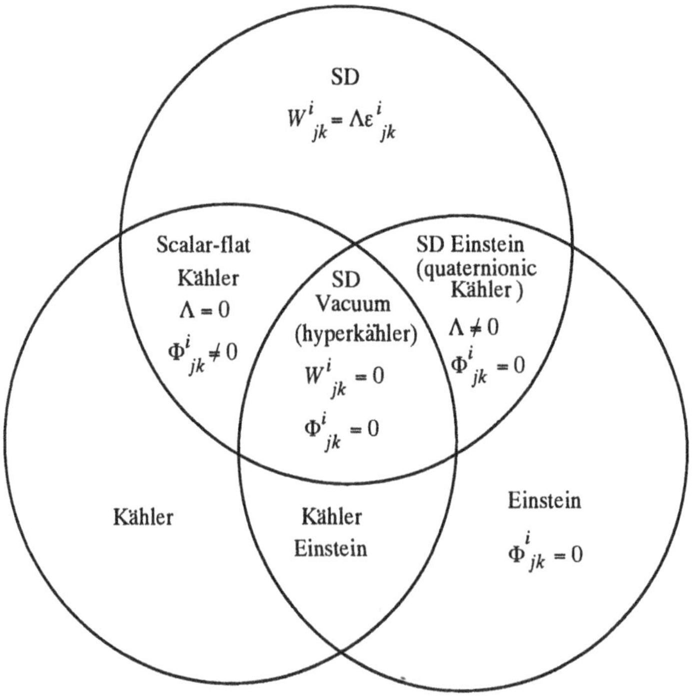

Figure 3.1: The various field equations

One may require the metric to be Einstein or to be Kähler or to have self- dual (SD) Weyl tensor. The last condition is conformally-invariant and the resulting equations will have a corresponding gauge freedom. To fix the gauge it is convenient to insist as well that the scalar curvature is zero and this we will often do. LeBrun (1991a) suggests the term 'scalar-flat' for metrics with zero scalar curvature.

Scalar-flat Kähler metrics with the standard orientation defined by the complex structure automatically have *anti*-self dual Weyl tensor. However, rather than change the relevant set of signs in the equations in Section 4, we will use the "wrong" orientation and think of these metrics as SD. This is one place where one needs to be careful about the distinction between SD and ASD Weyl tensors: **CP2** with the Fubini-Study metric is Kähler and has SD Weyl tensor with the "right" orientation (and is Einstein) but is not scalar-flat.

In four dimensions, Einstein metrics with SD Weyl tensor, but not scalar-flat, are sometimes known as quaternionic-Kähler metrics (this is a rather degenerate usage; usually quaternionic-Kähler manifolds are defined in dimension 4k with $k > 1$ Besse 1987). Note that this case is not Kähler.

On the other hand, vacuum metrics with SD Weyl tensor are known as hyper- Kähler and are 'very' Kählerian: any covariantly-constant spinor dyad, of which there are as many as possible, defines a Kähler form.

The different cases may be classified in terms of the connection on ASD 2-forms. If $\{\phi^i{}_-, i = 1, 2, 3\}$ is a normalised basis of ASD 2-forms then the connection 1-forms $\alpha^i{}_j$ are defined by

$$d\phi^i{}_- = \alpha^i{}_j \wedge \phi^j{}_- \tag{3.1}$$

If $\{\phi^i{}_+, i = 1, 2, 3\}$ is now a basis of SD 2-forms, then the curvature on ASD 2-forms is defined by the equations

$$d\alpha^i{}_j - \alpha^i{}_k \wedge \alpha^k{}_j = W^i{}_{jk}\phi^k{}_- + \Phi^i{}_{jk}\phi^k{}_+ \tag{3.2}$$

Here $W^i{}_{jk}$ corresponds to the ASD Weyl tensor together with the scalar curvature, and $\Phi^i{}_{jk}$ corresponds to the trace-free part of the Ricci tensor. In terms of these quantities, an Einstein metric has vanishing $\Phi^i{}_{jk}$ and a metric with SD Weyl tensor has $W^i{}_{jk} = \Lambda\epsilon^i{}_{jk}$ where Λ is a multiple of the scalar curvature; a scalar-flat metric with SD Weyl tensor has vanishing $W^i{}_{jk}$ and a vacuum metric with SD Weyl tensor has both $\Phi^i{}_{jk}$ and $W^i{}_{jk}$ vanishing, so that the connection on ASD 2-forms, from (3.2), is flat (see figure 3.1).

The framework of the Venn diagram as in figure 3.1 may also be used to pigeon-hole known Bianchi-type IX metrics as follows:

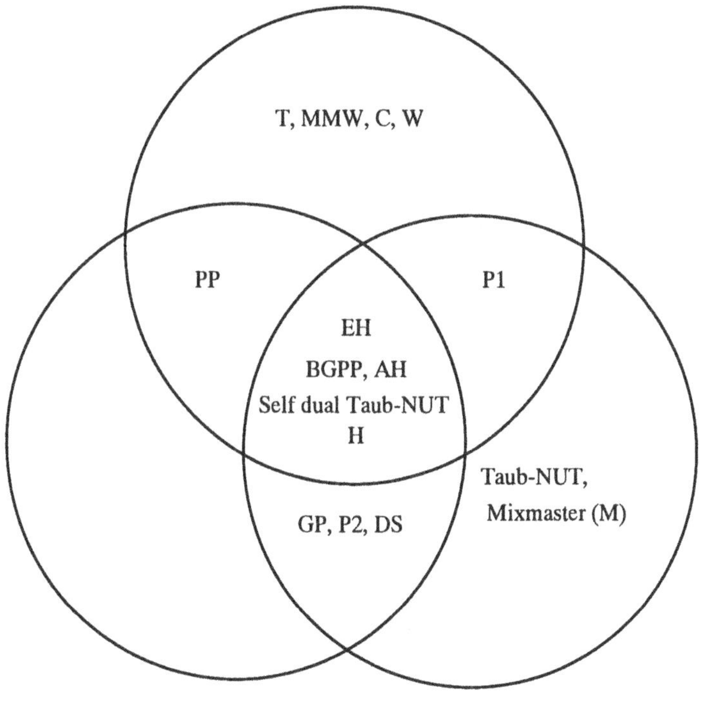

Key:

AH: Atiyah & Hitchin 1985, 1988.
BGPP: Belinski et al 1978.
C: Chakravarty 1993.
DS: Dancer and Strachan 1993a,b.
EH: Eguchi & Hansen 1978.
GP: Gibbons and Pope 1978, 1979.
H: Hawking 1977.
M: Misner 1969.
MMW: Maszczyk et al 1993.
P1: Pedersen 1986.
P2: Pedersen 1985.
PP: Pederson & Poon 1990.
Taub-NUT: see e.g. Hawking & Ellis 1973.
T: Tod 1991.
W: see Woodhouse, ch. 14, this volume.

Figure 3.2: The known Bianchi-IX metrics

One may check this figure against the integrability meta-theorem; everything inside the top circle is integrable, including the non-diagonal cases thanks to Maszczyk et al. (1993) (see the article by Woodhouse in these Proceedings); for Einstein but outside the top circle, the Lorentzian case includes the Mixmaster universe which is widely believed to be chaotic; for Kähler-Einstein but outside the top circle Dancer and Strachan (1993a,b) obtain a system which fails the Painlevé test. (However, with an extra symmetry, $U(2)$ instead of $SU(2)$, both the Kähler-Einstein and the Lorentzian vacuum cases can be solved in elementary functions.)

Our next step is to write down the Class A metric suitably parametrised and perform the calculation behind (3.1) and (3.2) in order to impose field equations.

4 IMPOSING THE FIELD EQUATIONS IN CLASS A

In this section we choose a parametrisation of the metric and impose the condition that $W^i{}_{jk}$ vanishes i.e. vanishing ASD Weyl tensor and vanishing scalar curvature. We solve the resulting equations, and then impose the Einstein condition by conformally rescaling the metrics obtained. In this way we find the desired class of Bianchi-type IX Einstein spaces, as well as a number of other, more special, cases. Finally, we consider imposing instead the Kähler condition.

With the benefit of experience, we shall take the metric to be

$$g = w_1 w_2 w_3 dt^2 + \frac{w_2 w_3}{w_1}\sigma_1{}^2 + \frac{w_3 w_1}{w_2}\sigma_2{}^2 + \frac{w_1 w_2}{w_3}\sigma_3{}^2 \tag{4.1}$$

where the σ_i are chosen to satisfy (2.9A). In this way, we may deal with all diagonal Class A metrics at once.

For the basis of ASD 2-forms we take

$$\phi^1{}_- = w_2 w_3 dt \wedge \sigma_1 - w_1 \sigma_2 \wedge \sigma_3 \text{ and cyclic permutations} \tag{4.2}$$

It turns out that (3.1) can be satisfied by setting

$$\alpha^1{}_2 = \frac{A_3 \sigma_3}{w_3} \text{ and cyclic permutations} \tag{4.3}$$

then (3.1) reduces to the system

$$\dot{w}_1 = -n_1 w_2 w_3 + w_1(A_2 + A_3) \text{ and cyclic permutations} \tag{4.4}$$

where a dot denotes differentiation with respect to t.

Finally imposing $W^i{}_{jk} = 0$ on $\alpha^i{}_j$ via (3.2) results in the system

$$\dot{A}_1 = -A_2 A_3 + A_1(A_2 + A_3) \text{ and cyclic permutations} \tag{4.5}$$

The problem thus reduces to solving the system (4.5), using the solution of this in (4.4) and then solving that system. This is essentially the problem set in Tod (1991) except with the inclusion of the n_i in (4.4) which allows all types of Class A to be considered at once. The system (4.5) has a long history, having been studied and solved more than once in the nineteenth century. It is dealt with in some detail by Ablowitz and Clarkson (1991).

For the case of type IX, where $n_1 = n_2 = n_3 = 1$, one may note the following points about the two systems:

(i) if all the A_i are zero, then the connection on ASD 2-forms is clearly flat and the metric is vacuum. This is the case of Belinski et al. (1978) and Eguchi-Hansen (1978).

(ii) if, for each i, $w_i = A_i$, then (4.4) and (4.5) are identical. This turns out to be the case of Atiyah-Hitchin (1985,1988); the metric is vacuum but the invariant ASD 2-forms are not covariantly constant.

(iii) if one imposes as an ansatz that all the A_i are constant then two of them are necessarily zero. Following through one arrives at the Pedersen-Poon scalar-flat, Kähler metric (Pedersen and Poon 1990).

(iv) there is a conserved quantity

$$\frac{{w_1}^2}{(A_1-A_2)(A_1-A_3)}+\frac{{w_2}^2}{(A_2-A_1)(A_2-A_3)}+\frac{{w_3}^2}{(A_3-A_1)(A_3-A_2)} \tag{4.6}$$

This will prove significant shortly (something similar is constant for the other types.

(v) there is a covariance under fractional linear transformations in t, see Tod (1991). This means in particular that the solution of (4.5) with say $A_1 = A_2 \neq 0$ is conformally related to the Pedersen-Poon solution.

We proceed to follow Brioschi (1881) in solving (4.5). Introduce the new dependent variable

$$x = \frac{A_1 - A_2}{A_3 - A_2} \tag{4.7}$$

then it is not too hard to show that (4.5) reduces to the third-order ODE

$$\frac{\dddot{x}}{\dot{x}} = \frac{3}{2}\frac{\ddot{x}^2}{\dot{x}^2} - \frac{\dot{x}^2}{2}\left(\frac{1}{x^2}+\frac{1}{x(1-x)}+\frac{1}{(1-x)^2}\right) \tag{4.8}$$

Remarkably this ODE is satisfied by the reciprocal of the elliptic modular function. Thus the solutions have a natural boundary in the complex t-plane (see Ablowitz and Clarkson 1991, who follow Chazy 1910, for more on this point).

Having solved (4.5), we may substitute into (4.4). Here it is convenient to redefine the dependent variables as Ω_i where

$$w_1 = \Omega_1\dot{x}(x(1-x))^{-1/2}, \quad w_2 = \Omega_2\dot{x}(x^2(1-x))^{-1/2}, \quad w_3 = \Omega_3\dot{x}(x(1-x)^2)^{-1/2} \tag{4.9}$$

and switch to using x as independent variable. With these changes, (4.4) reduces to the system

$$\Omega_1' = \frac{-n_1\Omega_2\Omega_3}{x(1-x)}, \quad \Omega_2' = \frac{-n_2\Omega_3\Omega_1}{x}, \quad \Omega_3' = \frac{-n_3\Omega_1\Omega_2}{1-x} \tag{4.10}$$

where prime denotes differentiation with respect to x.

For the moment we shall consider only type IX, with $n_1 = n_2 = n_3 = 1$. The resulting version of (4.10) has arisen elsewhere (Fokas et al. 1986; Dubrovin 1990) and it is known to reduce to the Painlevé-VI equation. There is a first integral

$${\Omega_2}^2 + {\Omega_3}^2 - {\Omega_1}^2 = 2\gamma, \quad \text{constant} \tag{4.11}$$

which is in fact the first integral noted above in (4.6). The general Painlevé-VI equation contains four parameters usually labelled $(\alpha, \beta, \gamma, \delta)$ and in this case these are $(1/8, -1/8, \gamma, (1-2\gamma)/2)$. (There is more than one way of reducing (4.10) to Painlevé-VI and different routes may lead to different sets of values for these parameters. However, complicated transformations link solutions corresponding to the different sets; see Fokas and Yortsos 1981 or Fokas and Ablowitz 1982).

One way of reducing (4.10) to Painlevé-VI was given in my Twistor Newsletter article (Tod 1992); another way was found by Chakravarty (1993); of course it was independently known to Hitchin and to Mason & Woodhouse and Maszczyk et al. that Painlevé-VI lay behind the Bianchi-type IX metrics with SD Weyl tensor.

If desired, one can now return to the system (4.10) and deal with the other types. Type VIII is an analytic extension of type IX. For type I, all the Ω_i are constant; the metric is the Einstein static cylinder and is conformally flat, as in fact is any non-diagonal type I. For type II, two of the Ω_i are constant, the third is then readily found, and the metric is not conformally flat. Finally, types VI_0 and VII_0 are given by solutions of a second-order linear ODE which can be transformed to a hypergeometric equation. (These results were also found independently by Dancer and Strachan 1993b.)

Returning to the type IX case, the next step is to impose the Einstein condition. This could be done by going back to the metric (4.1) and 2-forms (4.2) and imposing (3.2) but now with $\Phi^i{}_{jk} = 0$ and $W^i{}_{jk} = \Lambda \epsilon^i{}_{jk}$. However there is a more direct way, which is to seek to conformally rescale the metric which we have, in which the ASD Weyl tensor is already zero, in order to eliminate the trace-free Ricci tensor. This method has the advantage of using the calculations which we have already done and so this is the method which we shall follow.

With the redefinitions (4.9) and the choice of x as independent variable, the metric (4.1) becomes

$$g = \Theta\left[\frac{dx^2}{x(1-x)} + \frac{{\sigma_1}^2}{{\Omega_1}^2} + \frac{(1-x){\sigma_2}^2}{{\Omega_2}^2} + \frac{x{\sigma_3}^2}{{\Omega_3}^2}\right] \tag{4.12}$$

where $\Theta = \dfrac{\Omega_1\Omega_2\Omega_3\dot{x}}{x(1-x)}$.

Note that the time-dependence of x is only significant for the conformal factor Θ. The conformal structure, in square brackets, is determined by a solution of Painlevé-VI. The equation on x, which brings with it the natural boundary in t, arises solely from the requirement of scalar-flatness and is therefore only a 'gauge-fixing' equation.

The idea now is to use (4.12) subject to the system (4.10) but to choose Θ differently, namely so as to eliminate the trace-free Ricci tensor. This is an elementary, if complicated, calculation. Along the way, one finds that the procedure is only possible if γ in (4.11) takes the value 1/8 (thus quaternionic-Kähler Bianchi-type IX metrics arise from Painlevé- VI with parameter values (1/8,-1/8,1/8,3/8); this fact was known to Hitchin on the basis of the isomonodromic approach; see Hitchin 1993). In this case it is possible: with $W^i{}_{jk} = \Lambda\epsilon^i{}_{jk}$ we find

$$\begin{aligned}
\Lambda\Theta &= ND^{-2} \quad \text{where} \\
N &= 2\Omega_1\Omega_2\Omega_3(4x\Omega_1\Omega_2\Omega_3 + P) && (4.13)\\
P &= x({\Omega_1}^2 + {\Omega_2}^2) - (1-4{\Omega_3}^2)({\Omega_2}^2 - (1-x){\Omega_1}^2) && (4.14)\\
D &= x\Omega_1\Omega_2 + 2\Omega_3({\Omega_2}^2 - (1-x){\Omega_1}^2)
\end{aligned}$$

These are the new Einstein metrics promised in the Introduction. They form a 2-parameter family (3-parameter if Λ is counted) since one chooses data for the system (4.10) subject to

$$\Omega_2{}^2 + \Omega_3{}^2 - \Omega_1{}^2 = 1/4 \tag{4.15}$$

(A particular solution for this case is given in terms of complete elliptic integrals by Dubrovin 1990. As a check on the accuracy of the complicated expressions in (4.13) one may use the Fubini-Study metric on **CP2** with the opposite orientation.)

One of the motivations of this study was the filling-in problem. What can we say about it now? First of all we must seek the surface at infinity in the metric (4.12) with Θ as in (4.13). It is easy to see that infinity can only be at $x = x_0$ where x_0 is a zero of the function D whose square is the denominator of Θ. If we choose Ω_1 at x_0 then (4.14) together with the vanishing of $D(x_0)$ determines Ω_2 and Ω_3 at x_0. Rescaling (4.12) by the factor $(x - x_0)^2$ and setting $x = x_0$, we obtain for the 3-metric at infinity a diagonal, tri-axial left-invariant metric

$$h = \lambda\sigma_1{}^2 + \mu\sigma_2{}^2 + \nu\sigma_3{}^2 \tag{4.16}$$

where λ, μ, ν are distinct constants expressed in terms of x_0 and $\Omega_1(x_0)$. Since the general, left-invariant metric can be diagonalised, we are indeed filling in the general such metric. Is it positive-frequency? Equivalently, do we obtain a complete metric on the ball in $\mathbf{R}^4$?

This is a more delicate question and the answers, at the time of writing, are tentative. We must evolve the system (4.10) away from the given data at x_0 and see what happens. Now the system (4.10) has movable poles of order one. At such a pole, the metric is singular but this singularity can be removed (i.e. it is like the singularity at the origin of spherical polar coordinates). Thus if the evolution inwards from infinity arrives at a movable pole, which seems to be the generic situation, then the metric is complete. Our conclusion is that the metric (4.15) is positive frequency in general, though the problem needs more study to be certain.

We may briefly return to the question of imposing the Einstein condition on the other types in Class A. For type I, we know that the metrics are conformally flat but there are still some Einstein metrics with this symmetry, for example de Sitter space. For type II, there are two cases; a triaxial vacuum metric and a metric with non-zero Λ but an extra rotational symmetry. For type VI_0 and VII_0 we may take a limit of type IX or again solve a linear second-order ODE. Finally type VIII is again an analytic extension of type IX. More details of these solutions may be published elsewhere (for some of them, see also Lorentz-Petzold 1983a,b).

Instead of imposing the Einstein condition on the metrics with SD Weyl tensor we could require them to be Kähler. This has been considered for type IX by Pedersen and Poon (1990), and for this and other types by Dancer and Strachan (1993a,b) so I will be brief. Note that in this case it is a definite restriction to require the metric to be diagonal. The non-diagonal scalar-flat, Kähler metrics must lie in the general classes found by Maszczyk et al. (1993).

Begin by assuming that the Kähler-form ω is a constant linear combination of the ASD 2-forms (4.2); then closedness of ω implies that the metric is hyperKähler or that $\omega = \phi^1{}_-$ and $A_2 = A_3 = 0$, $A_1 =$ constant or a cyclic permutation of this. System (4.5) is solved and system (4.4) simplifies. For type IX the system reduces eventually to a special case of Painlevé III and one has the Pedersen-Poon solution. For type VIII as usual one has an analytic extension of type IX. The other types can be solved explicitly. As usual type I is trivial, this time it comes out as flat space.

5 IMPOSING THE FIELD EQUATIONS IN CLASS B

There is less rationale for considering diagonal metrics in Class B because of the constraint (2.14). Indeed this is where the known, non-diagonal vacuum metrics, the solutions of Lukash (see e.g. Kramer et al. 1980), are. However it can be done quite simply so I will briefly summarise the results.

We take the metric and basis of ASD 2-forms as in (4.1) and (4.2) but now use (2.9B). The connection 1-forms are

$$\alpha^1{}_2 = \frac{aw_1\sigma_2}{w_2} + \frac{A_3\sigma_3}{w_3};\ \alpha^3{}_1 = \frac{A_2\sigma_2}{w_2} - \frac{aw_1\sigma_3}{w_3};\ \alpha^2{}_3 = \frac{A_1\sigma_1}{w_1} \tag{5.1}$$

subject again to (4.4) but with $n_1 = 0$. In place of (4.5) we have the system

$$\begin{aligned} \dot{A}_1 &= -A_2A_3 + A_1(A_2 + A_3) + a^2{w_1}^2 \\ \dot{A}_2 &= -A_3A_1 + A_2(A_3 + A_1) + a^2{w_1}^2 \\ \dot{A}_3 &= -A_1A_2 + A_3(A_1 + A_2) + a^2{w_1}^2 \end{aligned} \tag{5.2}$$

The important difference between Class A and Class B is that we have one more equation namely

$$aw_1(A_2 - A_3) = 0 \tag{5.3}$$

so that, if $a \neq 0$, then A_2 and A_3 are equal. This is a significant restriction which enables us to obtain a separate equation for w_1:

$$\ddot{w}_1 - \frac{3{\dot{w}_1}^2}{2w_1} + 2a^2{w_1}^3 = 0 \tag{5.4}$$

The solution of (5.4) is

$$w_1 = \frac{\alpha}{a(1 + \alpha^2(t - t_0)^2)} \tag{5.5}$$

where α and t_0 are arbitrary constants. Knowing w_1, we may go through and calculate the other unknowns in an elementary fashion and arrive at the Class B metrics. The type V metric turns out to be conformally flat, but the other types retain some conformal curvature.

If we impose the Einstein condition then we expect severe restrictions, following the discussion of the constraint (2.14). Sure enough, solutions do exist for all types, but they are conformally flat.

If we impose the Kähler condition then there are no genuine (Riemannian) solutions but there are some curious pseudo-Kählerian metrics with indefinite signature. Two for type V have signature $(+---)$, and so are conformally flat, but there are some for the other types of signature $(++--)$ which have non-zero Weyl tensor.

The interesting unsolved problem is to find the non-diagonal Class B metrics which are the self-dual counterparts of the Lukash metrics. These are presumably contained in the non-diagonal families of Maszczyk et al.

REFERENCES

Ablowitz, M.J. and P.A. Clarkson 1991 'Solitons, non-linear evolution equations and inverse scattering' LMS Lecture Note Series 149, CUP; Cambridge.

Ashtekar, A. and J. Samuel 1991 Class. Quant. Grav. **8**, 2191.

Atiyah, M.F. and N.J. Hitchin 1985 Phys. Lett. **107A**, 21.

Atiyah, M.F. and N.J. Hitchin 1988 'The geometry and dynamics of magnetic monopoles' PUP; Princton.

Belinski, V.A., G.W. Gibbons, D.W. Page and C.N. Pope 1978 Phys. Lett. **76B**, 433.

Besse, A. 1987 'Einstein Manifolds' Springer-Verlag; Berlin.

Bianchi, L. 1897 Soc. Ital. Sci. Mem. di Mat. **11**, 267.

Brioschi, F. 1881 C.R. Acad. Sci. tXCII 1389.

Chakravarty, S. 1993 private communication.

Chazy, J. 1910 C.R. Acad. Sci. **150**, 456.

Dancer, A.S. and I.A.B. Strachan 1993a 'Kähler-Einstein metrics with $SU(2)$ action' to appear in Math. Proc. Cam. Phil. Soc.

Dancer, A.S. and I.A.B. Strachan 1993b 'Cohomogeneity-one Kähler metrics' these proceedings.

Dubrovin, B.A. 1990 Func. An. & its Appl. **24**, 280.

Eguchi, T. and A.J. Hanson 1978 Phys. Lett. **237** 249.

Ellis, G.F.R. and M.A.H. MacCallum 1969 Comm. Math. Phys. **12**, 108.

Estabrook, F.B., H.D. Wahlquist and C.G. Behr 1968 J. Math. Phys. **9**, 497.

Fokas, A.S. and M.J. Ablowitz 1982 J. Math. Phys. **23** 2033.

Fokas, A.S., R.A. Leo, L. Martina and G. Soliani 1986 Phys. Lett. **A115**, 329.

Fokas, A.S. and Y.C. Yortsos 1981 Lett. Nuovo Cim. **30**, 539.

Fujiwara, Y., H. Ishihara and H. Kodama 1993 Class. Quant. Grav. **10**, 859.

Gibbons, G.W. and C.N. Pope 1978 Comm. Math. Phys. bf 61, 239.

Gibbons, G.W. and C.N. Pope 1979 Comm. Math. Phys. bf 66, 267.

Graham, C.R. and J.M. Lee 1991 Adv. Math. **87**, 186.

Hawking, S.W. 1977 Phys. Lett. **60A** 81.

Hitchin, N.J. 1993 'Twistor spaces and isomonodromic deformation' in preparation.

Kramer, D., H. Stephani, M. MacCallum and E. Herlt 1980 'Exact solutions of Einstein's field equations' CUP; Cambridge.

Lebrun, C.R. 1982 Proc. Roy. Soc. Lond. **A380**, 171-185.

Lebrun, C.R. 1991a J. reine angew. Math. **420**, 161.

Lebrun, C.R. 1991b Duke Math. Jour. **63**, 723-743.

Lorentz-Petzold, D. 1983a J. Math. Phys. **24**, 2632.

Lorentz-Petzold, D. 1983b Acta Phys. Pol. **B14**, 791.

Maszczyk, R., L.J. Mason and N.M.J. Woodhouse 1993 'Self-dual Bianchi metrics and the Painlevé transcendents' Oxford preprint.

Misner, C.W. 1969 Phys. Rev. Lett. **62**, 1071.

Pedersen, H. 1985 Class. Quant. Grav. **2** 579.

Pedersen, H. 1986 Math. Ann. **274**, 35-39.

Pedersen, H. and Y.-S. Poon 1990 Class. Quant. Grav. **7**, 1707.

Penrose, R. and W. Rindler 1986 'Spinors and space-time: Vol 2' CUP; Cambridge.

Scott, P. 1983 Bull. Lond. Math. Soc. **15**, 401.

Thurston, W.P. 1982 Bull. Amer. Math. Soc. **6**, 357.

Tod, K.P. 1991 Class. Quant. Grav. **8**, 1049.

Tod, K.P. 1992 Twistor Newsletter **35**, 5.

Ward, R.S. 1980 Comm. Math. Phys. **78**, 1-17.

14
Twistor Theory and Isomonodromy

N. M. J. Woodhouse Wadham College, Oxford OX1 3PN, United Kingdom

1 SELF-DUALITY AND INTEGRABILITY

In this paper, I shall describe some recent work that I have done in collaboration with Lionel Mason, and later also with Roman Maszczyk, on the relationship between two central ideas in the theory of integrable systems. Our work builds on ideas of Nigel Hitchin and suggests the possibility of new applications of twistor theory to problems in classical analysis.

The first idea is an old one, given prominence by Sofya Kowalevskaya. It is that there is an intimate connection between integrability and the Painlevé property for ordinary differential equations (the property that the only movable singularities are poles). The precise nature of this connection is the subject of a large literature, of which the book by Ablowitz and Clarkson (1991) contains an up-to-date survey. I shall consider only one aspect of the connection, through the *Painlevé ODE test*:

> every ODE that arises from an integrable system of equations by similarity reduction has the Painlevé property

(see Ablowitz, Ramani, and Segur 1978, 1980). To take an example from Ablowitz and Clarkson, the KdV equation

$$u_t + u_{xxx} - 6uu_x = 0$$

is integrable. The similarity ansatz $u(x,t) = y(z) - \lambda t$, where $z = x - 3\lambda t^2$, reduces it to

$$y''' = 6yy' + \lambda,$$

which is equivalent to the derivative of the first Painlevé equation.

The second idea is more recent: it is that there is a fundamental connection between self-duality and integrability. Again the precise formulation is not clear. There is an impressive list of integrable systems of equations in one space and one time variable that can be obtained by symmetry reduction of the self-dual Yang-Mills (SDYM) equation. Richard Ward found many of them (see Ward 1990, and papers cited there), and a few years ago, Lionel Mason and George Sparling (1989) made the important additions of the KdV and nonlinear Schrödinger equations. The $2+1$ equations, such as the KP and Davey-Stewartson equations, have been harder to deal with. Lionel Mason (1990) suggested that they might be reductions of the self-dual Einstein equations, and more recently, that they could be brought within the fold by interpreting 'self-duality' more broadly as 'solvable by a twistor construction'.

Before stating our new results, I shall first describe in a little more detail some relevant aspects of these two ideas.

2 THE PAINLEVÉ PROPERTY

Singularities of the solutions to a linear ODE with holomorphic coefficients appear only at the singularities of the coefficients, and they do not move with the constants of integration. The same is not true in the typical nonlinear case. For example, the general solution to

$$y' + y^3 = 0$$

is $y = 1/\sqrt{2(t-c)}$, which has a branch point which moves with the constant c. This raises the question of whether there exist nonlinear ODEs that behave like linear equations in only having fixed singularities. This form of the Painlevé property is too restrictive. But one can, instead, require that that the only movable singularities should be poles, which is the condition in the problem as it was posed by Picard more than 100 years ago. Then there do exist interesting examples. In the first order case, the generalized Riccati equations

$$y' + ay^2 + by + c = 0,$$

where a, b, c are holomorphic, satisfy the condition. But here the Painlevé property is a trivial consequence of the fact the equation is transformed into a *linear* equation

$$aw'' + (ab - a')w' + a^2cw = 0$$

by the substitution $y = w'/aw$.

Painlevé, and others, catalogued the second-order examples of the form

$$y'' = F(y, y', t),$$

where F is rational in y and y'. They showed that an equation of this form with the Painlevé property could transformed into one of fifty canonical types, of which six, the *Painlevé equations* P_I–P_{VI}, required new transcendental functions for their solution. These are given in Table 1 (α, β, γ, and δ are constants).

P_I	$y'' = 6y^2 + t$
P_{II}	$y'' = 2y^3 + yt + \alpha$
P_{III}	$y'' = \dfrac{y'^2}{y} - \dfrac{y'}{t} + \dfrac{\alpha y^2 + \beta}{t} + \gamma y^3 + \dfrac{\delta}{y}$
P_{IV}	$y'' = \dfrac{y'^2}{2y} + \dfrac{3y^3}{2} + 4ty^2 + 2(t^2 - \alpha)y + \dfrac{\beta}{y}$
P_V	$y'' = \left(\dfrac{1}{2y} + \dfrac{1}{y-1}\right) y'^2 - \dfrac{y'}{t} + \dfrac{(y-1)^2}{t^2}\left(\alpha y + \dfrac{\beta}{y}\right) + \dfrac{\gamma y}{t} + \dfrac{\delta y(y+1)}{y-1}$
P_{VI}	$y'' = \dfrac{1}{2}\left(\dfrac{1}{y} + \dfrac{1}{y-1} + \dfrac{1}{y-t}\right) y'^2 - \left(\dfrac{1}{t} + \dfrac{1}{t-1} + \dfrac{1}{y-t}\right) y' + \dfrac{y(y-1)(y-t)}{t^2(t-1)^2}\left(\alpha + \dfrac{\beta t}{y^2} + \dfrac{\gamma(t-1)}{(y-1)^2} + \dfrac{\delta t(t-1)}{(y-t)^2}\right)$

Table 1: The six Painlevé equations

3 REDUCTIONS OF THE SDYM EQUATION

Ablowitz and Clarkson (1991, pp 343–4) give a list of integrable equations that can be obtained by symmetry reduction of the SDYM equation, and which can themselves be further reduced by similarity reduction to one or other of the Painlevé equations. They also give the references to the original papers. The main items in the list are the following.

We write the metric on complex Minkowski space in the form

$$ds^2 = dw d\tilde{w} - dz d\tilde{z}$$

and reduce the SDYM equation by imposing symmetry under the isometry group generated by two Killing vectors X and Y.

(1) $X = \partial_{\tilde{w}}$, $Y = \partial_{\tilde{z}}$. The reduction is the N-wave equation (the value of N depending on the choice of gauge group). This has reductions to P_{III}, P_{IV}, P_V, and P_{VI}.

(2) $X = \partial_w$, $Y = \partial_{\tilde{w}}$. The reductions, via the Toda field equation or the extended Toda field equation, include the sinh-Gordon equation, the sine-Gordon equation, and the Liouville equation. These have similarity reductions to P_{III}.

(3) $X = \partial_{\tilde{w}}$, $Y = \partial_z + \partial_{\tilde{z}}$. The reduction is either to the KdV equation or to the NLS equation. The KdV equation has further reductions to P_I and P_{II}; the NLS equation to P_{II} and P_{IV}.

(4) $X = z\partial_z - \tilde{z}\partial_{\tilde{z}}$ and either $Y = \partial_w + \partial_{\tilde{w}}$ or $Y = w\partial_w - \tilde{w}\partial_{\tilde{w}}$. Here the reduction is the Ernst equation

$$r^{-1}\partial_r\left(rJ^{-1}\partial_r J\right) + \partial_z\left(J^{-1}\partial_z J\right) = 0$$

on a matrix-valued function of r and z. This has similarity reductions to P_{III} and P_V.

Thus all six Painlevé are known to be reductions of the SDYM equation.[1] In making such a statement, it may be useful to distinguish between 'weak' and 'strong' reduction of the SDYM equation. By the former, I mean that the reduced equation can be obtained by substituting into the SDYM equation a particular form of the Yang-Mills potential with the relevant symmetry. By 'strong' reduction, I mean that the symmetry requirement alone forces the reduction; that is that the solutions to the SDYM with the particular symmetry are parametrized, up to gauge equivalence, by the solutions to the reduced equation, perhaps in conjunction with some extra data such as free functions. It is easier to make this clear by giving an extreme example of a weak reduction than by trying to formulate 'strong reduction' in a completely precise way that does not exclude all the interesting cases. Burgers' equation

$$u_w = 2uu_z + u_{zz}$$

is, in the weak sense, a reduction of the SD Maxwell equation, in that it is the condition that

$$A = udz + (u^2 + u_z)dw$$

should be a SD Maxwell potential. But every Maxwell field generated in this way from a solution to Burgers' equation vanishes. Thus what is being reduced is not simply the self-duality equation, but the self-duality equation combined with a particular, and very restrictive, gauge condition.[2]

In (1) and (2), the reduction is 'weak', in that a particular ansatz must be made for the potential. It is in fact possible to obtain the Toda field equation as a 'strong' reduction by imposing in addition symmetry under a group of rotations in the $w, \tilde{w}$-plane. The continuous group $w \mapsto e^{i\theta}w$, $\tilde{w} \mapsto e^{-i\theta}\tilde{w}$ gives the basic Toda field equation. The discrete group $w \mapsto e^{2\pi i/N}w$, $\tilde{w} \mapsto e^{-2\pi i/N}\tilde{w}$ gives the extended Toda field equation. The details will appear in the forthcoming book by myself and Lionel Mason. In (3) and (4), as well as in the reductions to the Painlevé equations below, the reduction is strong.

4 REDUCTIONS OF THE SELF-DUALITY CONDITION

A conformal structure in four dimensions is self-dual if its Weyl spinor $\psi_{A'B'C'D'}$ vanishes. Paul Tod (1992) has shown how a class of metrics with this property, together with vanishing scalar curvature and Bianchi-IX symmetry, can be obtained from a system of ODEs, which in turn comes down to P_{VI}, with $\alpha = -\beta = \frac{1}{8}$, $\delta = \frac{1}{2} - \gamma$. He explains the significance of these metrics elsewhere in this volume. Thus, in certain cases, the self-duality condition on a conformal structure passes the Painlevé ODE test.

[1] Not all possible values of the parameters appear in every case in the list. For example, the reductions of the Ernst equation to P_{III} known to Persides and Xanthopoulos (1988) have $\alpha = \beta = 0$, $\gamma = -\delta = 4$.

[2] This example was suggested to me by a preprint of A. D. Popov's.

5 STATEMENT OF THE RESULTS

Our results extend Paul Tod's and those in Ablowitz and Clarkson's list, and explain them by setting them within a natural geometric framework. We have shown the following by using twistor methods (which follow on from ideas of Nigel Hitchin's).

(1) The six Painlevé equations are the nontrivial reductions of the $SL(2,\mathbb{C})$ SDYM equation by a nondegenerate 3-dimensional abelian group of conformal transformations (Mason and Woodhouse 1993a).[3]

(2) There is a correspondence between the Bianchi IX conformal structures with $\psi_{A'B'C'D'} = 0$ and the generic reductions of the $SL(2,\mathbb{C})$ SDYM equation by three commuting Killing vectors (Maszczyk, Mason, and Woodhouse 1993).

The essential step in making these connections is to exploit the long-established relationship between the Painlevé equations and the isomonodromy property of families of ODEs (discussed in detail by Jimbo, Miwa, and Ueno 1981, and by Jimbo and Miwa 1981). We recall this next.

6 ISOMONODROMY

The isomonodromy problem concerns a family of linear differential equations of the form

$$\frac{d\psi}{d\zeta} = F(\zeta, x)\psi,$$

where F and ψ are $N \times N$ matrices, with F depending on $n+1$ complex parameters $x = (x_0, x_1, \ldots, x_n)$. For each fixed value of x, F is a rational function of the independent variable ζ, which we think of as taking values on the Riemann sphere. The problem is to determine the conditions under which the monodromy data of the equation remain constant as the parameters change.

When F has only simple poles $\zeta_0, \ldots, \zeta_n$, the monodromy data are encoded in the homomorphism

$$\pi_1(\mathbb{C}P_1 - \{\zeta_0, \ldots, \zeta_n\}) \to GL(N, \mathbb{C})$$

that determines the transformations of ψ under continuation around closed loops. The problem in this case is to determine conditions on the dependence of F on $\zeta_0, \ldots, \zeta_n$ that ensure that the homomorphism does not change when the poles move.

For poles of higher order, the data are more complicated. For example, in the case that $\zeta = 0$ is a pole of order r, and $\zeta^r F$ is diagonal at the origin, one can divide the complex plane into $2r$ sectors (with vertices at the pole), in each of which ψ has an asymptotic expansion of the form

$$\psi \sim \left(1 + c_1\zeta + \cdots\right) \exp\left(\sum_1^r T_j \zeta^j + T_0 \log\zeta\right),$$

as $\zeta \to 0$, where the Ts are diagonal and the cs are constants. The diagonal entries in T_0 are called the *exponents of formal monodromy*. The data are defined in a similar way at a pole

[3] The meaning of 'nondegenerate' is explained in Mason and Woodhouse (1993a). It excludes groups of translations.

at which the leading term in F is diagonalizable. The complete data here are the exponents of formal monodromy, the *Stokes' matrices* (which connect the expansions in neighbouring sectors), and other connection matrices that connect the expansions at different poles. All the data are required to remain constant as the parameters change.

This implies the more geometric condition that F should satisfy the deformation equations

$$d_x F = \frac{\partial \Omega}{\partial \zeta} + [\Omega, F], \quad d_x \Omega = \Omega \wedge \Omega$$

for some matrix-valued function $\Omega(\zeta, x)$, depending rationally on x. The deformation equations, in turn, are equivalent to the requirement that $\nabla = d - F d\zeta - \Omega$ should be a flat connection, on a trivial bundle over the product of $\mathbb{C}P_1$ with the parameter space, less the poles of F. In the simple case, the monodromy homomorphism is the holonomy of ∇ around the poles—the singularities in ∇. Its constancy in this case is an obvious consequence of the fact that the holonomy can depend only on the homotopy class of the closed loops in the complement of the poles in the product space.

When $N = 2$ and F has four simple poles, the deformation equations come down to P_{VI}. The other Painlevé equations are the degenerate cases that arise when various combinations of the four poles come into coincidence:

P_V: one double pole, two simple poles,

P_{IV}: one triple pole, one simple pole,

P_{III}: two double poles,

$P_{I,II}$: one quadruple pole.

The details can be found in Jimbo and Miwa (1981).

7 THE TWISTOR CONSTRUCTION

The first connection with the self-duality equations comes, via the Ward transform, from the correspondence between

isomonodromic families of ODEs with four poles

and

holomorphic bundles over a neighbourhood of a line in twistor space invariant under a three-dimensional abelian Lie algebra $\mathfrak{g}$ of infinitesimal projective transformations.

I shall explain how the correspondence is defined in one direction. The reverse construction for an arbitrary number of simple poles is given in Mason and Woodhouse (1993b). The reverse construction for the degenerate cases of four poles can be derived from the inversion theorem for the Ward transform (see Ward and Wells 1990).

Let $U \subset \mathbb{C}P_3$ be a neighbourhood of a line and let X_1, X_2, X_3 be commuting holomorphic vector fields on U that generate the action of an abelian subgroup of $PGL(4, \mathbb{C})$. Let $E \to U$ be a holomorphic $SL(2, \mathbb{C})$ vector bundle which is trivial on restriction to the lines in U. Suppose that E is invariant under the infinitesimal action of G, which means that the Xs lift to commuting vector fields on the total space. There are many possible choices for G, but in the generic case G is conjugate to the image of the diagonal subgroup under the projection $GL(4,\mathbb{C}) \to PGL(4,\mathbb{C})$. When this holds, the Xs are independent except on four planes[4] H_0, H_1, H_2, H_3. The action on E determines a flat connection ∇ on the restriction of E to the complement of the planes, with the horizontal subspaces spanned by the lifts of the Xs. In the special cases, some of the planes come into coincidence.

Let $L \subset U$ be a line. By triviality, $E|_L = L \times \mathbb{C}^N$. In this trivialization,

$$\nabla|_L = d\zeta\left(\frac{d}{d\zeta} - F\right),$$

where ζ is a stereographic coordinate and F has simple poles at the intersections of L with the four planes. Thus the action of G determines a family of linear ODEs. As L varies, the ODEs are isomonodromic because ∇ is flat. In the general case, therefore, an invariant bundle on U determines a solution to P_{VI}. As we continue F by moving L out of U, the fixed singularities of the solution arise when L passes through one of the intersections of the planes, and the movable singularities arise at the jumping lines, where the triviality condition fails.

There are many special cases in which G is not conjugate to the image of a diagonal subgroup. Some possible normal forms for the preimage in $GL(4,\mathbb{C})$ are shown in Table 2. The preimage is four-dimensional, and is parametrized by a, b, c, d; in the final case, the translation group, the normal form is five-dimensional, and G is the image of some four-dimensional subgroup. In the first four cases, the four planes come into coincidence in various combinations, and the isomonodromic families of ODEs give other Painlevé transcendents, as indicated. The remaining cases are degenerate in various ways, and the give, via the Ward transform, ODEs of rather different characters.

8 THE WARD TRANSFORM

We can now derive our first result by applying the Ward transform, which gives a correspondence between

> holomorphic G-invariant bundles on U

and

> SDYM fields on a neighbourhood in complex space-time, invariant under a three-dimensional abelian group of conformal isometries.

The three-dimensional conformal group is generated by three commuting Killing vectors, which we denote by X, Y, Z. In each of the Painlevé cases, it is possible to choose space-time coordinates x, y, z, t and the gauge of the Yang-Mills connection $d+\Phi$ so that $X = \partial_x$, $Y = \partial_y$, $Z = \partial_z$ and

$$\Phi = Pdx + Qdy + Rdz$$

where P, Q and R are 2×2 trace-free matrix-valued functions of t, the only non-ignorable space-time coordinate (they are the Higgs fields associated with the three generators of G).

[4]In the case of the diagonal subgroup itself, these are the coordinate planes $Z^\alpha = 0$, $\alpha = 0, 1, 2, 3$.

Painlevé cases	$P_{I,II}$ $\begin{pmatrix} a & b & c & d \\ 0 & a & b & c \\ 0 & 0 & a & b \\ 0 & 0 & 0 & a \end{pmatrix}$	P_{III} $\begin{pmatrix} a & b & 0 & 0 \\ 0 & a & 0 & 0 \\ 0 & 0 & c & d \\ 0 & 0 & 0 & c \end{pmatrix}$
	P_{IV} $\begin{pmatrix} a & b & c & 0 \\ 0 & a & b & 0 \\ 0 & 0 & a & 0 \\ 0 & 0 & 0 & d \end{pmatrix}$	P_V $\begin{pmatrix} a & b & 0 & 0 \\ 0 & a & 0 & 0 \\ 0 & 0 & c & 0 \\ 0 & 0 & 0 & d \end{pmatrix}$
	P_{VI} $\begin{pmatrix} a & 0 & 0 & 0 \\ 0 & b & 0 & 0 \\ 0 & 0 & c & 0 \\ 0 & 0 & 0 & d \end{pmatrix}$	
Degenerate cases	$\begin{pmatrix} a & b & c & 0 \\ 0 & a & 0 & 0 \\ 0 & 0 & a & 0 \\ 0 & 0 & 0 & d \end{pmatrix}$	$\begin{pmatrix} a & 0 & b & 0 \\ 0 & a & c & 0 \\ 0 & 0 & a & 0 \\ 0 & 0 & 0 & d \end{pmatrix}$
	$\begin{pmatrix} a & b & c & d \\ 0 & a & 0 & c \\ 0 & 0 & a & b \\ 0 & 0 & 0 & a \end{pmatrix}$	$\begin{pmatrix} a & b & c & d \\ 0 & a & b & 0 \\ 0 & 0 & a & 0 \\ 0 & 0 & 0 & a \end{pmatrix}$
	$\begin{pmatrix} a & b & c & d \\ 0 & a & 0 & 0 \\ 0 & 0 & a & 0 \\ 0 & 0 & 0 & a \end{pmatrix}$	$\begin{pmatrix} a & 0 & 0 & b \\ 0 & a & 0 & c \\ 0 & 0 & a & d \\ 0 & 0 & 0 & a \end{pmatrix}$
	$\begin{pmatrix} a & 0 & b & c \\ 0 & a & d & e \\ 0 & 0 & a & 0 \\ 0 & 0 & 0 & a \end{pmatrix}$	

Table 2: Conjugacy classes of G.

Table 3 gives X, Y, Z, and the fourth coordinate vector $T = \partial_t$ in terms the original space-time coordinates for one choice of G in each Painlevé conjugacy class, together with the SDYM equations on P, Q and R. One can check directly[5] that the reduced SDYM equations are either trivial (for special choices of the invariants of P, Q and R) or come down to the relevant Painlevé equation. In each case, the Painlevé transcendent y is an invariant of the three Higgs fields, and all possible values of the parameters appear. Since the Higgs fields—the contractions of the potential with the conformal Killing vectors in an invariant gauge—are themselves invariants of the connection (up to an overall conjugation by a matrix-valued function of t), the Painlevé transcendents are invariants of the SDYM fields. They are obtained explicitly from P, Q and R as follows. In the case of P_I, P_{II} and P_{IV}, y is related to a root of the invariant quadratic in λ

$$\det\big([P, Q - \lambda R]\big) = 0$$

by affine transformations of t and y. The reductions to P_I and P_{II} are distinguished by whether or not P is nilpotent. In the case of P_{III}, ty is a root of this quadratic. In the case of P_V, y is the cross-ratio of two eigenvectors of P, an eigenvector of Q, and an eigenvector of R, regarded as points of $\mathbb{C}P_1$. In the final case, P_{VI}, y is a root of

$$\det\Big([P + Q + R, t(1-y)P + (1-t)yQ]\Big) = 0.$$

Some of the reductions in Ablowitz and Clarkson's list are quite transparent in this framework. For example, in the $P_{I,II}$ case, reduction by X and Z gives either the KdV equation or the NLS equation (Mason and Sparling 1989). The imposition of the third symmetry along Y gives the further reduction of these equations to P_I and P_{II}. Similarly, in the case of P_{III}, reduction by X and Z gives the Toda equation, and reduction by $X + Z$ and Y gives the Ernst equation. Guy Calvert has used this last observation to obtain a list of real vacuum space-times with two commuting symmetries that can be constructed from solutions to P_{III}. Some of the metrics appear to be new.

Remark. The bundles on twistor space are defined only on the neighbourhood of a line. It might be thought that they could be extended by using their invariance under G. Unfortunately, there is a topological obstruction. In the P_{VI} case, for example, it arises because

$$\pi_1(\mathbb{C}P_1 - 4 \text{ points}) \not\simeq \pi_1(\mathbb{C}P_3 - 4 \text{ planes}).$$

9 AN INVOLUTION

The second of our results follows from a correspondence between

> SDYM fields in Minkowski space with gauge group $SL(2,\mathbb{C})$ and 3-dimensional abelian symmetry group G

and

> complex SD conformal metrics with $SL(2,\mathbb{C})$ symmetry.

[5]The calculations are not easy, even if one draws heavily on the work of Jimbo and Miwa (1981).

$P_{I,II}$	$X = \partial_{\tilde{w}}, \quad Y = w(\partial_z - \partial_{\tilde{z}}) + (\tilde{z} - z)\partial_{\tilde{w}} + \partial_w,$ $Z = \partial_z + \partial_{\tilde{z}}, \quad T = -\partial_{\tilde{z}} - w\partial_{\tilde{w}}$ $P' = 0, \quad Q' = [Q + tP, R], \quad R' = [Q, P]$
P_{III}	$X = \partial_{\tilde{w}}, \quad Y = z\partial_z - \tilde{z}\partial_{\tilde{z}},$ $Z = \partial_w, \quad T = 2\tilde{z}\partial_{\tilde{z}}$ $P' = 0, \quad Q' = 2t[P, R], \quad tR' = 2[R, Q]$
P_{IV}	$X = \partial_{\tilde{w}}, \quad Y = 2(z\partial_z + w\partial_w),$ $Z = w\partial_z + \tilde{z}\partial_{\tilde{w}} + \partial_{\tilde{z}}, \quad T = \partial_{\tilde{z}} + \tilde{z}\partial_{\tilde{w}}$ $P' = 0, \quad Q' = [Q, R], \quad R' = \frac{1}{2}[P, Q] + t[P, R]$
P_V	$X = \partial_{\tilde{w}}, \quad Y = w\partial_w + \tilde{z}\partial_{\tilde{z}},$ $Z = z\partial_z + w\partial_w, \quad T = wz^{-1}\partial_{\tilde{z}}$ $P' = 0, \quad tQ' = t[Q, P] - [Q, R], \quad tR' = [Q, R]$
P_{VI}	$X = -w\partial_w - z\partial_z, \quad Y = z\partial_z + \tilde{w}\partial_{\tilde{w}},$ $Z = w\partial_w + \tilde{z}\partial_{\tilde{z}}, \quad T = w\tilde{w}z^{-1}\partial_{\tilde{z}}$ $P' + Q' + R' = 0, \quad tP' + [P, R] = 0, \quad (t-1)Q' + [Q, R] = 0$

Table 3: The coordinate vectors and SDYM equations for the reductions to the Painlevé equations.

From the twistor point of view, the correspondence in the forward direction is obtained by using the Ward transform to encode the SDYM field in a holomorphic principal bundle over twistor space, and then by constructing a curved twistor space by taking the quotient by the action of G. The natural action of the gauge group on the total space of the principal bundle survives in the quotient and gives rise to the $SL(2,\mathbb{C})$ symmetry of the corresponding SD conformal structure.

From another point of view, the correspondence can be seen directly as a consequence of the existence of the following involution. It acts on objects constructed from the data:

a complex space-time M with SD conformal structure g,

a SDYM connection D on a principal $\tilde{G}$-bundle $\pi : P \to M$, and

a symmetry group G of g and D of the same dimension as $\tilde{G}$.

The involution interchanges the roles of G and $\tilde{G}$. To construct it, we regard D as being encoded in the spaces of horizontal tangent vectors on P. Both groups G and $\tilde{G}$ act on P and preserve the horizontal distribution: the action $\tilde{G}$ is part of the structure of P as a principal bundle, while the action of G lifts from its symmetry action on M. We have to make one assumption, which holds in the generic case, that

the orbits of G and $\tilde{G}$ are both transverse to the horizontal subspaces.

The new conformal metric $\tilde{g}$ is defined on $\tilde{M} = P/G$ by pulling g back to P, restricting it to the horizontal subspaces, and then projecting into $\tilde{M}$. That is, the distance between a nearby pair of G-orbits is measured by taking a horizontal connecting vector and finding its length with respect to g. If $\tilde{\pi} : P \to \tilde{M}$ is the projection, then at each $p \in P$, the horizontal subspace is identified with $T_{\pi(p)}M$ and with $T_{\tilde{\pi}(p)}\tilde{M}$ by π_* and $\tilde{\pi}_*$.

The conformal class of $\tilde{g}$ is independent of the various choices involved (the metric within the conformal class of g and the horizontal subspaces in which the connecting vectors are chosen) because g is invariant up to scale under the action of G. By taking G as the structure group, instead of $\tilde{G}$, P becomes a principal G-bundle over $\tilde{M}$, and the horizontal subspaces again determine a connection $\tilde{D}$. Both the new conformal structure and the new connection are self-dual. We see the self-duality of $\tilde{g}$ from the existence of β-surfaces. Any β surface in M can be lifted horizontally to a family of horizontal surfaces in P (since D is flat on β-surfaces). These then project under $P \to \tilde{M}$ onto β surfaces of $\tilde{g}$. The self-duality of $\tilde{D}$ follows from the fact that the restriction of its curvature 2-form on P to the horizontal subspaces coincides with that of the curvature 2-form of D, which is self-dual. Finally the action of $\tilde{G}$ on P descends to the quotient $\tilde{M} = P/G$, where it preserves $\tilde{g}$ and $\tilde{D}$.

It is clear that the construction is involutive: if we apply it to $\tilde{g}$, $\tilde{D}$ and $\tilde{M}$, we shall recover g, D and M.

We can, in particular, take M to be flat space-time and D to be a SDYM connection constructed from one of the Painlevé transcendents. Then $\tilde{G} = SL(2,\mathbb{C})$ and D is invariant under a three-dimensional abelian subgroup of the conformal group. The application of the involution gives a new self-dual conformal metric $\tilde{g}$ and three self-dual Maxwell fields, which are invariant under an action of $SL(2,\mathbb{C})$ with three-dimensional orbits. That is, $\tilde{g}$ is a (complexified) Bianchi IX SD conformal structure. In general $\tilde{g}$ is not flat. In fact, by applying the involution again, one can show that every such conformal structure arises in this way.

10 EXPLICIT FORM OF THE METRICS

I shall illustrate the construction by considering the generic case, in which the transcendent satisfies P_{VI}. The calculations are due to Roman Maszczyk. Suppose that we are given a SDYM potential Φ on flat space-time with the symmetries corresponding to P_{VI}. We first make a coordinate transformation

$$t = \frac{z\tilde{z}}{w\tilde{w}}, \quad x^1 = \tfrac{1}{2}\log\left(\frac{\tilde{w}\tilde{z}}{wz}\right), \quad x^2 = \tfrac{1}{2}\log\left(\frac{s\tilde{w}}{w}\right), \quad x^3 = \tfrac{1}{2}\log\left(\frac{s\tilde{z}}{z}\right),$$

where $s = z\tilde{z} - w\tilde{w}$, to bring the flat space-time metric into the form

$$ds^2 = \Omega^2\left(dt^2 + g_{ij}dx^i dx^j\right)$$

where

$$\left(g_{ij}\right) = 2t(t-1)\begin{pmatrix} 0 & -1 & t \\ -1 & 0 & 1-t \\ t & 1-t & 0 \end{pmatrix}.$$

We also make the gauge transformation $\Phi \mapsto q^{-1}\Phi q + q^{-1}q'dt$ of the SDYM potential on flat space-time, where[6]

$$q'q^{-1} = \frac{(t-1)P + tQ + (2t-1)R}{2t(t-1)}.$$

Then the P_{VI} conformal Killing vectors X, Y and Z are the coordinate vectors of the coordinates x^i and the new potential has no dt component.

If we take $G = \mathbb{C}^3 = \{(x^1, x^2, x^3)\}$, then the coordinate transformation splits flat space-time (locally) into a product $\mathbb{C} \times G$, where t is the coordinate on the first factor. The result of applying the involution is a self-dual conformal metric on $\mathbb{C}\times\tilde{G}$, where $\tilde{G} = SL(2,\mathbb{C})$ (the gauge group). If we choose a basis e_a for $sl(2,\mathbb{C})$, and let γ^a be the corresponding invariant 1-forms on $\tilde{G}$ ($a = 1,2,3$), then we can write

$$\Phi = q^{-1}(P\,dx^1 + Q\,dx^2 + R\,dx^3)q = \Pi_i^a e_a\,dx^i$$

where the matrix $\Pi = \left(\Pi_i^a\right)$ depends on t. The new metric is, up to a conformal factor,

$$dt^2 + g_{ij}\Pi_a^i\Pi_b^j\gamma^a\gamma^b,$$

where $\Pi_i^a\Pi_b^i = \delta_b^a$. *This is the generic complex Bianchi IX self-dual conformal structure.*

11 A SYMPLECTIC CONSTRUCTION

I remark, finally, that the construction in the preceding sections fits into a general symplectic framework. Suppose that we are given a Yang-Mills field on $\mathbb{M}$ (flat space-time) together with a symplectic manifold (C, σ) on which the gauge group has a Hamiltonian action. By applying the homomorphism of the Lie algebra of the gauge group into the Poisson algebra of functions on C, we can construct from the gauge potential a 1-form $\Phi_a dx^a$ on space-time with values in the functions on C. From this, we get a symplectic form

$$\omega = dp_a \wedge dx^a + d(\Phi_a dx^a) + \sigma$$

[6]This corrects the corresponding formula in Maszczyk *et. al.* (1993), which is not quite right.

on $S = T^*\mathbb{M} \times C$. The self-duality condition on Φ implies that

$$\{w - \zeta\tilde{z}, z - \zeta\tilde{w}\} = 0$$

for all constant values of the spectral parameter ζ. Here $\{\cdot,\cdot\}$ is the Poisson bracket. This in turn is equivalent to the commutation condition

$$[X_w - \zeta X_{\tilde{z}}, X_z - \zeta X_{\tilde{w}}] = 0$$

where the Xs are the Hamiltonian vector fields of the coordinate functions on $\mathbb{M}$. Apart from the fact that they are defined on a space of larger dimension, this is the same as Mason and Newman's (1989) form of the self-dual vacuum equations on a four-dimensional metric. They reduce the equations to the condition

$$[U - \zeta\tilde{V}, V - \zeta\tilde{U}] = 0$$

on four volume preserving vector fields $U, V, \tilde{U}, \tilde{V}$ on $\mathbb{C}^4$.

By choosing C to be two-dimensional, and then taking the quotient of S by a two-dimensional translation group, together with the vector fields $\partial/\partial p_a$, one can project X_w, X_z, $X_{\tilde{w}}$, $X_{\tilde{z}}$ down to volume preserving vector fields in four dimensions. In this way, one can construct SD vacuum metrics from solutions to KdV and NLS equations, as Lionel Mason observed in his Twistor Newsletter article (Mason 1990). By taking $C = T^*\tilde{G}$, where $\tilde{G} = SL(2, \mathbb{C})$, and then taking the quotient by a three-dimensional abelian symmetry group G, together with Hamiltonian flows on S of the coordinate functions on $\tilde{G}$ and $\mathbb{M}$, we can construct the space-time $\tilde{M}$. The Xs project at each point onto a null tetrad for the conformal metric on $\tilde{M}$. It is likely that more families of SD metrics can be constructed within this general framework from reductions of the SDYM equation.

REFERENCES

1. Ablowitz, M. J. and Clarkson, P. A. (1991). *Solitons, nonlinear evolution equations and inverse scattering.* LMS Lecture Note Series **149**. Cambridge University Press, Cambridge.
2. Ablowitz, M. J. , Ramani, A. and Segur, H. (1978). Nonlinear evolution equations and ordinary differential equations of Painlevé type. *Lett. Nuovo Cim.* **23**, 333–8.
3. Ablowitz, M. J. , Ramani, A. and Segur, H. (1980). A connection between nonlinear evolution equations and ordinary differential equations of P-type. I. *J. Math. Phys.* **21**, 715–21.
4. Jimbo, M., Miwa, T. and Ueno, K. (1981). Monodromy preserving deformation of linear ordinary differential equations with rational coefficients. I. *Physica* **2D**, 306–52.
5. Jimbo, M. and Miwa, T. (1981). Monodromy preserving deformation of linear ordinary differential equations with rational coefficients. II. *Physica* **2D**, 407–48.
6. Mason, L. J. (1990) $\mathcal{H}$-space—a universal integrable system? *Twistor Newsletter* **30**, 14–17.
7. Mason L. J., and Newman, E. T. (1989). A connection between the Einstein and Yang-Mills equations. *Commun. Math. Phys.* **121**, 659–68.

8. Mason, L. J. and Sparling, G. A. J. (1989). Nonlinear Schrödinger and Korteweg-de Vries are reductions of self-dual Yang-Mills. *Phys. Lett.* **137A**, 29–33.
9. Mason, L. J. and Woodhouse, N. M. J. (1993a). Self-duality and the Painlevé transcendents. *Nonlinearity* **6**, 569–81.
10. Mason, L. J. and Woodhouse, N. M. J. (1993b). Twistor theory and the Schlesinger equations. *Proceedings NATO Advanced Research Workshop*, ed. P. A. Clarkson, Kluwer, Dordrecht.
11. Maszczyk, R., Mason, L. J. and Woodhouse, N. M. J. (1993) Self-dual Bianchi metrics and the Painlevé transcendents. To appear in *Class. Quantum Grav.*
12. Persides, S. and Xanthopoulos, B. C. (1988). Some new stationary axisymmetric asymptotically flat space-times obtained from Painlevé transcendents. *J. Math. Phys.* **29**, 674–680.
13. Tod K. P. (1992). Metrics with SD Weyl tensor from Painlevé-VI *Twistor Newsletter* **35**, 5–10.
14. Ward, R. S. (1990). Integrable systems in twistor theory. In *Twistors in Mathematics and Physics*, eds. T.N. Bailey and R.J. Baston. LMS Lecture Note Series **156**. Cambridge University Press, Cambridge.
15. Ward, R. S. and Wells, R. O. (1990). *Twistor geometry and field theory.* Cambridge University Press, Cambridge.

Appendix: Geometry of Relative Deformations. I

Sergey A. Merkulov
School of Mathematics and Statistics, University of Plymouth
Plymouth, Devon PL4 8AA, United Kingdom

1 Introduction

Let Y and M be complex manifolds and let $\pi_1 : Y \times M \longrightarrow Y$ and $\pi_2 : Y \times M \longrightarrow M$ be the natural projections. An analytic family of compact submanifolds of the complex manifold Y with the moduli space M is a complex submanifold $F \hookrightarrow Y \times M$ such that the restriction of the projection π_2 on F is a proper regular map (regularity means that the rank of the differential of $\nu \equiv \pi_2 \mid_F$: $F \longrightarrow M$ is equal at every point to $\dim M$). Thus the family F has the structure of a double fibration

$$\begin{array}{ccc} & F & \\ {}^{\mu}\swarrow & & \searrow^{\nu} \\ Y & & M \end{array}$$

where $\mu \equiv \pi_1 \mid_F$. For each $t \in M$ the compact complex submanifold $X_t = \mu \circ \nu^{-1}(t) \hookrightarrow Y$ is said to belong to the family M.

Let us denote the normal bundle $TA|_B / TB$ of a complex submanifold B of a complex manifold A by $N_{B|A}$.

If $F \hookrightarrow Y \times M$ is an analytic family of compact submanifolds, then, for any $t \in M$, there is a natural linear map (Kodaira 1962),

$$k_t : T_t M \longrightarrow H^0(X_t, N_{X_t|Y}),$$

from the tangent space at t to the vector space of global holomorphic sections of the normal bundle $N_{X_t|Y} = TY|_{X_t} / TX_t$ to the submanifold $X_t \hookrightarrow Y$, which can be described as follows. First note that the normal bundle of the embedding $\nu^{-1}(t) \hookrightarrow F$ is trivial and thus there is a canonical map $\rho_t : T_t M \longrightarrow H^0(\nu^{-1}(t), N_{\nu^{-1}(t)|F})$. Then a composition $d\nu \circ \rho_t$ gives the desired map k_t for the differential of ν maps global sections of $N_{\nu^{-1}(t)|F}$ to global sections of $N_{X_t|Y}$.

An analytic family $F \hookrightarrow Y \times M$ of compact submanifolds is called *complete* if the

Kodaira map $\boldsymbol{k}_t$ is an isomorphism at each point t in the moduli space M. It is called *maximal* if for any other analytic family $\tilde{F} \hookrightarrow Y \times \tilde{M}$ such that $\boldsymbol{\nu}^{-1}(t) = \tilde{\boldsymbol{\nu}}^{-1}(\tilde{t})$ for some points $t \in M$ and $\tilde{t} \in \tilde{M}$ there is a neighbourhood $\tilde{U} \subset \tilde{M}$ of the point $\tilde{t}$ and a holomorphic map $f : \tilde{U} \longrightarrow M$ such that $\tilde{\boldsymbol{\nu}}^{-1}(\tilde{t}') = \boldsymbol{\nu}^{-1}\left(f(\tilde{t}')\right)$ for every $\tilde{t}' \in \tilde{U}$. Here the equality $\boldsymbol{\nu}^{-1}(t) = \tilde{\boldsymbol{\nu}}^{-1}(\tilde{t})$ means that $\boldsymbol{\mu} \circ \boldsymbol{\nu}^{-1}(t)$ and $\tilde{\boldsymbol{\mu}} \circ \tilde{\boldsymbol{\nu}}^{-1}(\tilde{t})$ are the same submanifolds of Y.

In 1962 Kodaira proved the following theorem.

Theorem (Kodaira) *If $X \hookrightarrow Y$ is a compact complex submanifold with normal bundle N such that $H^1(X, N) = 0$, then X belongs to the complete analytic family $\{X_t : t \in M\}$ of compact submanifolds X_t of Y. The family is maximal and its moduli space is of complex dimension* $\dim_{\mathbb{C}} H^0(X, N)$.

After the profound paper by Penrose (1976) on the non-linear graviton construction the Kodaira relative deformation theory was applied to find a number of remarkable correspondences between a compact complex submanifold X of a complex manifold Y and solution spaces of a number of non-linear differential equations which are expressible in terms of (equivalence classes of) torsion-free affine connections (see the books by Baston and Eastwood 1989, Besse 1987, Manin 1988, Penrose and Rindler 1986 and Ward and Wells 1990 and references cited therein). Though all known (to the author) examples involve only rather special embeddings of rational curves and quadrics, the resulting picture suggests that there should exist a general theory relating a pair (X, Y), X being a compact complex submanifold of a complex manifold Y with normal bundle N satisfying some restrictions, to an equivalence class of affine connections on the associated complete moduli space of relative deformations of X which satisfy some integrability conditions. In this paper we prove several general theorems which support this conjecture and demonstrate that a number of successful "curved twistor" constructions (such as, for example , Penrose's (1976) solution of the self-dual Einstein's equations, the twistor descriptions of self-dual space-times with non-vanishing cosmological constant (Hitchin 1982, Ward 1980), and of quaternionic Kähler manifolds (Bailey and Eastwood 1991, LeBrun 1989, Pedersen and Poon 1989, Salamon 1982), and Hitchin's (1982) interpretation of several classes of second-order differential equations) are in fact different faces of one and the same general phenomenon.

Let $X \hookrightarrow Y$ be a pair of complex manifolds satisfying the conditions of the Kodaira theorem and let M be (a domain in) the associated complete moduli space. After the pioneer work by Ward (1977) on instanton solutions of the Yang-Mills equations holomorphic vector bundles E on Y which are trivial on submanifolds X_t for all $t \in M$ have been intensively studied and their geometric role has been clarified — such bundles on Y generate vector bundles on M together with connections satisfying some integrability conditions (see, e.g., Manin 1988). In this paper we find a geometric meaning of holomorphic vector bundles on Y which, when restricted to X_t, $t \in M$, are isomorphic to the normal bundle of the embedding $X_t \hookrightarrow Y$. We show that such bundles often generate torsion-free affine connections on the moduli space M automatically satisfying some integrability conditions. It is also worth pointing out that the method used to prove main theorems of the paper also yields practical tools for computing such connections.

The paper is organized as follows. In Sect. 2 we recall some useful facts about affine connections and then introduce equivalence classes of affine connections on a manifold M which are associated with some fixed subbundles $\Lambda \subset TM \otimes \odot^2 \Omega^1 M$ and $\Xi \subset \mathrm{End}(TM)$ and are called Λ-connections and Ξ-connections respectively. In the case when Ξ is the structure sheaf $\mathcal{O}_M$ with its natural action on TM, a Ξ-connection is the same thing as a projective connection. The main reason behind our interest in such creatures is that, for any compact complex rigid submanifold $X \subset Y$ with rigid normal bundle N such that $H^1(X, N) = 0$, the associated complete moduli space, M, of relative deformations always comes equipped with subbundles, $\Lambda_{H^0(X, N \otimes \odot^2 N^*)} \subset TM \otimes \odot^2 \Omega^1 M$ and $\Xi_{H^0(X, N \otimes N^*)} \subset TM \otimes \Omega^1 M$, defined by the natural "pointwise contraction" actions of the groups $H^0(X, N \otimes \odot^2 N^*)$ and $H^0(X, N \otimes N^*)$ on the vector spaces $H^0(X, N \odot N)$ and $H^0(X, N)$ respectively.

In Sect. 3 we prove a theorem which states that if there is a holomorphic vector bundle E on Y such that $E|_{X_t}$ is isomorphic to $N_{X_t|Y}$ for all $t \in M$, then the moduli space M comes equipped canonically with a torsion-free $\Xi_{H^0(X, N \otimes N^*)}$-connection such that, for every point $y \in Y$, the associated submanifold $P_y \subseteq \{t \in M : y \in X_t\}$ of M is totally geodesic. Another theorem shows that if in addition the cohomology group $H^1(X, N \otimes \odot^2 N^*)$ vanishes, then there is a distinguished torsion-free $\Lambda_{H^0(X, N \otimes \odot^2 N^*)}$-connection on M which is also totally geodesic on the submanifolds P_y, $y \in Y$. In particular, when the group $H^0(X, N \otimes \odot^2 N^*)$ is zero, this canonical $\Lambda_{H^0(X, N \otimes \odot^2 N^*)}$-connection degenerates into a single affine connection satisfying the natural integrability conditions. It is this theorem which gives a unified treatment of Penrose's construction of all local Ricci-flat self-dual metrics on 4-folds and the twistor description of quaternionic Kähler manifolds with non-zero scalar curvature .

In section 4 we prove Theorem 4.3 which strengthens this result by replacing the condition $H^1(X, N \otimes \odot^2 N^*) = 0$ by a vanishing condition on some cohomology class in $H^1(X, N \otimes \odot^2 N^*)$ naturally associated with the triple (X, Y, E). Moreover this theorem presents a clear geometric means allowing one to pick out a single affine connection from the $\Lambda_{H^0(X, N \otimes \odot^2 N^*)}$-equivalence class induced canonically on the moduli space. If the compact complex submanifold $X \hookrightarrow Y$ is not a *rigid* complex manifold, then local versions of all the results described above still hold.

We conclude the paper by discussing a number of examples.

2 Jet bundles and torsion-free connections

2.1 Jet bundles

Let M be a complex manifold and let $\triangle \hookrightarrow M \times M$ be a diagonal. Denote by $p_{1,2} : M \times M \to M$ the two natural projections and set

$$\mathcal{O}_M^{(m)} = \mathcal{O}_{M\times M}/J_\triangle^{m+1},$$

where $J_\triangle$ is the ideal subsheaf of $\mathcal{O}_{M\times M}$ consisting of germs of functions vanishing on $\triangle$. The space $(\triangle, \mathcal{O}_M^{(m)})$ is the mth-order infinitesimal neighbourhood of the diagonal.

Consider two maps,

$$\partial_1 : \mathcal{O}_M \to \mathcal{O}_M^{(1)}, \quad f \to (p_1^* f)/ \bmod J_\triangle^2,$$

and

$$\partial_2 : \mathcal{O}_M \to \mathcal{O}_M^{(1)}, \quad f \to (p_2^* f)/ \bmod J_\triangle^2,$$

which provide two identifications of $\triangle$ with M. One may check that the map

$$d = \partial_1 - \partial_2$$

is a differentiation, $d : \mathcal{O}_M \to \Omega^1 M$, identifying the holomorphic cotangent sheaf $\Omega^1 M$ with the kernel of the natural reduction $\bmod J_\triangle : \mathcal{O}_M^{(1)} \to \mathcal{O}_M$. Thus we have an exact sequence

$$0 \longrightarrow \Omega^1 M \longrightarrow \mathcal{O}_M^{(1)} \longrightarrow \mathcal{O}_M \longrightarrow 0.$$

Another natural map

$$\partial = \partial_1 + \partial_2$$

is actually an $\mathcal{O}_M$-linear morphism of sheaves providing a canonical splitting of the above extension

$$\mathcal{O}_M^{(1)} = \mathcal{O}_M \oplus \Omega^1 M.$$

The sheaf of mth-order jets of $\Omega^1 M$ is defined by the formula

$$J^m(\Omega^1 M) = p_{2*}\left(p_1^*\Omega^1 M \,/\, J_\triangle^{m+1}\, p_1^*(\Omega^1 M)\right).$$

The above discussion implies the existence of an exact sequence

$$0 \longrightarrow \Omega^1 M \otimes \Omega^1 M \xrightarrow{\alpha} J^1(\Omega^1 M) \xrightarrow{\beta} \Omega^1 M \longrightarrow 0.$$

An affine connection on M can be defined as a splitting of this exact sequence, i.e. as an $\mathcal{O}_M$-linear map

$$\gamma : J^1(\Omega^1 M) \to \Omega^1 M \otimes \Omega^1 M$$

such that $\gamma \circ \alpha = id_{\Omega^1 M \otimes \Omega^1 M}$. A more familiar description of an affine connection as a covariant differential operator $\nabla_\gamma : \Omega^1 M \to \Omega^1 M \otimes \Omega^1 M$ can be reconstructed from γ as follows. There exists a universal operator

$$\begin{aligned} u : \Omega^1 M &\longrightarrow J^1(\Omega^1 M) \\ \omega &\longrightarrow p_{2*}\left(p_1^*\omega \,/\, J_\Delta^2 p_1^*(\Omega^1 M)\right) \end{aligned}$$

such that $\beta \circ u = id_{\Omega^1 M}$. Then $\nabla_\gamma = \gamma \circ u$. If D_γ is the composition

$$D_\gamma : \Omega^1 M \xrightarrow{u} J^1(\Omega^1 M) \xrightarrow{\gamma} \Omega^1 M \otimes \Omega^1 M \xrightarrow{\text{antisymmetrization}} \Omega^2 M,$$

and $d : \Omega^1 M \longrightarrow \Omega^2 M$ is the outer differential, then the difference

$$T_\gamma = D_\gamma - d$$

is an $\mathcal{O}_M$-linear, morphism of locally free sheaves,

$$T_\gamma : \Omega^1 M \longrightarrow \Omega^2 M$$

which is called the *torsion* of the connection γ. It is clear that $T_\gamma \in H^0(M, TM \otimes \Omega^2 M)$. A connection γ is called *torsion-free* if T_γ vanishes.

A choice of a local coordinate chart $\{t^\alpha,\ \alpha = 1, \ldots, m = \dim M\}$ on a domain $U \subset M$ determines the associated splitting of $J^1(\Omega^1 M)$ over U,

$$\begin{aligned} J^1(\Omega^1 M)\mid_U &\simeq \Omega^1 M \oplus \Omega^1 M \otimes \Omega^1 M \mid_U, \\ \omega^{(1)} &\simeq (\omega_\alpha, \omega_{\alpha\beta}). \end{aligned}$$

Then a connection $\gamma : J^1(\Omega^1 M) \to \Omega^1 M \otimes \Omega^1 M$ is given by

$$\gamma\left(\omega^{(1)}\right) \simeq \omega_{\alpha\beta} - \Gamma^\delta_{\alpha\beta}\,\omega_\delta$$

for some holomorphic functions $\Gamma^\delta_{\alpha\beta}$. The covariant differential ∇_γ has the coordinate representation

$$\nabla\omega \simeq \partial_\alpha \omega_\beta - \Gamma^\delta_{\alpha\beta}\,\omega_\delta,$$

for some 1-form $\omega = \omega_\alpha\, dt^\alpha$, while the torsion tensor T_γ is represented by $\Gamma^\delta_{[\alpha\beta]}$. The functions $\Gamma^\delta_{\alpha\beta}$ are called the coefficients of the connection γ in the coordinate chart $\{t^\alpha\}$.

Since

$$\Omega^1 M \otimes \Omega^1 M = \Omega^1 M \odot \Omega^1 M \oplus \Omega^2 M$$

the quotient sheaf $J^1(\Omega^1 M)/\alpha\,(\Omega^2 M)$ fits into the exact sequence

$$0 \longrightarrow \Omega^1 M \odot \Omega^1 M \longrightarrow J^1(\Omega^1 M)/\alpha\left(\Omega^2 M\right) \xrightarrow{\beta} \Omega^1 M \longrightarrow 0.$$

We shall often use its dual vector bundle

$$T^{[2]}M \equiv \left(J^1(\Omega^1 M)/\alpha\left(\Omega^2 M\right)\right)^*,$$

which has canonically the extension structure

$$0 \longrightarrow TM \xrightarrow{\beta^*} T^{[2]}M \longrightarrow TM \odot TM \longrightarrow 0.$$

and is called a second-order tangent bundle. Global splittings of the latter exact sequence, i.e. $\mathcal{O}_M$-linear morphisms $\gamma : T^{[2]}M \longrightarrow TM$ such that $\gamma \circ \beta^* = id_{TM}$, are in one-to-one correspondence with torsion-free affine connections on M. In a local coordinate chart a torsion-free connection

$$\gamma : T^{[2]}M \longrightarrow TM,$$

is given by

$$\gamma : (V^{\alpha\beta}, V^\alpha) \longrightarrow V^\alpha + \Gamma^\alpha_{\beta\gamma} V^{\beta\gamma}.$$

Any section $v^{[2]}$ of $T^{[2]}M$ over a domain $U \in M$ gives a second-order differential operator

$$\begin{array}{rccl} v^{[2]}: & \mathcal{O}_U & \longrightarrow & \mathcal{O}_U \\ (V^{\alpha\beta}, V^\alpha): & f & \longrightarrow & V^{\alpha\beta}\partial_\alpha\partial_\beta f + V^\alpha \partial_\alpha f. \end{array}$$

To see that this operator is well-defined, i. e. that it does not depend on the choice of the local coordinate system used in the construction, we give another but equivalent definition of $T^{[2]}M$. For any point $t \in M$ and any germ f of a holomorphic function at t define the "2-jet" $f^{[2]}$ as the quotient

$$f^{[2]} = (f - f(t))\,/J_t^3,$$

where $f(t) \in \mathbb{C}$ is the value of the germ at the point t and J_t is the ideal of germs of holomorphic functions vanishing at t. The set of germs of all such 2-jets can be given a natural topology which makes it a locally free sheaf of $\mathcal{O}_M$-modules which we denote $\mathcal{O}_M^{[2]}$. This sheaf is canonically isomorphic to $J^1(\Omega^1 M)/\alpha\,(\Omega^2 M)$ and thus $T^{[2]}M$ is isomorphic to the dual bundle $\left(\mathcal{O}_M^{[2]}\right)^*$. If

$$\begin{array}{rccl} pr^{[2]}: & \mathcal{O}_M & \longrightarrow & \mathcal{O}_M^{[2]} \\ & f & \longrightarrow & f^{[2]} \end{array}$$

is the natural projection, then

$$v^{[2]}(f) =< v^{[2]}, pr^{[2]}(f) >,$$

where the angular brackets denote the natural pairing between $\mathcal{O}_M^{[2]}$ and its dual $T^{[2]}M$.

2.2 Some equivalence classes of torsion-free connections

An affine connection on M determines a family of geodesic curves. We say that two connections γ and $\hat{\gamma}$ are projectively equivalent if they have the same geodesics considered as unparameterized paths. In a local coordinate chart this condition reads (Hitchin 1982)

$$\hat{\Gamma}^{\alpha}_{(\beta\gamma)} = \Gamma^{\alpha}_{(\beta\gamma)} + b_\beta \delta^\alpha_\gamma + b_\gamma \delta^\alpha_\beta$$

for some 1-form $b = b_\alpha dt^\alpha$. We shall need a natural generalization of such an equivalence relation.

If $\gamma_1 : T^{[2]}M \longrightarrow TM$ and $\gamma_2 : T^{[2]}M \longrightarrow TM$ are two torsion-free connections on M, then their difference vanishes on the subsheaf $TM \hookrightarrow T^{[2]}M$ and thus induces an $\mathcal{O}_M$-linear morphism of locally free sheaves

$$\gamma_1 - \gamma_2 : T^{[2]}M/TM \cong TM \odot TM \longrightarrow TM$$

Thus $\gamma_1 - \gamma_2 \in H^0(M, TM \otimes \odot^2\Omega^1 M)$.

Let Λ be a fixed (not necessary locally free) subsheaf of $\mathcal{O}_M$-modules of the locally free sheaf $TM \otimes \odot^2\Omega^1 M$. We call such a datum a Λ-structure on M and say that torsion-free connections γ_1 and γ_2 are Λ-equivalent if

$$\gamma_1 - \gamma_2 \in H^0(M, \Lambda).$$

A corresponding equivalence class is called a *Λ-equivalence class of torsion-free connections* on M. There is a closely related concept of *Λ-connection* on M which is, by definition, a collection of ordinary torsion-free connections $\{\gamma_i \mid i \in I\}$ on an open covering $\{U_i \mid i \in I\}$ of M which, on overlaps $U_{ij} = U_i \cap U_j$, have their differences in $H^0(U_{ij}, \Lambda)$. Locally a Λ-connection is the same thing as a Λ-equivalence class of torsion-free connections, but globally they are different — the obstruction for the existence of a Λ-connection on M lies in $H^1(M, TM \otimes \Omega^1 M \otimes \Omega^1 M/\Lambda)$, while the obstruction for the existence of a Λ-equivalence class of torsion-free connections is an element of $H^1(M, TM \otimes \odot^2\Omega^1 M)$.

We shall be interested also in a subclass of Λ-connections associated with Λ-structures on M of a special type. If $\Xi \subset TM \otimes \Omega^1 M$ is a subsheaf of $\mathcal{O}_M$-modules, we define a *Ξ-equivalence class of torsion-free connections* on M as a Λ-equivalence class of torsion-free connections with $\Lambda = \boldsymbol{sym}(\Xi \otimes \Omega^1 M)$, where $\boldsymbol{sym}$ is the natural projection

$$\boldsymbol{sym} : TM \otimes \Omega^1 M \otimes \Omega^1 M \to TM \otimes \Omega^1 M \odot \Omega^1 M.$$

Analogously, a Ξ-*connection* on M is, by definition, a $\boldsymbol{sym}(\Xi \otimes \Omega^1 M)$-connection.

Note that the structure sheaf $\mathcal{O}_M$ can be identified with a rank-1 locally free subsheaf $[\mathcal{O}_M] \subset \mathrm{End}(TM) \cong TM \otimes \Omega^1 M$, and a $[\mathcal{O}_M]$-connection on M is exactly a torsion-free projective connection.

2.3 Totally geodesic submanifolds

Let M be a manifold with a specified subsheaf of $\mathcal{O}_M$-modules $\Lambda \subset TM \otimes \Omega^1 M \odot \Omega^1 M$. We say that a submanifold $P \subset M$ is *consistent with the* Λ*-structure* on M if

$$\boldsymbol{f}_P(\Lambda) \subseteq TP \otimes \Omega^1 P \odot \Omega^1 P,$$

where

$$\boldsymbol{f}_P : TM \otimes \Omega^1 M \odot \Omega^1 M\Big|_P \xrightarrow{\boldsymbol{id} \otimes \boldsymbol{res} \odot \boldsymbol{res}} TM|_P \otimes \Omega^1 P \odot \Omega^1 P,$$

is a natural map. Here $\boldsymbol{res}$ denotes the restriction of 1-forms from $\Omega^1 M \mid_P$ to vector fields tangent to P. If Λ is of the form $\boldsymbol{sym}(\Xi \otimes \Omega^1 M)$ for some $\mathcal{O}_M$-submodule $\Xi \subset TM \otimes \Omega^1 M$, then the consistency condition is equivalent to

$$\Xi|_P \subseteq \mathrm{End}(TP).$$

We say in this case that the submanifold $P \hookrightarrow M$ is *consistent with the* Ξ*-structure.*

Let $\{U_i \mid i \in I\}$ be an open covering of M such that for some subset $I' \subset I$ the intersections $\{U_i' \equiv U_i \cap P \mid i \in I'\}$ provide an open covering of P. Let $\hat{\gamma}$ be a Λ-connection on M represented in this open covering by a collection of ordinary connections $\{\gamma_i \mid i \in I\}$. If the submanifold P is consistent with the Λ-structure on M, then, for all $i \in I'$, the compositions

$$\boldsymbol{\Phi}_P^{\hat{\gamma}}\Big|_{U_i'} : TP|_{U_i'} \xrightarrow{\nabla_{\gamma_i}} TM \otimes \Omega^1 M\Big|_{U_i'} \xrightarrow{\boldsymbol{id} \otimes \boldsymbol{res}} TM \otimes \Omega^1 P\Big|_{U_i'} \longrightarrow (TM/TP) \otimes \Omega^1 P\Big|_{U_i'},$$

glue to form a global $\mathcal{O}_P$-linear operator on TP which does not depend on the particular choices of both the covering $\{U_i \mid i \in I\}$ and the local lifts of the Λ-connection $\hat{\gamma}$ to the ordinary affine connections γ_i used in the construction.

A submanifold P of a manifold M equipped with a subsheaf of $\mathcal{O}_M$-modules $\Lambda \subset TM \otimes \odot^2 \Omega^1 M$ and a Λ-connection $\hat{\gamma}$ is called *totally geodesic relative to* $\hat{\gamma}$ if

(i) P is consistent with the Λ-structure on M;

(ii) the form $\boldsymbol{\Phi}_P^{\hat{\gamma}}$ vanishes.

In the case of a projective connection the condition (i) above is redundant.

3 Canonical geometric structures on moduli spaces

3.1 Analytic subspaces of moduli spaces

Let Y be a complex manifold and $X \hookrightarrow Y$ a complex submanifold with normal bundle N such that $H^1(X, N) = 0$. By the Kodaira theorem, there is a complete analytic family $F \hookrightarrow Y \times M$ of compact submanifolds which contains X. If N_F is the normal bundle of the embedding $F \hookrightarrow Y \times M$, then, for any point $t \in M$, we have

$$N_F|_{\nu^{-1}(t)} = N_{X_t|Y},$$

where $N_{X_t|Y}$ is the normal bundle of the embedding $\mu \circ \nu^{-1}(t) = X_t \hookrightarrow Y$. The completeness of the family implies the isomorphism of $\mathcal{O}_M$-modules

$$k : TM \xrightarrow{\simeq} \nu_*^0(N_F) \tag{1}$$

on the moduli space M.

Let us denote the point in the moduli space M corresponding to X by t_0, i.e. $X = \mu \circ \nu^{-1}(t_0)$. From the local coordinate description of this family given in Kodaira's (1962) paper it easily follows that there is a sufficiently small neighbourhood $M_0 \subseteq M$ of the point t_0 such that for each $y \in Y' \equiv \cup_{t \in M_0} X_t$ the set $\nu \circ \mu^{-1}(y) \cap M_0$ is a complex analytic subspace of M_0. We denote by P_y the manifold content of $\nu \circ \mu^{-1}(y) \cap M_0$, i.e. $P_y = \nu \circ \mu^{-1}(y) \cap M_0 \setminus \{\text{singular points}\}$. If the natural evaluation map

$$\begin{array}{rcl} H^0(X_t, N_{X_t|Y}) & \longrightarrow & N_z \\ \phi & \longrightarrow & \phi(z), \end{array}$$

where N_z is the fibre of N at a point $z \in X_t$ and $\phi(z)$ is the value of the global section $\phi \in H^0(X_t, N_{X_t|Y})$ at z, is surjective at all points $z \in X_t$ and for all $t \in M_0$, then $P_y = \nu \circ \mu^{-1}(y) \cap M_0$, i.e. the alpha-surface is a complex submanifold of M_0.

3.2 Distinguished Λ-structures

Let $F \hookrightarrow Y \times M$ be a complete analytic family of compact submanifolds. There exists a natural morphism of sheaves of $\mathcal{O}_M$-modules

$$\phi : \nu_*^0\left(N_F \otimes \odot^2 N_F^*\right) \longrightarrow TM \otimes \odot^2 \Omega^1 M.$$

Indeed, if χ is a germ of $\nu^0_*(N_F \otimes \odot^2 N^*_F)$ at a point $t \in M$, then the action

$$\phi(\chi) : TM \odot TM \longrightarrow TM$$

of the corresponding germ $\phi(\chi) \in TM \otimes \Omega^1 M \odot \Omega^1 M$ on $TM \odot TM$ may be described as follows. First we note that there is a natural morphism of $\nu^{-1}(\mathcal{O}_M)$-modules

$$\begin{aligned} \lambda : \nu^0_* N_F \odot_{\nu^{-1}(\mathcal{O}_M)} \nu^0_* N_F &\longrightarrow \nu^0_*(N_F \odot_{\mathcal{O}_F} N_F), \\ \sigma_1 \odot_{\nu^{-1}(\mathcal{O}_M)} \sigma_2 &\longrightarrow \sigma_1 \odot_{\mathcal{O}_F} \sigma_2, \end{aligned}$$

defined by pointwise symmetric tensor product of germs of global sections of N_F over the submanifold $\nu^{-1}(t) \subset F, t \in M$. Combining this map with the Kodaira map $\boldsymbol{k}$, we obtain a natural composition

$$TM \odot TM \xrightarrow{\boldsymbol{k}\odot\boldsymbol{k}} \nu^0_* N_F \odot_{\nu^{-1}(\mathcal{O}_M)} \nu^0_* N_F \xrightarrow{\lambda} \nu^0_*(N_F \odot N_F) \xrightarrow{\chi} \nu^0_*(N_F) \xrightarrow{\boldsymbol{k}^{-1}} TM$$

which explains the action of $\phi(\chi)$ on $TM \odot TM$. The image $\phi(\nu^0_*(N_F \otimes \odot^2 N^*_F)) \subset TM \otimes \odot^2 \Omega^1 M$ is a subsheaf of $\mathcal{O}_M$-modules which we denote by $\Lambda_{H^0(X, N \otimes \odot^2 N^*)}$.

Analogously one defines a natural morphism of $\mathcal{O}_M$-modules

$$\nu^0_*(N_F \otimes N^*_F) \longrightarrow TM \otimes \Omega^1 M,$$

whose image we denote by $\Xi_{H^0(X, N \otimes N^*)}$. Note that the $\mathcal{O}_M$-modules $\Lambda_{H^0(X, N \otimes \odot^2 N^*)}$ and $\Xi_{H^0(X, N \otimes N^*)}$ may fail to be locally free in general.

3.3 Canonical Ξ-connections

We want to distinguish in this paper between "equivalence" and "isomorphism" of two vector bundles, E and V, on a complex manifold X. We say that E and V are equivalent and use the notation $E \simeq V$, if these vector bundles are represented in the set $H^1(X, GL(n, \mathcal{O}_X))$ by one and the same cohomology class. We say that E and V are isomorphic and use the notation $E = V$ if there is a *fixed* vector bundle isomorphism $\phi : E \longrightarrow V$.

Theorem 3.1 *Let $X \hookrightarrow Y$ be a compact complex submanifold with normal bundle N such that*

$$H^1(X, N) = H^1(X, N \otimes N^*) = 0$$

and let $F \hookrightarrow Y \times M$ be the associated complete analytic family of compact submanifolds of Y. If there is a holomorphic vector bundle E on Y such that $\pi_1^*(E)|_{\nu^{-1}(M_0)} = N_F|_{\nu^{-1}(M_0)}$ for some open neighbourhood $M_0 \subseteq M$ of the point t_0 corresponding to X, then a (possibly smaller) open neighbourhood $U \subseteq M_0$ of this point comes equipped canonically with *a $\Xi_{H^0(X,N\otimes N^*)}$-connection such that, for every point $y \in Y' \equiv \cup_{t\in U} X_t$, the associated submanifold $P_y \subseteq \nu \circ \mu^{-1}(y) \cap U$ is totally geodesic.*

Proof. We suppose for simplicity that M_0 is a ball in $\mathbb{C}^m$, $m = H^0(X, N)$. Functoriality of the construction of the canonical class of torsion-free connections on $U \subseteq M_0$ given below makes it obvious how to deal with a more general case.

An open neighbourhood of the submanifold X in Y can always be covered by a finite number of sufficiently small Stein coordinate charts $\{W_{\rm i} \mid {\rm i} \in I\}$ with local coordinate functions[1], $(w_{\rm i}^A, z_{\rm i}^a)$, on each neighbourhood $W_{\rm i}$ such that the intersection $X \cap W_{\rm i}$ coincides with the subspace of $W_{\rm i}$ determined by the equations $w_{\rm i}^A = 0$. On the intersection $W_{\rm i} \cap W_{\rm j}$ the coordinates $w_{\rm i}^A, z_{\rm i}^a$ are holomorphic functions of $w_{\rm j}^B$ and $z_{\rm j}^b$,

$$w_{\rm i}^A = f_{\rm ij}^A(w_{\rm j}^B, z_{\rm j}^b), \qquad z_{\rm i}^a = g_{\rm ij}^a(w_{\rm j}^B, z_{\rm j}^b),$$

with $f_{\rm ij}^A(0, z_{\rm j}^b) = 0$. Let $t^\alpha, \alpha = 1, \ldots, m$, be a global coordinate system on M_0. Then the coordinate domains $W_{\rm i} \times M_0$ with coordinate functions $\left(w_{\rm i}^A, z_{\rm i}^a, t^\alpha\right)$ cover an open neighbourhood of $X \times M_0$ in the manifold $Y \times M_0$. Shrinking M_0 as necessary and using the implicit function theorem we conclude that the submanifold $F_0 = \nu^{-1}(M_0)$ is described in each coordinate chart $W_{\rm i} \times M_0$ by simultaneous equations of the form (Kodaira 1962)

$$w_{\rm i}^A = \phi_{\rm i}^A(z_{\rm i}, t),$$

where $\phi_{\rm i}^A(z_{\rm i}, t)$ are holomorphic functions of $z_{\rm i} = (z_{\rm i}^1, \ldots, z_{\rm i}^n)$ and $t = (t^1, \ldots, t^m)$ which satisfy the conditions $\phi_{\rm i}^A(z_{\rm i}, t_0) = 0$. For each fixed $t \in M_0$ the equation $w_{\rm i}^A = phi_{\rm i}^A(z_{\rm i}, t)$ defines a submanifold $X_t \cap W_{\rm i} \hookrightarrow W_{\rm i}$.

Thus F_0 is covered by coordinate neighbourhoods $\{V_{\rm i} \equiv F_0 \cap (W_{\rm i} \times M_0)\}$ with local coordinate functions $(z_{\rm i}^a, t^\alpha)$ which are related to each other on the overlaps $V_{\rm i} \cap V_{\rm j}$ as follows

$$z_{\rm i}^a = g_{\rm ij}^a\left(\phi_{\rm j}(z_{\rm j}, t), z_{\rm j}\right),$$

where $\phi_{\rm j}(z_{\rm j}, t) = \left(\phi_{\rm j}^1(z_{\rm j}, t), \ldots, \phi_{\rm j}^p(z_{\rm j}, t)\right)$. Obviously we have

$$\phi_{\rm i}^A\left(g_{\rm ij}\left(\phi_{\rm j}(z_{\rm j}, t), z_{\rm j}\right), t\right) = f_{\rm ij}^A\left(\phi_{\rm j}(z_{\rm j}, t), z_{\rm j}\right).$$

Define the functions

$$w_{t{\rm i}}^A = w_{\rm i}^A - \phi_{\rm i}^A(z_{\rm i}, t)$$

and note that the collections of functions $(w_{t{\rm i}}^A, z_{\rm i}^a, t^\alpha)$ form a coordinate system on $Y \times M_0$ with the property that the submanifold $F_0 \hookrightarrow Y \times M_0$ is described in each chart $W_{\rm i} \times M_0$

[1]Here and throughout the paper Capital Latin indices take values $1, \ldots, p = \dim Y - \dim X$, small Latin indices take values $1, \ldots, n = \dim X$, and Greek indices take values $1, \ldots, m = \dim M$.

by the equations $w_{t\mathrm{i}}^A = 0$. On the intersection $(W_\mathrm{i} \times W_\mathrm{j}) \cap M_0$ we have

$$\begin{aligned} w_{t\mathrm{i}}^A &= f_{\mathrm{ij}}^A\left(w_{t\mathrm{j}}^B + \phi_\mathrm{j}^B(z_\mathrm{j}, t),\, z_\mathrm{j}^b\right) - \phi_\mathrm{i}^A\left(g_{\mathrm{ij}}^a\left(w_{t\mathrm{j}}^B + \phi_\mathrm{j}^B(z_\mathrm{j}, t),\, z_\mathrm{j}^b\right), t^\beta\right), \\ z_\mathrm{i}^a &= g_{\mathrm{ij}}^a\left(w_{t\mathrm{j}}^B + \phi_\mathrm{j}^B(z_\mathrm{j}, t),\, z_\mathrm{j}^b\right). \end{aligned}$$

The sheaf J_{F_0} of functions on $Y \times M_0$ which vanish on F_0 is generated on $W_\mathrm{i} \times M_0$ by functions $w_{t\mathrm{i}}^1, \ldots, w_{t\mathrm{i}}^p$. Therefore, a general element $\sigma \in H^0(F_0, N_F)$, $N_F \equiv (J_F/J_F^2)^*$, is represented by a 0-cochain, $\{\sigma_\mathrm{i}^A(z_\mathrm{i}, t)\}$, of vector-valued holomorphic functions defined respectively on V_i which are pasted together according to the rule

$$\sigma_\mathrm{i}^A = F_{\mathrm{ij}B}^{\;A}\, \sigma_\mathrm{j}^B, \tag{2}$$

where the functions,

$$F_{\mathrm{ij}B}^{\;A} \equiv \left.\frac{\partial w_{t\mathrm{i}}^A}{\partial w_{t\mathrm{j}}^B}\right|_{w_t=0} = \left.\frac{\partial f_{\mathrm{ij}}^A}{\partial w_\mathrm{j}^B}\right|_{w_\mathrm{j}=\phi(z_\mathrm{j},t)} - \left.\frac{\partial \phi_\mathrm{i}^A}{\partial z_\mathrm{i}^a}\right|_{z_\mathrm{i}=g_\mathrm{ij}(\phi_\mathrm{j}^A, z_\mathrm{j})} \left.\frac{\partial g_{\mathrm{ij}}^a}{\partial w_\mathrm{j}^B}\right|_{w_\mathrm{j}=\phi_\mathrm{j}(z_\mathrm{j},t)}, \tag{3}$$

form a Čech 1-cocycle representing the cohomology class in $H^1(F_0, GL(p, \mathcal{O}_F))$ associated with the normal bundle $N_{F_0} \equiv N_F|_{F_0}$.

The Kodaira isomorphism

$$k : TM_0 \longrightarrow \nu_*^0(N_{F_0})$$

can now be described very explicitly: take any vector field v on M_0 and apply the associated 1st-order differential operator $V^\alpha \partial_\alpha$ to each function $\phi_\mathrm{i}^A(z_\mathrm{i}, t)$, where $p_\alpha = \partial/\partial t^\alpha$. The result is a collection of vector-valued holomorphic functions

$$\sigma_\mathrm{i}^A = V^\alpha\, \partial_\alpha \phi_\mathrm{i}^A(z_\mathrm{i}, t)$$

defined respectively on V_i. On the overlap $V_\mathrm{i} \cap V_\mathrm{j}$ they are related to each other according to formula (2) which implies that the 0-cochain

$$\left\{ V^\alpha \frac{\partial \phi_\mathrm{i}^A(z_\mathrm{i}, t)}{\partial t^\alpha} \right\}$$

is the Čech 0-cocycle representing the global section $k(v) \in H^0(F_0, N_F)$.

Now let us take any section, $v^{[2]}$, of $T^{[2]}M$ over M_0 and apply the associated 2nd-order differential operator $V^{\alpha\beta}\partial_\alpha\partial_\beta + V^\alpha\partial_\alpha$ to each function $\phi_\mathrm{i}^A(z_\mathrm{i}, t)$. The result is a 0-cochain of vector-valued holomorphic functions

$$\left\{ V^{\alpha\beta} \frac{\partial^2 \phi_\mathrm{i}^A(z_\mathrm{i}, t)}{\partial t^\alpha \partial t^\beta} + V^\alpha \frac{\partial \phi_\mathrm{i}^A(z_\mathrm{i}, t)}{\partial t^\alpha} \right\}$$

defined respectively on V_i. Let us investigate how these functions are related to each other on the intersection $V_i \cap V_j$.

Since

$$\left.\frac{\partial \phi_i^A(z_i,t)}{\partial t^\alpha}\right|_{z_i=g_{ij}(\phi_j,z_j)} = F_{ij}{}^A_B \frac{\partial \phi_j^B(z_j,t)}{\partial t^\alpha}$$

we find

$$\left.\frac{\partial^2 \phi_i^A}{\partial t^\alpha \partial t^\beta}\right|_{z_i=g_{ij}(\phi_j,z_j)} = F_{ij}{}^A_B \frac{\partial^2 \phi_j^B}{\partial t^\alpha \partial t^\beta} + F_{ij}{}^A_{BC} \frac{\partial \phi_j^B}{\partial t^\alpha}\frac{\partial \phi_j^C}{\partial t^\beta} - G_{ij}{}^A_{\alpha B}\frac{\partial \phi_j^B}{\partial t^\beta} - G_{ij}{}^A_{\beta B}\frac{\partial \phi_j^B}{\partial t^\alpha}, \tag{4}$$

where

$$\begin{aligned} F_{ij}{}^A_{BC} = {} & \left.\frac{\partial^2 f_{ij}^A}{\partial w_j^B \partial w_j^C}\right|_{w_j=\phi_j(z_j,t)} - \left.\frac{\partial \phi_i^A}{\partial z_i^a}\right|_{z_i=g_{ij}(\phi_j,z_j)} \left.\frac{\partial^2 g_{ij}^a}{\partial w_j^B \partial w_j^C}\right|_{w_j=\phi_j(z_j,t)} \\ & - \left.\frac{\partial^2 \phi_i^A}{\partial z_i^a \partial z_i^b}\right|_{z_i=g_{ij}(\phi_j,z_j)} \left.\left(\frac{\partial g_{ij}^a}{\partial w_j^B}\frac{\partial g_{ij}^b}{\partial w_j^C}\right)\right|_{w_j=\phi_j(z_j,t)}, \end{aligned} \tag{5}$$

and

$$G_{ij}{}^A_{\alpha B} = \left.\frac{\partial^2 \phi_i^A}{\partial z_i^a \partial t^\alpha}\right|_{z_i=g_{ij}(\phi_j,z_j)} \left.\frac{\partial g_{ij}^a}{\partial w_j^B}\right|_{w_j=\phi_j(z_j,t)}. \tag{6}$$

The collections $\left\{F_{ij}{}^A_{BC}\right\}$ and $\left\{G_{ij}{}^A_{\alpha B}\right\}$ form 1-cochains with coefficients in, respectively, $N_F \otimes N_F^* \odot N_F^*$ and $N_F \otimes N_F^* \otimes \nu^*(\Omega^1 M)$ relative the covering $\{V_i\}$. Straightforward (though very tedious) calculations reveal obstructions for these two 1-cochains to be 1-cocycles,

$$\delta\left\{F_{ik}{}^A_{BC}\right\} = \left.\frac{\partial F_{ij}{}^A_D(z_j,t)}{\partial z_j^a}\right|_{z_j=g_{jk}(\phi_k,z_k)} \frac{\partial g_{jk}^a}{\partial w_k^C} F_{jk,B}{}^D + \left.\frac{\partial F_{ij}{}^A_D(z_j,t)}{\partial z_j^a}\right|_{z_j=g_{jk}(\phi_k,z_k)} \frac{\partial g_{jk}^a}{\partial w_k^B} F_{jk}{}^D_C \tag{7}$$

and

$$\delta\left\{G_{ik}{}^A_{\alpha B}\right\} = \left.\frac{\partial F_{ij}{}^A_C(z_j,t)}{\partial z_j^a}\right|_{z_j=g_{jk}(\phi_k,z_k)} \frac{\partial g_{jk}^a}{\partial w_k^B}\frac{\partial \phi_j^C}{\partial t^\alpha}, \tag{8}$$

which are defined on triple intersections $V_i \cap V_j \cap V_k$. Here δ stands for the coboundary operator.

By assumption, there is a holomorphic vector bundle E on Y such that $\pi_1^*(E)|_{F_0} = N_{F_0}$. If $\left\{E_{ij}{}^A_B(w_j,z_j)\right\}$ is a 1-cocycle on Y representing E as an element of $H^1(Y, GL(p, \mathcal{O}_Y))$ relative the covering $\{W_i\}$, then

$$F_{ij}{}^A_B(z_i,t) = H_i{}^A_C(z_i,t)\Big|_{z_i=g_{ij}(\phi_j,z_j)} E_{ij}{}^C_D(w_j,z_j)\Big|_{w_j=\phi_j(z_j,t)} (H_j(z_j,t)^{-1})^D_B$$

for some 0-cochain $\{H_{\rm i}(z_{\rm i},t)\}$ of invertible holomorphic $p \times p$ matrices. From the latter equation and the cocycle condition

$$E_{\rm ik} = E_{\rm ij}\, E_{\rm jk}$$

one finds that the obstruction for the following 1-cochain

$$\begin{aligned} \tau_{{\rm ij}\,BC}^{\ A} \equiv\ & H_{{\rm i}\,K}^{A}(z_{\rm i},t)\Big|_{z_{\rm i}=g_{\rm ij}(\phi_{\rm j},z_{\rm j})} \frac{\partial E_{{\rm ij}\,D}^{\ K}}{\partial w_{\rm j}^{C}}\Big|_{w_{\rm j}=\phi_{\rm j}(z_{\rm j},t)} (H_{\rm j}(z_{\rm j},t)^{-1})_{B}^{D} \\ & +\Big(\frac{\partial H_{\rm i}}{\partial z_{\rm i}^{a}} H_{\rm i}^{-1}\Big)_{K}^{A}\Big|_{z_{\rm i}=g_{\rm ij}(\phi_{\rm j},z_{\rm j}^{a})} \frac{\partial g_{\rm ij}^{a}}{\partial w_{\rm j}^{C}}\, F_{{\rm ij}\,B}^{\ K}, \end{aligned} \tag{9}$$

to be a 1-cocycle with coefficients in $N_F \otimes \odot^2 N_F^*$, is :

$$\delta\{\tau_{{\rm ik}\,BC}^{\ A}\} = \frac{\partial F_{{\rm ij}\,D}^{\ A}(z_{\rm j},t)}{\partial z_{\rm j}^{a}}\Big|_{z_{\rm j}=g_{\rm jk}(\phi_{\rm k},z_{\rm k})} \frac{\partial g_{\rm jk}^{a}}{\partial w_{\rm k}^{C}}\, F_{{\rm jk}\,B}^{\ D}. \tag{10}$$

From equations (7) and (10) we infer that the combination

$$\left\{F_{{\rm ij}\,BC}^{\ A} - \tau_{{\rm ij}\,BC}^{\ A} - \tau_{{\rm ij}\,CB}^{\ A}\right\}$$

is a 1-cocycle on F_0 with values in the sheaf $N_F \otimes N_F^* \odot N_F^*$. Therefore, the combination

$$\left\{\left(F_{{\rm ij}\,BC}^{\ A} - \tau_{{\rm ij}\,BC}^{\ A} - \tau_{{\rm ij}\,CB}^{\ A}\right) \frac{\partial \phi_{\rm j}^{B}}{\partial t^{\alpha}}\right\} \tag{11}$$

is also a 1-cocycle, on F_0, this time with coefficients in the sheaf $N_F \otimes N_F^* \otimes \nu^*(\Omega^1 M)$. By assumption, the group $H^1(X, N\otimes N^*)$ vanishes. Then the semi-continuity principle (Kodaira 1986, Wells 1980) implies that $H^1(X_t, N_{X_t|Y} \otimes N_{X_t|Y}^*) = 0$ for all t in some open Stein neighbourhood $U \subseteq M_0$. Hence, by the Leray spectral sequence, the cohomology group $H^1(\nu^{-1}(U), N_F \otimes N_F^*)$ vanishes also. Therefore $H^1(\nu^{-1}(U), N_F \otimes N_F^* \otimes \nu^*(\Omega^1 M)) = 0$, and so the 1-cocycle (11) is actually a coboundary on $\nu^{-1}(U)$,

$$\left\{\left(F_{{\rm ij}\,BC}^{\ A} - \tau_{{\rm ij}\,BC}^{\ A} - \tau_{{\rm ij}\,CB}^{\ A}\right) \frac{\partial \phi_{\rm j}^{B}}{\partial t^{\alpha}}\right\} = \delta\left\{\sigma_{{\rm i}\,\alpha C}^{\ A}\right\},$$

or, more explicitly,

$$\left(F_{{\rm ij}\,BC}^{\ A} - \tau_{{\rm ij}\,BC}^{\ A} - \tau_{{\rm ij}\,CB}^{\ A}\right) \frac{\partial \phi_{\rm j}^{B}}{\partial t^{\alpha}} = -\sigma_{{\rm i}\,\alpha E}^{\ A}\, F_{{\rm ij}\,C}^{\ E} + F_{{\rm ij}\,E}^{\ A}\, \sigma_{{\rm j}\,\alpha C}^{\ E}, \tag{12}$$

where $\left\{\sigma_{{\rm i}\,\alpha C}^{\ A}\right\}$ is a 0-cochain with coefficients in the sheaf $N_F \otimes N_F^* \otimes \nu^*(\Omega^1 M)$ which splits this cocycle.

From equations (8) and (10) it follows that the combination

$$\left\{\tau_{\mathrm{ij}}{}_{CB}^{A}\,\frac{\partial\phi_{\mathrm{j}}^{C}}{\partial t^{\alpha}}-G_{\mathrm{ij}}{}_{\alpha B}^{A}\right\}\tag{13}$$

is also a 1-*cocycle* with coefficients in $N_F\otimes N_F^*\otimes\nu^*(\Omega^1M)$. Therefore

$$\delta\left\{\tau_{\mathrm{ij}}{}_{CB}^{A}\,\frac{\partial\phi_{\mathrm{j}}^{C}}{\partial t^{\alpha}}-G_{\mathrm{ij}}{}_{\alpha B}^{A}\right\}=\delta\left\{\theta_{\mathrm{i}}{}_{\alpha B}^{A}\right\},\tag{14}$$

for some 0-cochain $\left\{\theta_{\mathrm{i}}{}_{\alpha B}^{A}\right\}$ with coefficients in $N_F\otimes N_F^*\otimes\nu^*(\Omega^1M)$.

However, the 0-cochains $\left\{\sigma_{\mathrm{i}}{}_{\alpha B}^{A}\right\}$ and $\left\{\theta_{\mathrm{i}}{}_{\alpha B}^{A}\right\}$ are defined non-uniquely. For any global sections ξ and ζ of $N_F\otimes N_F^*\otimes\nu^*(\Omega^1M)$ over F_0, the 0-cochains

$$\begin{aligned}\tilde{\sigma}_{\mathrm{i}}{}_{\alpha B}^{A}&=\sigma_{\mathrm{i}}{}_{\alpha B}^{A}+\zeta_{\alpha B}^{A}\big|_{V_{\mathrm{i}}}\\ \tilde{\theta}_{\mathrm{i}}{}_{\alpha B}^{A}&=\theta_{\mathrm{i}}{}_{\alpha B}^{A}+\xi_{\alpha B}^{A}\big|_{V_{\mathrm{i}}}\end{aligned}\tag{15}$$

split the same 1-cocycles (11) and (13) respectively. As we shall see later, this non-uniqueness has an important geometric meaning.

If we rewrite equation (4) in the form

$$\begin{aligned}\left.\frac{\partial^2\phi_{\mathrm{i}}^{A}}{\partial t^{\alpha}\partial t^{\beta}}\right|_{z_{\mathrm{i}}=g_{\mathrm{ij}}(\phi_{\mathrm{j}},z_{\mathrm{j}})}&=F_{\mathrm{ij}}{}_{B}^{A}\,\frac{\partial^2\phi_{\mathrm{j}}^{B}}{\partial t^{\alpha}\partial t^{\beta}}+\left(F_{\mathrm{ij}}{}_{BC}^{A}-\tau_{\mathrm{ij}}{}_{BC}^{A}-\tau_{\mathrm{ij}}{}_{CB}^{A}\right)\frac{\partial\phi_{\mathrm{j}}^{B}}{\partial t^{(\alpha}}\frac{\partial\phi_{\mathrm{j}}^{C}}{\partial t^{\beta)}}\\&+\left(\tau_{\mathrm{ij}}{}_{CB}^{A}\,\frac{\partial\phi_{\mathrm{j}}^{C}}{\partial t^{\alpha}}-G_{\mathrm{ij}}{}_{\alpha B}^{A}\right)\frac{\partial\phi_{\mathrm{j}}^{B}}{\partial t^{\beta}}+\left(\tau_{\mathrm{ij}}{}_{CB}^{A}\,\frac{\partial\phi_{\mathrm{j}}^{C}}{\partial t^{\beta}}-G_{\mathrm{ij}}{}_{\beta B}^{A}\right)\frac{\partial\phi_{\mathrm{j}}^{B}}{\partial t^{\alpha}},\end{aligned}$$

and take into account equations (12) and (14), we obtain the equality

$$\begin{aligned}&\left.\left(\frac{\partial^2\phi_{\mathrm{i}}^{A}}{\partial t^{\alpha}\partial t^{\beta}}+\chi_{\mathrm{i}}{}_{\alpha B}^{A}\,\frac{\partial\phi_{\mathrm{i}}^{B}}{\partial t^{\beta}}+\chi_{\mathrm{i}}{}_{\beta B}^{A}\,\frac{\partial\phi_{\mathrm{i}}^{B}}{\partial t^{\alpha}}\right)\right|_{z_{\mathrm{i}}=g_{\mathrm{ij}}(\phi_{\mathrm{j}},z_{\mathrm{j}})}=\\&=F_{\mathrm{ij}}{}_{C}^{A}\left(\frac{\partial^2\phi_{\mathrm{j}}^{C}}{\partial t^{\alpha}\partial t^{\beta}}+\chi_{\mathrm{j}}{}_{\alpha B}^{C}\,\frac{\partial\phi_{\mathrm{j}}^{B}}{\partial t^{\beta}}+\chi_{\mathrm{j}}{}_{\beta B}^{C}\,\frac{\partial\phi_{\mathrm{j}}^{B}}{\partial t^{\alpha}}\right),\end{aligned}$$

where

$$\chi_{\mathrm{i}}{}_{\alpha B}^{C}\equiv\frac{1}{2}\sigma_{\mathrm{i}}{}_{\alpha B}^{C}+\theta_{\mathrm{i}}{}_{\alpha B}^{C}.$$

This equality implies the key observation that, for each value of α and β, the vector-valued functions,

$$\Phi_{\mathrm{i}}{}_{\alpha\beta}^{A}\equiv\frac{\partial^2\phi_{\mathrm{i}}^{A}}{\partial t^{\alpha}\partial t^{\beta}}+\chi_{\mathrm{i}}{}_{\alpha B}^{A}\,\frac{\partial\phi_{\mathrm{i}}^{B}}{\partial t^{\beta}}+\chi_{\mathrm{i}}{}_{\beta B}^{A}\,\frac{\partial\phi_{\mathrm{i}}^{B}}{\partial t^{\alpha}},$$

represent a *global section* of the bundle N_F over $\nu^{-1}(U)$. Hence, for any section

$$v^{[2]} = V^{\alpha\beta}\partial_\alpha \odot \partial_\beta + V^\alpha \partial_\alpha$$

of $T^{[2]}M$ over U, there is an associated global section,

$$V^{\alpha\beta}\, \Phi_{i\,\alpha\beta}^{\,A} + V^\alpha\, \partial_\alpha \phi_i^A(z_i, t)$$

of N_F and hence an associated section of $TM|_U$. Thus we have constructed a map

$$\begin{array}{rccc} \gamma: & T^{[2]}M\big|_U & \longrightarrow & TM|_U \\ & \left(V^{\alpha\beta}, V^\alpha\right) & \longrightarrow & k^{-1}\left\{V^{\alpha\beta}\, \Phi_{i\,\alpha\beta}^{\,A} + V^\alpha\, \partial_\alpha \phi_i^A\right\} \end{array}$$

which obviously splits the exact sequence

$$0 \longrightarrow TM \longrightarrow T^{[2]}M \longrightarrow TM \odot TM \longrightarrow 0$$

over the domain $U \subseteq M_0$ and thus gives a torsion-free affine connection on U. Since the functions $\partial_\alpha \phi_i^A(z_i, t)$ represent a basis of the free $\mathcal{O}_M$-module of global sections of N_{F_0}, the coefficients $\Gamma^\gamma_{\alpha\beta}(t)$ of the connection γ may be found from the equation

$$\Phi_{i\,\alpha\beta}^{\,A}(z_i, t) = \Gamma^\gamma_{\alpha\beta}(t)\, \partial_\gamma \phi_i{}^A(z_i, t). \tag{16}$$

It is the compactness of the submanifold $X_t \hookrightarrow Y$ for each $t \in U$ that makes the solution $\Gamma^\gamma_{\alpha\beta}(t)$ of these equations to be independent of the coordinates z_i.

However the map γ is not a canonical one — it depends on the choices of particular splittings (15) of the 1-cocycles (11) and (13). Forgetting for a moment about this arbitrariness we check by straightforward computations that γ is invariant under

(i) general coordinate transformations

$$\begin{array}{rcl} w_i^A & \longrightarrow & \tilde{w}_i^A = f_i^A(w_i^B, z_i^b) \\ z_i^a & \longrightarrow & \tilde{z}_i^a = g_i^a(w_i^B, z_i^b) \end{array}$$

on the manifold Y with $\det\left(\frac{\partial f_i^A}{\partial w_i^B}\right) \neq 0$, and

(ii) arbitrary changes of fibre coordinates in E,

$$\begin{array}{rcl} E_{ij\,B}^{\,A}(w_j, z_j) & \longrightarrow & [B_i^{-1}(w_i, z_i)]^A_C\, E_{ij\,D}^{\,C}(w_j, z_j)\, B_{j\,B}^{\,D}(w_j, z_j), \\ H_{i\,B}^{\,A}(z_i, t) & \longrightarrow & H_{i\,C}^{\,A}(z_i, t)\; B_{i\,B}^{\,C}(w_i, z_i)\Big|_{w_i = \phi_i(z_i, t)}, \\ H_{j\,B}^{\,A}(z_j, t) & \longrightarrow & H_{j\,C}^{\,A}(z_j, t)\; B_{j\,B}^{\,C}(w_j, z_j)\Big|_{w_j = \phi_j(z_j, t)}, \end{array}$$

where $\left\{B_{iB}^{A}(w_i, z_i)\right\}$ is an arbitrary 0-cochain of holomorphic invertible matrices on Y.

Invariance under transformations (i) and (ii) is not a surprise — the final expression (16) for the connection depends only on the choice of coordinates t^α on M_0.

Thus the connection γ is well-defined except for the arbitrariness in its construction described by transformations (15) which show that what is *canonically* induced on U is not a connection but the whole family of torsion-free connections with the property that the difference of any two connections of the family is a section of the bundle $\boldsymbol{sym}(\Xi_{H^0(X,N\otimes N^*)} \otimes \Omega^1 M) \subset TM \otimes \odot^2\Omega^1 M$. Therefore, we conclude that the complete moduli space M of relative deformations of X in Y comes equipped with a distinguished local $\Xi_{H^0(X,N\otimes N^*)}$-connection. We call it the *canonical* $\Xi_{H^0(X,N\otimes N^*)}$-connection.

Now we investigate how the submanifolds $P_y \subseteq \boldsymbol{\nu} \circ \boldsymbol{\mu}^{-1}(y) \cap U$, $y \in Y'$, of $U \subseteq M_0$ are related to the canonical $\Xi_{H^0(X,N\otimes N^*)}$-connection.

Let us first show that, for any $y_0 \in Y'$, the submanifold $P_{y_0} \subset M$ is consistent with the $\Xi_{H^0(X,N\otimes N^*)}$-structure on the moduli space M. Suppose that $y_0 \in W_i$ for some label $i \in I$. Then $y_0 = (w_{i0}^A, z_{i0}^a)$ and the submanifold P_{y_0} is given locally by the equations

$$w_{i0}^A - \phi_i^A(z_{i0}, t) = 0,$$

where $t \in \boldsymbol{\nu} \circ \boldsymbol{\mu}^{-1}(y_0)\backslash\{\text{singular points}\}$. So a vector field $v(t) = V^\alpha \partial_\alpha|_{P_{y_0}}$ is tangent to P_{y_0} if and only if it satisfies the simultaneous equations

$$V^\alpha\, \partial_\alpha \phi_i{}^A(z_{i0}, t) = 0. \tag{17}$$

On the other hand the vector field $v(t)$ on U is represented by a global section of N_F over $\boldsymbol{\nu}^{-1}(U)$ which in turn is represented by a 0-cocycle

$$\left\{\sigma_i^A = V^\alpha\, \partial_\alpha \phi_i{}^A(z_i, t)\right\}.$$

Since any $\xi \in \Xi_{H^0(X,N\otimes N^*)}$ is also represented by a global section of $N_F \otimes N_F^*$ over $\boldsymbol{\nu}^{-1}(U)$,

$$\xi = \left\{\xi_{iB}^A\right\},$$

the action of the endomorphism ξ on the vector field v can be described as a map

$$\begin{array}{ccc} v & \longrightarrow & \xi(v) \\ \left\{\sigma_i^A\right\} & \longrightarrow & \left\{\xi_{iB}^A \sigma_i^B\right\}. \end{array}$$

Now it is evident that if a vector v at $t \in M$ satisfies equation (17), then for any $\xi \in \Xi_{H^0(X,N\otimes N^*)}$ the vector $\xi(v)$ also satisfies (17). Therefore, $\Xi_{H^0(X,N\otimes N^*)}\,|_{TP} \subset \mathrm{End}(TP)$, and any submanifold $P_y \subset U$, $y \in Y'$, is consistent with the canonical $\Xi_{H^0(X,N\otimes N^*)}$-structure on $U \subseteq M$.

In order to prove that the submanifold P_{y_0} for arbitrary $y_0 \in Y'$ is totally geodesic relative to the canonical $\Xi_{H^0(X,N\otimes N^*)}$-connection $\hat{\gamma}$ on M, we have to show that, for any vector fields $v(t)$ and $w(t)$ on P_{y_0}, the quantity $< \Phi^{\hat{\gamma}}_{P_{y_0}}(v), w >$ vanishes (see subsection 2.3). If $\{t^\alpha\}$ is a local coordinate chart on U and $\Gamma^\gamma_{\alpha\beta}$ are coefficients of any lift of $\hat{\gamma}$ to a torsion-free affine connection, then the equation $< \Phi^{\hat{\gamma}}_{P_{y_0}}(v), w > = 0$ reads

$$\left(W^\beta \partial_\beta V^\alpha + \Gamma^\alpha_{\beta\gamma} V^\gamma W^\beta\right) \bmod TP_{y_0} = 0. \tag{18}$$

If $v(t) = V^\alpha(t)\partial_\alpha$ and $w(t) = W^\beta(t)\partial_\beta$ are tangent to $P_{y_0} \subset U$, then the equation

$$W^\beta(t)\frac{\partial}{\partial t^\beta}\left(V^\alpha(t)\frac{\partial \phi_{\mathrm{i}}^A(z_{\mathrm{i}\,0}, t)}{\partial t^\alpha}\right) = 0 \tag{19}$$

must hold. By construction, the coefficients, $\Gamma^\gamma_{\alpha\beta}$, of the lift of $\hat{\gamma}$ to an affine connection satisfy the equation

$$\Phi_{\mathrm{i}\,\alpha\beta}^{\ A}(z_{\mathrm{i}}, t) = \Gamma^\gamma_{\alpha\beta}(t)\,\partial_\gamma \phi_{\mathrm{i}}{}^A(z_{\mathrm{i}}, t). \tag{20}$$

where

$$\Phi_{\mathrm{i}\,\alpha\beta}^{\ A} \equiv \frac{\partial^2 \phi_{\mathrm{i}}^A}{\partial t^\alpha \partial t^\beta} + \chi_{\mathrm{i}\,\alpha B}^{\ A}\frac{\partial \phi_{\mathrm{i}}^B}{\partial t^\beta} + \chi_{\mathrm{i}\,\beta B}^{\ A}\frac{\partial \phi_{\mathrm{i}}^B}{\partial t^\alpha}.$$

Since

$$V^\alpha\, \partial_\alpha \phi_{\mathrm{i}}{}^A(z_{\mathrm{i}\,0}, t) = 0, \qquad W^\beta\, \partial_\beta \phi_{\mathrm{i}}{}^A(z_{\mathrm{i}\,0}, t) = 0,$$

the equation (20) implies

$$V^\alpha W^\beta \frac{\partial^2 \phi_{\mathrm{i}}^A(z_{\mathrm{i}\,0}, t)}{\partial t^\alpha \partial t^\beta} = \Gamma^\gamma_{\alpha\beta}\frac{\partial \phi_{\mathrm{i}}^A(z_{\mathrm{i}\,0}, t)}{\partial t^\gamma} V^\alpha W^\beta.$$

From the latter equation and the equation (19) it follows that

$$\left(W^\beta \partial_\beta V^\alpha + \Gamma^\alpha_{\beta\gamma} V^\gamma W^\beta\right)\frac{\partial \phi_{\mathrm{i}}^A(z_{\mathrm{i}\,0}, t)}{\partial t^\alpha} = 0.$$

By (17) this means that $\left(W^\beta \partial_\beta V^\alpha + \Gamma^\alpha_{\beta\gamma} V^\gamma W^\beta\right)\partial_\alpha \in TP_{y_0}$, and thus the equation (18) holds. The proof is completed. ■

Remark. If X is a hypersurface in Y, then N is a line bundle. Hence $H^0(X, N\otimes N^*) = H^0(X, \mathcal{O}_X) = \mathbb{C}$ and $\Xi_{H^0(X,N\otimes N^*)} = [\mathcal{O}_M]$. As was noted in subsection 2.2, an $[\mathcal{O}_M]$-connection is the same thing as a projective connection, and Theorem 3.1 gives thus a prescription of how to compute the local canonical projective connection on the moduli space M.

If the submanifold $X \hookrightarrow Y$ is rigid (i.e. $H^1(X, TX) = 0$), then $F \longrightarrow M$ is locally trivial (Kodaira 1986), i.e. for any point $t \in M$ there is a domain $U \subseteq M$ such that $\nu^{-1}(U) \simeq U \times X$. For each $t \in M$ the complex submanifold $X_t = \mu \circ \nu^{-1}(t)$ is isomorphic to X. If the normal bundle N is also rigid (i.e. $H^1(X, N \otimes N^*) = 0$), then the normal bundle N_t of the embedding $X_t \hookrightarrow Y$ is isomorphic to N for all $t \in M$. This implies that both $\mathcal{O}_M$-modules $\Xi_{H^0(X,N\otimes N^*)} \subset TM \otimes \Omega^1 M$ and $\Lambda_{H^0(X,N\otimes\odot^2 N^*)} \subset TM \otimes \Omega^1 M \odot \Omega^1 M$ are locally free on M. For each $y \in Y' \equiv \cup_{t\in M} X_t$ the set $\nu \circ \mu^{-1}(y)$ is a complex analytic subspace of M. Again we let $P_y = \nu \circ \mu^{-1}(y) \setminus \{\text{singular points}\}$.

Corollary 3.2 *Let Y be a complex manifold and $X \hookrightarrow Y$ a compact complex rigid submanifold such that*

$$H^1(X, N) = H^1(X, N \otimes N^*) = 0,$$

and let $F \hookrightarrow Y \times M$ be the associated complete analytic family of compact submanifolds of Y. If there is a holomorphic vector bundle E on Y such that $\pi_1^*(E)|_{\nu^{-1}(M_0)} = N_F|_{\nu^{-1}(M_0)}$ for some domain M_0 in the moduli space M, then M_0 comes equipped canonically with a $\Xi_{H^0(X,N\otimes N^*)}$-connection such that, for every point $y \in Y'$, the associated submanifold $P_y \cap M_0$ is totally geodesic.

Proof. Let us cover the domain M_0 by sufficiently small coordinate neighbourhoods $\{U_i\}$ and use Theorem 3.1 to define a canonical $\Xi_{H^0(X,N\otimes N^*)}$-equivalence class of affine connections in each U_i. Functorial properties of the construction described in the proof of Theorem 3.1 immediately imply that these local families of affine connections glue to form a global $\Xi_{H^0(X,N\otimes N^*)}$-connection on M_0 which does not depend on the particular covering of M_0 used in the construction. ■

There is an important case when the requirement $E|_{X_t} = N_t$ for $t \in M_0$ can be replaced by a weaker condition $E|_{X_t} \simeq N_t$ without altering the main result.

Proposition 3.3 *Let $X \hookrightarrow Y$ be a compact complex rigid hypersurface with normal bundle N such that $H^1(X, N) = 0$ and let $F \hookrightarrow Y \times M$ be the associated complete analytic family of compact submanifolds of Y containing X. If $H^1(X, \mathcal{O}_X) = 0$ and there is a holomorphic line bundle E on Y such that $\pi_1^*(E)|_{\nu^{-1}(M_0)} \simeq N_F$ for some domain M_0 in the moduli space M, then M_0 comes equipped canonically with a torsion-free projective connection such that, for every point $y \in Y'$, the associated submanifold $P_y \cap M_0$ is totally geodesic.*

Proof. Since the vector bundles $\pi_1^*(E)|_{F_0}$ and N_{F_0} are known to be only equivalent (rather than isomorphic) to each other, the projective connection on M_0 constructed along the lines suggested in the proof of Theorem 3.1 is well-defined if it is invariant under additional transformations of the form

$$\begin{aligned} E_{ij}(w_j, z_j) &\longrightarrow E_{ij}(w_j, z_j) \\ H_i(z_i, t) &\longrightarrow H_i(z_i, t)\, S_i(z_i, t), \\ H_j(z_j, t) &\longrightarrow H_j(z_j, t)\, S_j(z_j, t), \end{aligned}$$

where the 0-cocycle $\{S_i\}$ represents an arbitrary global holomorphic section,

$$E_{ij}|_{F_0} = S_i \; E_{ij}|_{F_0} \; S_j^{-1}.$$

of $\pi_1^*(E)|_{F_0}$. Since E has rank 1 and X is compact,

$$\frac{\partial S_i(z_i, t)}{\partial z_i} = 0,$$

and the 1-cochain (9) remains invariant. Therefore the canonical projective connection on the moduli space is invariant under such transformations. ■

3.4 Canonical Λ-connections

In this subsection we study $\Lambda_{H^0(X, N\otimes N^*\odot N^*)}$-connections induced canonically on moduli spaces M of Kodaira's complete analytic families of compact submanifolds. In contrast to the distinguished $\Xi_{H^0(X, N\otimes N^*)}$-connections studied in the previous subsection, these $\Lambda_{H^0(X, N\otimes\odot^2 N^*)}$-connections degenerate sometimes into ordinary torsion-free connections on M providing thereby a link with many results of twistor theory.

Theorem 3.4 *Let $X \hookrightarrow Y$ be a compact complex submanifold with normal bundle N such that*

$$H^1(X, N) = H^1(X, N \otimes \odot^2 N^*) = 0$$

and let $F \hookrightarrow Y \times M$ be the associated complete analytic family of compact submanifolds of Y containing X. Suppose that there is a domain $M_0 \subseteq M$ containing the point t_0 corresponding to X and a holomorphic vector bundle E on Y such that $\pi_1^*(E)|_{\nu^{-1}(M_0)} = N_F|_{\nu^{-1}(M_0)}$. Then there is an open neighbourhood $U \subseteq M_0$ of t_0 which comes equipped canonically with a $\Lambda_{H^0(X, N\otimes\odot^2 N^*)}$-connection such that, for every point $y \in Y'$, the associated submanifold $P_y \subseteq \nu \circ \mu^{-1}(y) \cap U$ is totally geodesic.

Proof. We prove this theorem by explicit calculations in coordinate charts $\{V_i\}$ on $F_0 \equiv \nu^{-1}(M_0)$ introduced in the proof of Theorem 3.1. As before, the isomorphism

$$\pi_1^*(E)|_{F_0} = N_F|_{F_0}$$

implies the equality

$$F_{ij}{}^A_B(z_j,t) = H^A_{i\,C}(z_i,t)\Big|_{z_i=g_{ij}(\phi_j,z_j)} E_{ij}{}^C_D(w_j,z_j)\Big|_{w_j=\phi_j(z_j,t)} (H_j(z_j,t)^{-1})^D_B \tag{21}$$

for some 0-cochain $\{H_i(z_i,t)\}$ of invertible holomorphic $p \times p$ matrices (see the proof of Theorem 3.1). Then

$$\frac{\partial F_{ij}{}^A_B}{\partial t^\alpha} = (\frac{\partial H_i}{\partial t^\alpha} H_i^{-1})^A_C F_{ij}{}^C_B - F_{ij}{}^A_C(\frac{\partial H_j}{\partial t^\alpha} H_j^{-1})^C_B + \tau_{ij}{}^A_{BC}\frac{\partial \phi_j^C}{\partial t^\alpha},$$

where the functions $\tau_{ij}{}^A_{BC}$ are given by (9). On the other hand, the explicit expression (3) yields

$$\frac{\partial F_{ij}{}^A_B}{\partial t^\alpha} = F_{ij}{}^A_{BC}\frac{\partial \phi_j^C}{\partial t^\alpha} - G_{ij}{}^A_{\alpha B},$$

where $F_{ij}{}^A_{BC}$ and $G_{ij}{}^A_{\alpha B}$ are given by (5) and (6) respectively.

¿From the latter two equalities one finds

$$\begin{aligned} G_{ij}{}^A_{\alpha B}\frac{\partial \phi_j^B}{\partial t^\beta} &= \left(F_{ij}{}^A_{BC} - \tau_{ij}{}^A_{BC}\right)\frac{\partial \phi_j^B}{\partial t^\beta}\frac{\partial \phi_j^C}{\partial t^\alpha} + \\ &+ F_{ij}{}^A_C(\frac{\partial H_j}{\partial t^\alpha} H_j^{-1})^C_B\frac{\partial \phi_j^B}{\partial t^\beta} - (\frac{\partial H_i}{\partial t^\alpha} H_i^{-1})^A_B\frac{\partial \phi_i^B}{\partial t^\beta}, \end{aligned}$$

and substituting this expression into (4) one obtains

$$\begin{aligned} \frac{\partial^2 \phi_i^A}{\partial t^\alpha \partial t^\beta} &= F_{ij}{}^A_B\frac{\partial^2 \phi_j^B}{\partial t^\alpha \partial t^\beta} + \left(\tau_{ij}{}^A_{BC} + \tau_{ij}{}^A_{CB} - F_{ij}{}^A_{BC}\right)\frac{\partial \phi_j^B}{\partial t^\alpha}\frac{\partial \phi_j^C}{\partial t^\beta} + \\ &+ (\frac{\partial H_i}{\partial t^\alpha} H_i^{-1})^A_B\frac{\partial \phi_i^B}{\partial t^\beta} + (\frac{\partial H_i}{\partial t^\beta} H_i^{-1})^A_B\frac{\partial \phi_i^B}{\partial t^\alpha} - \\ &- F_{ij}{}^A_C(\frac{\partial H_j}{\partial t^\alpha} H_j^{-1})^C_B\frac{\partial \phi_j^B}{\partial t^\beta} - F_{ij}{}^A_C(\frac{\partial H_j}{\partial t^\beta} H_j^{-1})^C_B\frac{\partial \phi_j^B}{\partial t^\alpha}. \end{aligned} \tag{22}$$

We have met once again the combination

$$\left\{F_{ij}{}^A_{BC} - \tau_{ij}{}^A_{BC} - \tau_{ij}{}^A_{CB}\right\},$$

which was shown in the proof of Theorem 3.1 to be a 1-cocycle representing some cohomology class in $H^1(F_0, N_F \otimes \odot^2 N_F^*)$. Since $H^1(X, N \otimes \odot^2 N^*) = 0$ the semi-continuity principle implies $H^1(X_t, N_t \otimes \odot^2 N_t^*) = 0$ for all points t in some open Stein neighbourhood $U \subseteq M_0$ of t_0. Hence, by the Leray spectral sequence, $H^1(\nu^{-1}(U), N_F \otimes \odot^2 N_F^*) = 0$. Thus the 1-cocycle

$$\left\{\tau_{ij}{}^A_{BC} + \tau_{ij}{}^A_{CB} - F_{ij}{}^A_{BC}\right\} \tag{23}$$

is actually a coboundary,

$$\left\{\tau_{ij}{}_{BC}^{A} + \tau_{ij}{}_{CB}^{A} - F_{ij}{}_{BC}^{A}\right\} = \delta\left\{\sigma_{i}{}_{BC}^{A}\right\},$$

or, more explicitly,

$$\tau_{ij}{}_{BC}^{A} + \tau_{ij}{}_{CB}^{A} - F_{ij}{}_{BC}^{A} = -\sigma_{i}{}_{FE}^{A}\, F_{ij}{}_{B}^{F}\, F_{ij}{}_{C}^{E} + F_{ij}{}_{E}^{A}\, \sigma_{j}{}_{BC}^{E}, \tag{24}$$

where $\left\{\sigma_{ij}{}_{BC}^{A}\right\}$ is a 0-cochain with coefficients in $N_F \otimes \odot^2 N_F^*$ which splits this 1-cocycle. However this splitting is non-unique — for any global section $\left\{\xi_{i}{}_{BC}^{A}\right\}$ of $N_F \otimes \odot^2 N_F^*$ over $\nu^{-1}(U)$ the 0-cochain

$$\tilde{\sigma}_{i}{}_{BC}^{A} = \sigma_{i}{}_{BC}^{A} + \xi_{i}{}_{BC}^{A} \tag{25}$$

splits the same 1-cocycle (23).

Therefore, choosing some 0-cochain $\{\sigma_{i}{}_{BC}^{0}\}$ which splits the 1-cocycle (23), we rewrite the equation (22) in the form

$$\Psi_{i}{}_{\alpha\beta}^{A} = F_{ij}{}_{B}^{A}\, \Psi_{j}{}_{\alpha\beta}^{B},$$

where

$$\begin{aligned}\Psi_{i}{}_{\alpha\beta}^{A}(z_i, t) \;\equiv\; & \frac{\partial^2 \phi_i^A}{\partial t^\alpha \partial t^\beta} + \sigma_{i}{}_{BC}^{A}\, \frac{\partial \phi_i^B}{\partial t^\alpha} \frac{\partial \phi_i^C}{\partial t^\beta} - \\ & - \left(\frac{\partial H_i}{\partial t^\alpha} H_i^{-1}\right)_B^A \frac{\partial \phi_i^B}{\partial t^\beta} - \left(\frac{\partial H_i}{\partial t^\beta} H_i^{-1}\right)_B^A \frac{\partial \phi_i^B}{\partial t^\alpha}.\end{aligned}$$

Therefore for each value of α and β the 0-cochain $\left\{\Psi_{i}{}_{\alpha\beta}^{A}(z_i, t)\right\}$ is actually a 0-cocycle representing some global section of N_F over $\nu^{-1}(U)$. Thus we have constructed a map

$$\begin{aligned} T^{[2]}M\big|_U \;&\longrightarrow\; TM|_U \\ V^{\alpha\beta}(t)\partial_\alpha\partial_\beta + V^\alpha(t)\partial_\alpha \;&\longrightarrow\; k^{-1}\left(V^{\alpha\beta}\, \Psi_{i}{}_{\alpha\beta}^{A} + V^\alpha\, \partial_\alpha \phi_i^A\right)\end{aligned}$$

which obviously splits the exact sequence

$$0 \longrightarrow TM \longrightarrow T^{[2]}M \longrightarrow TM \odot TM \longrightarrow 0$$

over U and thus gives a torsion-free affine connection ∇ on U. It is straightforward to check that this connection is independent of the particular choices of coordinates used in the construction (i.e. it is invariant under the transformations (i) and (ii) described in the proof of Theorem 3.1) and thus is well-defined. The coefficients of this connection in a coordinate chart $\{t^\alpha\}$ on U can be described as follows. We know that the collection, $\left\{\Psi_{i\alpha\beta}^{A}\right\}$ of vector-valued functions represents a global section of N_F over $\nu^{-1}(U)$. Since the 0-cocycles of holomorphic vector-valued functions,

$$\left\{\frac{\partial \phi_{\mathrm{i}}^{A}(z_{\mathrm{i}},t)}{\partial t^{\alpha}}\right\}, \quad \alpha = 1,\dots,m,$$

represent a basis of the free $\mathcal{O}_U$-module $H^0(\boldsymbol{\nu}^{-1}(U), N_F) \simeq H^0(U, TM)$, we have

$$\Psi_{\mathrm{i}\alpha\beta}^{A}(z_{\mathrm{i}},t) = \Gamma_{\alpha\beta}^{\gamma}(t)\,\partial_{\gamma}\phi_{\mathrm{i}}^{A}(z_{\mathrm{i}},t) \tag{26}$$

for some holomorphic functions $\Gamma_{\alpha\beta}^{\gamma}(t)$ on $\boldsymbol{\nu}^{-1}(U)$. Since X_t is compact for each $t \in M$, these functions are independent of the coordinates z_{i} and thus give an explicit coordinate description of the torsion-free affine connection ∇.

Thus the only arbitrariness in the construction of the connection γ is described by the transformations (25) which show that what is *canonically* induced on the domain U is not a single connection but the whole family of torsion-free connections with the property that the difference of any two connections of the family is a section of the sheaf $\Lambda_{H^0(X,N\otimes\odot^2N^*)} = \phi(\boldsymbol{\nu}_*^0(N_F \otimes \odot^2 N_F^*)) \subset TM \otimes \odot^2\Omega^1 M$ over U. Therefore, we conclude that a sufficiently small neighbourhood U of the point t_0 in the moduli space M of the complete family $F \hookrightarrow Y \times M$ of relative deformations of X in Y comes equipped canonically with a distinguished $\Lambda_{H^0(X,N\otimes\odot^2N^*)}$-equivalence class of affine connections. We call it the canonical $\Lambda_{H^0(X,N\otimes\odot^2N^*)}$-connection.

Let us now show that for any $y_0 \in Y' = \cup_{t\in U} X_t$ the submanifold $P_{y_0} \subseteq \boldsymbol{\nu} \circ \boldsymbol{\mu}^{-1}(y_0)$ is consistent with the canonical $\Lambda_{H^0(X,N\otimes\odot^2N^*)}$-structure on M.

Suppose that $y_0 \in W_{\mathrm{i}}$ for some label $\mathrm{i} \in I$. Then $y_0 = (w_{\mathrm{i}0}^A, z_{\mathrm{i}0}^a)$ and the submanifold P_{y_0} is given locally by the equations

$$w_{\mathrm{i}0}^{A} - \phi_{\mathrm{i}}^{A}(z_{\mathrm{i}0},t) = 0,$$

where $t \in \boldsymbol{\nu}\circ\boldsymbol{\mu}^{-1}(y_0)\backslash\{\text{singular points}\}$. A vector field $v(t) = V^{\alpha}\partial_{\alpha}|_{P_{y_0}}$ can be represented by a global section of N_F over $\boldsymbol{\nu}^{-1}(U)$ which in turn is represented in the covering $\{V_{\mathrm{i}}\}$ by a 0-cocycle

$$\left\{v_{\mathrm{i}}^{A}(z_{\mathrm{i}},t) \equiv V^{\alpha}\,\partial_{\alpha}\phi_{\mathrm{i}}^{A}(z_{\mathrm{i}},t)\right\}.$$

The vector field $v(t)$ is tangent to P_{y_0} if and only if

$$v_{\mathrm{i}}^{A}(z_{\mathrm{i}0},t) = 0.$$

Let $v(t)$ and $w(t)$ be two vector fields on P_{y_0} represented by

$$\left\{v_{\mathrm{i}}^{A}(z_{\mathrm{i}},t) \equiv V^{\alpha}\,\partial_{\alpha}\phi_{\mathrm{i}}^{A}(z_{\mathrm{i}},t)\right\}$$

and, respectively,

$$\left\{w_i^A(z_i,t) \equiv W^\alpha\, \partial_\alpha \phi_i^A(z_i,t)\right\}$$

and let ξ be a global section of $\Lambda_{H^0(X,N\otimes\odot^2N^*)}$ represented on $\nu^{-1}(M_0)$ by the 0-cocycle $\left\{\xi_{i\,BC}^{\ A}\right\}$. Then the action of ξ on $v \odot w$ is given by

$$u = \xi(v \odot w) \Longleftrightarrow \left\{u_i^A = \xi_{i\,BC}^{\ A}\, v_i^B\, w_i^C\right\}.$$

It is clear that

$$u_i^A(z_{i0}, t) = 0,$$

and thus $\xi(v \odot w) \in TP_{y_0}$. This means that

$$f_P\left(\Lambda_{H^0(X,N\otimes\odot^2N^*)}\right) \subset TP_{y_0} \otimes \odot^2\Omega^1 P_{y_0}$$

and thus any submanifold $P_y \cap U$, $y \in Y'$, is consistent with the canonical $\Lambda_{H^0(X,N\otimes\odot^2N^*)}$-structure on U.

Therefore , in order to prove that P_{y_0}, $y_0 \in Y'$, is totally geodesic relative to the local canonical $\Lambda_{H^0(X,L_X\otimes\odot^2N^*)}$-connection $\hat{\gamma}$ it remains to show that the form $\Phi^{\hat{\gamma}}_{P_{y_0}}$ vanishes. This part of the proof relies on exactly the same arguments as given in the proof of Theorem 3.1 and we omit them. The conclusion is that, for any point $y \in Y'$, the associated submanifold $P_y \subseteq \nu \circ \mu^{-1}(y) \cap U$ is totally geodesic relative to the connection $\hat{\gamma}$. The proof is completed. ■

Corollary 3.5 *Let $X \hookrightarrow Y$ be a compact complex rigid submanifold with normal bundle N such that*

$$H^1(X,N) = H^1(X, N \otimes N^*) = H^1(X, N \otimes \odot^2 N^*) = 0$$

and let $F \hookrightarrow Y \times M$ be the associated complete analytic family of compact submanifolds of Y containing X. If there is a holomorphic vector bundle E on Y such that $\pi_1^*(E)|_{\nu^{-1}(M_0)} = N_F|_{\nu^{-1}(M_0)}$ for some domain M_0 in the moduli space M, then M_0 comes equipped canonically with a torsion-free $\Lambda_{H^0(X,N\otimes\odot^2N^*)}$-connection such that, for every point $y \in Y'$, the associated submanifold $P_y \subseteq \nu \circ \mu^{-1}(y) \cap M_0$ is totally geodesic.

The next proposition presents the conditions which ensure that the complete moduli space of a relative deformation problem $X \hookrightarrow Y$ has a distinguished affine connection rather than some equivalence class of connections.

Proposition 3.6 *Let $X \hookrightarrow Y$ be a compact complex rigid submanifold with normal bundle N such that*

$$H^1(X,N) = H^1(X, N \otimes N^*) = H^1(X, N \otimes \odot^2 N^*) = H^0(X, N \otimes \odot^2 N^*) = 0$$

and let $F \hookrightarrow Y \times M$ be the associated complete analytic family of compact submanifolds of Y containing X. If there is holomorphic vector bundle E on Y such that $\pi_1^*(E)|_{\nu^{-1}(M_0)} = N_F|_{\nu^{-1}(M_0)}$ for some domain M_0 in the moduli space M, then M_0 comes equipped canonically with a torsion-free connection such that, for every point $y \in Y'$, the associated submanifold $P_y \subseteq \nu \circ \mu^{-1}(y) \cap M_0$ is totally geodesic.

Proof. Since X is a rigid complex manifold, the family $F \hookrightarrow Y \times M$ is locally trivial. The vanishing of the cohomology group $H^1(X, N \otimes N^*)$ implies that, for each $t \in M$ the normal bundle of the embedding $X_t = \mu \circ \nu^{-1}(t) \hookrightarrow Y$ is isomorphic to N. Then the vanishing of the cohomology group $H^1(X, N \otimes \odot^2 N^*)$, the local triviality of F and the Künneth theorem (Burns 1979) imply that the $\mathcal{O}_M$-module $\nu_*^0(N_F \otimes \odot^2 N_F^*)$ is zero. Therefore the distinguished $\Lambda_{H^0(X,N\otimes\odot^2 N^*)}$-connection induced on M_0 in accordance with Corollary 3.5 degenerates into an ordinary torsion-free affine connection. This fact completes the proof. ∎

Remark. Let a triple (X, Y, E), X being a compact hypersurface in Y, satisfy the conditions of Proposition 3.6. Let us study how the connection ∇ predicted by Proposition 3.6 changes under the transformations (cf. the proof of Proposition 3.3)

$$\begin{aligned} E_{ij}(w_j, z_j) &\longrightarrow E_{ij}(w_j, z_j) \\ H_i(z_i, t) &\longrightarrow H_i(z_i, t)\, S(t), \\ H_j(z_j, t) &\longrightarrow H_j(z_j, t)\, S(t), \end{aligned}$$

where $S(t)$ is an arbitrary nowhere vanishing holomorphic function. One finds that these transformations change the 0-cocycle $\Psi_{i\alpha\beta}$ as follows

$$\Psi_{i\alpha\beta} \longrightarrow \Psi_{i\alpha\beta} - \Lambda_\alpha \frac{\partial \phi_i}{\partial t^\beta} - \Lambda_\beta \frac{\partial \phi_i}{\partial t^\alpha},$$

where

$$\Lambda_\alpha = \frac{\partial S}{\partial t^\alpha} S^{-1}.$$

Then the equation (26) implies that the coefficients of the connection ∇ change as follows

$$\hat{\Gamma}^\alpha_{(\beta\gamma)} = \Gamma^\alpha_{(\beta\gamma)} + \Lambda_\beta\, \delta^\alpha_\gamma + \Lambda_\gamma\, \delta^\alpha_\beta.$$

Thus if we had asked in Proposition 3.6 only for the equivalence $\pi_1^*(E)|_{F_0} \simeq N_{F_0}$ rather than for the isomorphism $\pi_1^*(E)|_{F_0} = N_{F_0}$, then our construction would have resulted merely in the somewhat different approach to Proposition 3.3. Thus it is the isomorphism $\pi_1^*(E)|_{F_0} = N_{F_0}$ rather than equivalence $\pi_1^*(E)|_{F_0} \simeq N_{F_0}$ that allows us to pick out a single affine connection on M.

Remark. The proof of Theorem 3.4 shows that, without altering the main result, the condition $\pi_1^*(E)|_F = N_F$ can be replaced by a weaker condition that there is a monomorphism realizing N_F as a direct summand in $\pi_1^*(E)|_F$.

3.5 Some applications

Proposition 3.7 *Let Y be a complex manifold equipped with a holomorphic distribution $D \subset TY$, and let $X \hookrightarrow Y$ be a $(\dim Y - \operatorname{rank} D)$-dimensional compact complex submanifold which is transverse to D and has normal bundle N such that*

$$H^1(X, N) = H^1(X, N \otimes \odot^2 N^*) = H^0(X, N \otimes \odot^2 N^*) = 0.$$

If $F \hookrightarrow Y \times M$ is the associated complete analytic family of compact submanifolds containing X, then a sufficiently small open neighbourhood $U \subseteq M$ of the point t_0 corresponding to X, comes equipped canonically with a torsion-free affine connection such that, for every point $y \in Y'$, the associated submanifold $P_y \subseteq \nu \circ \mu^{-1}(y) \cap U$ is totally geodesic.

Proof. Since X is transverse to D, there is an open neighbourhood $M_0 \subseteq M$ of the associated point t_0 in M such that the submanifolds $X_t = \mu \circ \nu^{-1}(t)$ are transverse to D for all $t \in U$. Then the triple (X, Y, D) satisfies all the conditions of Theorem 3.4 which implies that there is an open neighbourhood $W \subseteq M_0$ of the point t_0 which comes equipped canonically with a torsion free $\Lambda_{H^0(X, N \otimes \odot^2 N^*)}$-connection $\hat{\gamma}$. By the semi-continuity principle (Kodaira 1986), the vanishing of the cohomology group $H^0(X, N \otimes \odot^2 N^*)$ implies the vanishing of $H^0(X_t, N_t \otimes \odot^2 N_t^*)$ for each t in some open Stein neighbourhood $U \subseteq W$. Then, by the Leray spectral sequence, $\nu_*^0(N_F \otimes \odot^2 N_F^*)|_U = 0$ and the connection $\hat{\gamma}$ is nothing but an ordinary torsion-free affine connection on U. ■

Corollary 3.8 *Let $X \hookrightarrow Y$ be a compact complex submanifold with normal bundle such that*

$$H^1(X, N) = H^0(X, N \otimes \odot^2 N^*) = H^1(X, N \otimes \odot^2 N^*) = 0,$$

and let $F \hookrightarrow Y \times M$ be the associated complete analytic family of compact submanifolds containing X. If Y has the structure of holomorphic fibration over X then a sufficiently small neighbourhood $U \subseteq M$ of the point t_0 corresponding to X, comes equipped canonically with a torsion-free affine connection such that, for any $y \in Y'$, the submanifold $P_y \subseteq \nu \circ \mu^{-1}(y) \cap U$ is totally geodesic.

Proof. By assumption, the manifold Y has the structure of a holomorphic fibration,

$$\pi : Y \longrightarrow X$$

over its submanifold X. If τ is the distribution of π-vertical vector fields, then the triple (X, Y, τ) satisfies the conditions of Proposition 3.7 and the result follows. ■

This corollary gives a new interpretation of the famous Penrose (1976) non-linear graviton construction which provides a one-to-one correspondence between "small" 4-dimensional complex Riemannian manifolds with self-dual Riemann tensor and 3-dimensional complex manifolds Y equipped with the following data: (i) a submanifold $X \simeq \mathbb{C}P^1$ with normal bundle $N \simeq \mathbb{C}^2 \otimes \mathcal{O}_{\mathbb{C}P^1}(1)$; (ii) a fibration $\pi : Y \longrightarrow X$; (iii) a global non-vanishing section of the "twisted" determinant bundle $Det(\tau) \otimes \pi^*(\mathcal{O}(2))$, where τ is the sheaf of π-vertical vector fields on Y. It is only the data (i) and (ii) that is used in the construction the associated torsion-free connection along the lines suggested in the proof of Theorem 3.4. (The datum (iii) ensures that the resulting connection is consistent with the metric induced on M by the Penrose construction, see Merkulov 1992.)

Remark. In fact the Corollary 3.8 is one of the manifestations of a general phenomenon peculiar to complete analytic families $F \hookrightarrow Y \times M$ generated by relative deformations of a compact complex submanifold $X \hookrightarrow Y$ such that the ambient manifold Y has the structure of holomorphic fibration over X (and hence over all X_t for t sufficiently close to t_0, $X = \mu \circ \nu^{-1}(t_0)$). Denote by J_{t_0} the ideal sheaf of the inclusion

$$t_0 \longrightarrow M$$

and by J_X the ideal sheaf of the corresponding embedding

$$X \longrightarrow Y.$$

With these two embeddings there are associated two towers of infinitesimal conormal sheaves $\tilde{N}^{(m)}_{t_0|M} \equiv J_{t_0}/J_{t_0}^{m+2}$ and $\tilde{N}^{(m)}_{X|Y} \equiv J_X/J_X^{m+2}$, where $m = 0, 1, 2, \ldots$. The sheaf $\tilde{N}^{(0)}_{X|Y}$ is always locally free on X and is exactly the conormal vector bundle. The point is that when Y has the structure of a holomorphic fibration over X all the higher order conormal sheaves $\tilde{N}^{(m)}_{X|Y}$, $m = 1, 2, \ldots$, are also locally free on X and one can interpret their $\mathcal{O}_M$-duals $N^{(m)}_{X|Y} \equiv \left(\tilde{N}^{(m)}_{X|Y}\right)^*$ as the mth-order normal bundles of the embedding $X \hookrightarrow Y$ (Merkulov 1992). Thus we obtain two towers in this case,

$$N^{(0)}_{t_0|M} \subset N^{(1)}_{t_0|M} \subset N^{(2)}_{t_0|M} \subset \ldots,$$

and

$$N^{(0)}_{X|Y} \subset N^{(1)}_{X|Y} \subset N^{(2)}_{X|Y} \subset \ldots.$$

The Kodaira theorem states that $N^{(0)}_{t_0|M} = H^0(X, N^{(0)}_{X|Y})$ and thus clarifies the interplay between the ground floors of these towers. What happens at higher floors? The answer is delightfully simple (Merkulov 1992) — for any number m there is a canonical linear map

$$\boldsymbol{k}_{t_0}^{(m)} : N_{t_0|M}^{(m)} \longrightarrow H^0\left(X, N_{X|Y}^{(m)}\right)$$

from the mth-order normal sheaf, $N_{t_0|M}^{(m)}$, of the point $t_0 \in M$, to the vector space of global holomorphic sections of the mth-order normal bundle, $N_{X|Y}^{(m)}$, of the associated submanifold X of M. When $m = 0$, $N_{t_0|M}^{(0)} \simeq T_{t_0}M$ and the map $\boldsymbol{k}_{t_0}^{(0)}$ coincides precisely with the Kodaira map $\boldsymbol{k}_{t_0}$. Since such basic differential geometric operations on the manifold M as a covariant derivative of a vector field, the Lie bracket of two vector fields or the Frobenius map $\wedge^2 D \to TM/D$ associated with a distribution $D \subset TM$, depend not only the values of vector fields at a point $t \in M$ but also on their representatives in the first order neighborhood $t^{(1)}$, it is clear that the first order map $\boldsymbol{k}_t^{(1)}$ should be useful in encoding (or decoding) information about the above mentioned differential operators on M into (or from) the holomorphic data of the embedding $X \hookrightarrow Y$. We refer to Merkulov (1992) for details where it is shown in particular that such statements as Corollary 3.8 are indeed rooted in the first order correspondence $\boldsymbol{k}^{(1)}$.

Let us now consider a relative deformation problem $X \hookrightarrow Y$ in the case when the ambient $(2n+1)$-dimensional manifold Y has a complex contact structure, meaning a holomorphic distribution of $2n$-planes $D \subset TY$ such that the Frobenius form

$$\begin{aligned} \boldsymbol{\Phi} : D \times D &\longrightarrow TY/D \\ (v, w) &\longrightarrow [v, w] \bmod D \end{aligned}$$

is non-degenerate Arnold (1978). Proposition 3.7 implies

Corollary 3.9 *Let X be a compact Riemann surface embedded into a complex contact manifold Y transversely to the contact structure with normal bundle N such that*

$$H^1(X, N) = H^0(X, N \otimes \odot^2 N^*) = H^1(X, N \otimes \odot^2 N^*) = 0,$$

and let $F \hookrightarrow Y \times M$ be the associated complete analytic family of compact submanifolds containing X. Then there is a neighbourhood $U \subseteq M$ of the point t_0 corresponding to X, which comes equipped canonically with a torsion-free affine connection such that, for any $y \in Y'$, the submanifold $P_y \subseteq \boldsymbol{\nu} \circ \boldsymbol{\mu}^{-1}(y) \cap U$ is totally geodesic.

Let X be a rational curve in a $(2n+1)$-dimensional contact manifold which is transverse to the contact structure and has normal bundle isomorphic to $\mathbb{C}^{2n} \otimes \mathcal{O}(1)$. Then Corollary 3.9 implies that the associated $4n$-dimensional moduli space comes equipped with a torsion-free connection ∇ such that every $2n$-dimensional submanifold $P_y = \boldsymbol{\nu} \circ \boldsymbol{\mu}^{-1}(y)$, $y \in Y$, is horizontal relative to ∇. This statement reproduces a well-known result in twistor geometry. The case $n = 1$ has been considered by Hitchin (1982) and Ward (1980) who showed that the moduli space comes equipped with a complex Riemannian metric satisfying self-dual Einstein equations with non-zero scalar curvature. The case $n \geq 2$ has been investigated by Bailey and Eastwood (1991), LeBrun (1989) and Pedersen and Poon (1989) who proved that the moduli space M comes equipped canonically with a torsion-free connection compatible with the induced (complexified) quaternionic Kähler structure on M. In Section 5 we shall study the sphere S^4 along the lines suggested in the proof of Theorem 3.4 and show that the canonical connection predicted by Corollary 3.9 is exactly the Levi-Civita connection of the symmetric metric on S^4 thus demonstrating the equivalence of two different twistor methods.

4 Infinitesimal neighbourhoods and affine connections

Let Y be a complex manifold and X a complex submanifold of Y. Let $\mathcal{O}_Y$ and $\mathcal{O}_X$ denote the structure sheaves of manifolds Y and X respectively, and let J denote the ideal of functions on Y which vanish on X,

$$0 \longrightarrow J \longrightarrow \mathcal{O}_Y \longrightarrow \mathcal{O}_X \longrightarrow 0. \tag{27}$$

Then the mth-order infinitesimal neighborhood of X in Y is (Griffiths 1966) the ringed space $X^{(m)} = (X, \mathcal{O}_X^{(m)})$ with the structure sheaf $\mathcal{O}_X^{(m)}$ defined by the following exact sequence

$$0 \longrightarrow J^{m+1} \longrightarrow \mathcal{O}_Y \longrightarrow \mathcal{O}_X^{(m)} \longrightarrow 0.$$

With the $(m+1)$th-order infinitesimal neighborhood there is naturally associated an $\mathcal{O}_X^{(m)}$-module $\mathcal{O}_X^{(m)}(N^*)$ defined by

$$0 \longrightarrow J^{m+2} \longrightarrow J \longrightarrow \mathcal{O}_X^{(m)}(N^*) \longrightarrow 0. \tag{28}$$

The exact sequences (27) and (28) imply that $\mathcal{O}_X^{(m)}(N^*)$ is an ideal subsheaf of $\mathcal{O}_X^{(m+1)}$,

$$0 \longrightarrow \mathcal{O}_X^{(m)}(N^*) \longrightarrow \mathcal{O}_X^{(m+1)} \longrightarrow \mathcal{O}_X \longrightarrow 0,$$

consisting of all nilpotent elements. Motivated by the fact that in the case $m = 0$ this sheaf coincides precisely with the conormal bundle N^* of the embedding $X \hookrightarrow Y$ we call the sheaf $\mathcal{O}_X^{(m)}(N^*)$ the *mth-order conormal sheaf* of the embedding $X \hookrightarrow Y$ (cf. Eastwood and LeBrun 1986, 1992).

We are interested in this paper in first order conormal sheaves. As follows from (28), such a sheaf fits into the exact sequence of $\mathcal{O}_X^{(1)}$-modules

$$0 \longrightarrow \odot^2 N^* \longrightarrow \mathcal{O}_X^{(1)}(N^*) \longrightarrow N^* \longrightarrow 0.$$

Let E be a holomorphic vector bundle on Y. Then its restriction $E_X^{(1)} = E|_{X^{(1)}}$ to the first order infinitesimal neighbourhood of X in Y is an $\mathcal{O}_X^{(1)}$-module which fits into the exact sequence (Manin 1988)

$$0 \longrightarrow N^* \otimes E|_X \xrightarrow{i} E_X^{(1)} \longrightarrow E|_X \longrightarrow 0.$$

If in addition

$$E|_X = N^*,$$

one can define the quotient $\mathcal{O}_X^{(1)}$-module

$$\mathcal{O}_X^{(1)}(E) \equiv E_X^{(1)}/i(\wedge^2 N^*),$$

which fits into the exact sequence of $\mathcal{O}_X^{(1)}$-modules

$$0 \longrightarrow \odot^2 N^* \longrightarrow \mathcal{O}_X^{(1)}(E) \longrightarrow N^* \longrightarrow 0,$$

of the same type as $\mathcal{O}_X^{(1)}(N^*)$.

Lemma 4.1 *Let $X \hookrightarrow Y$ be a complex submanifold and E a holomorphic bundle on Y such that $E|_X = N^*$. Then the difference $\left[\mathcal{O}_X^{(1)}(N^*) - \mathcal{O}_X^{(1)}(E)\right] \in \mathrm{Ext}_{\mathcal{O}_X^{(1)}}(N^*, \odot^2 N^*)$ is a locally free $\mathcal{O}_X$-module.*

Proof. Let us cover the manifold Y by coordinate charts $Y = \cup W_i$ with local coordinate functions (w_i^A, z_i^a) on each neighbourhood W_i such that $X \cap W_i$ coincides with the subspace of W_i determined by $w_i^A = 0$. On the intersection $W_i \cap W_j$ the coordinates w_i^A, z_i^a are holomorphic functions of w_j^B and z_j^b,

$$w_i^A = f_{ij}^A(w_j^B, z_j^b), \qquad z_i^a = g_{ij}^a(w_j^B, z_j^b),$$

with $f_{ij}^A(0, z_j^b) = 0$. The sheaf $\mathcal{O}_X^{(1)}(N^*) = J/J^3$ is generated on W_i by functions w_i^A and $w_i^B w_i^C \ (\mathrm{mod}(w_i)^3)$. On the intersection $W_i \cap W_j$ we have

$$w_i^A \bmod (w_j)^3 = F_{ij}{}_B^A\Big|_{t=0} w_j^B + \frac{1}{2} F_{ij}{}_{BC}^A\Big|_{t=0} w_j^B w_j^C,$$

where $F_{ij}{}_B^A$ and $F_{ij}{}_{BC}^A$ are given by (3) and (5) respectively. Hence

$$\begin{pmatrix} w_i^A \\ w_i^B w_i^C \end{pmatrix} \bmod (w_j)^3 = \begin{pmatrix} F_{ij}{}_D^A|_{t=0} & \frac{1}{2} F_{ij}{}_{EF}^A|_{t=0} \\ 0 & F_{ij}{}_E^B|_{t=0} F_{ij}{}_F^C|_{t=0} \end{pmatrix} \begin{pmatrix} w_j^D \\ w_j^E w_j^F \end{pmatrix}.$$

Note that in general these transformation matrices do not match up on the triple intersections $X \cap W_i \cap W_j \cap W_k$.

An analogous "gluing" mapping for the sheaf $\mathcal{O}_X^{(1)}(E)$ is given by the matrix

$$\begin{pmatrix} F_{ij}{}^A_D \,|_{t=0} & \frac{1}{2}(\tau_{ij}{}^A_{EF} + \tau_{ij}{}^A_{FE}) \,|_{t=0} \\ 0 & F_{ij}{}^B_E \,|_{t=0} \; F_{ij}{}^C_F \,|_{t=0} \end{pmatrix}.$$

Therefore, by definition of the group structure in $Ext_{\mathcal{O}_X^{(1)}}(N^*, \odot^2 N^*)$, the local generators of $\left[\mathcal{O}_X^{(1)}(N^*) - \mathcal{O}_X^{(1)}(E)\right]$ are glued on the intersection $W_i \cap W_j$ with the help of the matrix

$$G_{ij} = \begin{pmatrix} F_{ij}{}^A_D \,|_{t=0} & \frac{1}{2}(F_{ij}{}^A_{EF} - \tau_{ij}{}^A_{EF} - \tau_{ij}{}^A_{FE}) \,|_{t=0} \\ 0 & F_{ij}{}^B_E \,|_{t=0} \; F_{ij}{}^C_F \,|_{t=0} \end{pmatrix}.$$

The point is that for any holomorphic bundle E such that $E|_X = N^*$, the combination $\left\{F_{ij}{}^A_{EF} - \tau_{ij}{}^A_{EF} - \tau_{ij}{}^A_{FE}\right\}$ is always a 1-cocycle with coefficients in $N \otimes \odot^2 N^*$ relative to the covering $\{W_i\}$ (see the proof of Theorem 3.1). This immediately implies that the matrices G_{ij} satisfy on triple intersections $W_i \cap W_j \cap W_k$ the condition

$$G_{ik} = G_{ij}\, G_{jk}$$

which is a necessary and sufficient condition for $\left[\mathcal{O}_X^{(1)}(N^*) - \mathcal{O}_X^{(1)}(E)\right]$ be a locally free $\mathcal{O}_X$-module. ■

Corollary 4.2 $\left[\mathcal{O}_X^{(1)}(N^*) - \mathcal{O}_X^{(1)}(E)\right] \in H^1(X, N \otimes \odot^2 N^*)$.

Theorem 4.3 *Let $X \hookrightarrow Y$ be a compact complex submanifold with normal bundle N such that $H^1(X, N) = 0$ and let $F \hookrightarrow Y \times M$ be the associated complete analytic family of compact submanifolds containing X. Suppose there is a holomorphic vector bundle E on Y such that $\pi_1^*(E)|_{\nu^{-1}(M_0)} = N_F^*|_{\nu^{-1}(M_0)}$ and the cohomology class $\left[\mathcal{O}_F^{(1)}(N_F^*) - \mathcal{O}_F^{(1)}(\pi_1^*(E))\right]\big|_{\nu^{-1}(M_0)} \in H^1(\nu^{-1}(M_0), N_F \otimes N_F^* \odot N_F^*)$ vanishes for some neighbourhood $M_0 \subseteq M$ of the point t_0 corresponding to X. Then any splitting of the exact sequence of the locally free $\mathcal{O}_F$-modules*

$$0 \longrightarrow \odot^2 N_F^* \longrightarrow \left[\mathcal{O}_F^{(1)}(N_F^*) - \mathcal{O}_F^{(1)}(\pi_1^*(E))\right] \longrightarrow N_F^* \longrightarrow 0 \qquad (29)$$

over $\nu^{-1}(M_0)$ induces canonically a torsion-free affine connection on M_0 such that, for every point $y \in Y'$, the associated submanifold $P_y \subseteq \nu \circ \mu^{-1}(y) \cap M_0$ is totally geodesic.

Proof. In the notation introduced in the proof of Theorem 3.1 the cohomology class $\left[\mathcal{O}_F^{(1)}(N_F^*) - \mathcal{O}_F^{(1)}(\pi_1^*(E))\right]\big|_{\nu^{-1}(M_0)} \in H^1(\nu^{-1}(M_0), N_F \otimes \odot^2 N_F^*)$ is represented by the 1-cocycle $\left\{F_{ij}{}^A_{EF} - \tau_{ij}{}^A_{EF} - \tau_{ij}{}^A_{FE}\right\}$. Therefore this 1-cocycle is a coboundary. Any splitting of the exact sequence (29) over $\nu^{-1}(M_0)$ determines an associated 0-cochain $\left\{\sigma_{\mathrm{i}}{}^{A}_{BC}\right\}$ such that

$$\left\{F_{\mathrm{ij}}{}^{A}_{EF} - \tau_{\mathrm{ij}}{}^{A}_{EF} - \tau_{\mathrm{ij}}{}^{A}_{FE}\right\} = \delta\left\{\sigma_{\mathrm{i}}{}^{A}_{BC}\right\}.$$

Then one proceeds with the proof of Theorem 4.3 exactly in the same way as in the case of Theorem 3.4. ■

5 Examples

Example 1. If $X = \mathbb{CP}^{n-1}$ is a hyperplane in $Y = \mathbb{CP}^n$, then the normal bundle of X is equivalent to $E|_X$, where E is the hyperplane section bundle $\mathcal{O}_{\mathbb{CP}^n}(1)$. Thus all the conditions of Proposition 3.3 are satisfied. The moduli space, M, of relative deformations of $\mathbb{CP}^{n-1}$ in $\mathbb{CP}^n$ is isomorphic to $\mathbb{CP}^n$. Geodesics of the canonical projective connection induced canonically on $M = \mathbb{CP}^n$ are just projective lines. For any $z \in X$ the corresponding surface $P_z = \nu \circ \mu^{-1}(z)$ is a hyperplane $\mathbb{CP}^{n-1}$ in M. It is clear that each hyperplane P_z is totally geodesic relative to the induced projective connection.

Example 2. Let the pair $X \hookrightarrow Y$ be the same as in the previous example. If y_0 is a point in $Y \setminus X$, then the manifold $Y_0 \equiv Y \setminus y_0$ has a natural structure of a holomorphic fibration

$$\begin{aligned} \pi : Y_0 &\longrightarrow X \\ y &\longrightarrow \overline{(y, y_0)} \cap X \end{aligned}$$

over X, where $\overline{(y, y_0)}$ stands for the projective line through points y and y_0. By Corollary 3.8, the complete moduli space $M_0 \simeq \mathbb{C}^n$ associated with the pair $X \hookrightarrow Y^I$ comes equipped canonically with a torsion free affine connection. It is easy to check that this connection is exactly the flat connection on $\mathbb{C}^n$.

Example 3. Let X be a rational curve in a 2-dimensional complex manifold Y. Any line bundle on $X = \mathbb{CP}^1$ is of the form $\mathcal{O}(m) = \otimes^m \mathcal{O}(1)$ for some $m \in \mathbb{Z}$. If κ is the canonical bundle on Y, then the bundle

$$E = \kappa^{-\frac{m}{m+2}}$$

is such that

$$E|_X \simeq \mathcal{O}(m)$$

in some open neighbourhood of X. Thus, according to Proposition 3.3, rational curves in a 2-dimensional manifold Y with normal bundle $N \simeq \mathcal{O}(m)$ for *any* non-negative integer m generate a canonical projective connection on the moduli space M. For $m \leq 2$ such relative deformation problems have been considered by Hitchin (1982) who proved that

- The case $N \simeq \mathcal{O}$ is trivial in the sense that a neighbourhood of $\mathbb{CP}^1$ with a trivial normal bundle is biholomorphically equivalent to the product $\mathbb{CP}^1 \times \mathbb{C}$.
- There is a one-to-one correspondence between pairs $\{(X, Y)$ consisting of a 2-dimensional manifold Y and a rational curve X embedded into Y with a normal bundle $N \simeq \mathcal{O}(1)\}$ and {equivalence classes under coordinate transformation of differential equations of the form

$$\frac{d^2y}{dx^2} = a_3\left(\frac{dy}{dx}\right)^3 + a_2\left(\frac{dy}{dx}\right)^2 + a_1\left(\frac{dy}{dx}\right) + a_0, \tag{30}$$

  where the coefficients a_k are holomorphic functions of x and y}.
- There is a one-to-one correspondence between {pairs (X, Y) consisting of a 2-dimensional manifold Y and a rational curve X embedded into Y with normal bundle $N \simeq \mathcal{O}(2)\}$ and {solutions of the 3-dimensional Einstein-Weyl equations}.

In proving the second statement Hitchin first used Cartan's (1925) idea to associate equation (30) to a projective connection and then showed that the moduli space M of rational curves in Y with normal bundle $\mathcal{O}(1)$ comes equipped with a projective connection.

Example 4. Let X be a non-singular quadric in $Y = \mathbb{CP}^3$. Then its normal sheaf N fits into the exact sequence

$$0 \longrightarrow \mathcal{O}_Y \longrightarrow \mathcal{O}_Y(2) \longrightarrow N \longrightarrow 0. \tag{31}$$

Hence $H^1(X, N) = 0$ and the Kodaira theorem applies. The moduli space, M, of non-singular quadrics in $\mathbb{CP}^3$ is easily found to be $\mathbb{CP}^9 \setminus Q$, where Q is a non-singular quartic. Since $X \simeq \mathbb{CP}^1 \times \mathbb{CP}^1$, we have $H^1(X, N \otimes N^*) = H^1(X, \mathcal{O}_X) = 0$. Hence Proposition 3.3 applies, and the moduli space $M \simeq \mathbb{CP}^9 \setminus Q$ comes equipped with a canonical projective connection. This connection is easy to describe — its geodesics are just projective lines lying in $\mathbb{CP}^9 \setminus Q$, while totally geodesic submanifolds P_y are hyperplanes in $\mathbb{CP}^9 \setminus Q$.

Example 5. Consider $Y = \mathbb{CP}^3$ and let $X = \mathbb{CP}^1$ be the intersection of two generic hyperplanes in Y. We can always choose homogeneous coordinates (Z_0, Z_1, Z_2, Z_3) on Y in such a way that $X \hookrightarrow Y$ is given by

$$Z_0 = Z_1 = 0.$$

Since the normal bundle N of $X \hookrightarrow Y$ is equivalent to $\mathbb{C}^2 \otimes \mathcal{O}(1)$, the cohomology groups $H^1(X, N)$, $H^1(X, N \otimes N^* \odot N^*)$ and $H^0(X, N \otimes \odot^2 N^*)$ vanish. By the Kodaira theorem, there is a 4-dimensional moduli space $M \cong \mathbb{CP}^3$ of rational curves in Y. Covering an open neighbourhood of X in Y by two coordinate charts

$$W = \left(w^0 = \frac{Z_2}{Z_0}, w^1 = \frac{Z_3}{Z_0}, z = \frac{Z_1}{Z_0}\right), \qquad \hat{W} = \left(\hat{w}^0 = \frac{Z_2}{Z_1}, \hat{w}^1 = \frac{Z_3}{Z_1}, \hat{z} = \frac{Z_0}{Z_1}\right),$$

one easily finds the explicit description,

$$X_t \cap W \Longleftrightarrow \begin{cases} w^0 &= t_0 + t_1 z \\ w^1 &= -\tilde{t}_1 + \tilde{t}_0 z \end{cases},$$

$$X_t \cap \hat{W} \Longleftrightarrow \begin{cases} \hat{w}^0 &= t_0 \hat{z} + t_1 \\ \hat{w}^1 &= -\tilde{t}_1 \hat{z} + \tilde{t}_0 \end{cases},$$

of submanifolds $X_t \hookrightarrow Y$ for all points t in a big cell $M_0 \cong \mathbb{C}^4$ of M which is covered by one coordinate chart $\{t^\alpha\} = \{t_0, t_1, \tilde{t}_0, \tilde{t}_1\}$.

Suppose now that Y is given a complex contact structure described by the "twisted" one form (Besse 1987)

$$\theta = Z_0\, dZ_1 - Z_1\, dZ_0 + Z_2\, dZ_3 - Z_3\, dZ_2.$$

Then the contact distribution $D = \{v \in TY : < v, \theta > = 0\}$ is given explicitly by

$$D|_W = \text{span}\left(D_0 \equiv \frac{\partial}{\partial w^0} + w^1 \frac{\partial}{\partial z}, D_1 \equiv \frac{\partial}{\partial w^1} - w^0 \frac{\partial}{\partial z}\right),$$

$$\hat{D}|_{\hat{W}} = \text{span}\left(\hat{D}_0 \equiv \frac{\partial}{\partial \hat{w}^0} - \hat{w}^1 \frac{\partial}{\partial \hat{z}}, \hat{D}_1 \equiv \frac{\partial}{\partial \hat{w}^1} + \hat{w}^0 \frac{\partial}{\partial \hat{z}}\right).$$

All curves X_t, $t \in M_0$, are transverse to D and , in accordance with Corollary 3.9, the moduli space M_0 comes equipped canonically with a torsion-free affine connection ∇. For pedagogical reasons we shall include a detailed computation of this connection along the lines suggested in the proof of Theorem 3.4.

Step 1. Let us first find explicit expressions for all matrices which enter the basic equality (21). In the present notation this equality takes the form

$$F^A_B(z,t) = \hat{H}^A_C(\hat{z},t)\; E^C_D(w^0, w^1, z)\Big|_{X_t}\; (H(z,t)^{-1})^D_B. \tag{32}$$

Since

$$\begin{aligned} \hat{w}^0 &= z^{-1} w^0, \\ \hat{w}^1 &= z^{-1} w^0, \\ \hat{z} &= z^{-1}, \end{aligned}$$

one finds

$$\begin{aligned} D_0 &= z^{-2}(z - w^0 w^1)\hat{D}_0 - z^{-2}(w^1)^2 \hat{D}_1, \\ D_1 &= z^{-2}(w^0)^2 \hat{D}_0 + z^{-2}(z + w^0 w^1)\hat{D}_1. \end{aligned}$$

Therefore, the coordinate patching matrix $E(w^0, w^1, z)$ of the bundle D has the form

$$E^A_B = z^{-2}\begin{pmatrix} z - w^0 w^1 & (w^0)^2 \\ -(w^1)^2 & z + w^0 w^1 \end{pmatrix}.$$

Define a new coordinate system

$$\begin{aligned} w^0_t &= w^0 - t_0 - t_1 z \\ w^1_t &= w^1 + \tilde{t}_1 - \tilde{t}_0 z, \end{aligned}$$

and note that the tangent bundle to X_t is spanned in these coordinates by $\partial/\partial z$. Therefore

$$e^t_0 \equiv \frac{\partial}{\partial w^0_t} \bmod \frac{\partial}{\partial z}$$

and

$$e^t_1 \equiv \frac{\partial}{\partial w^1_t} \bmod \frac{\partial}{\partial z}$$

form a basis of the normal bundle $N_t = TY|_{X_t} \bmod TX_t$ over $W \cap X_t$. There is an analogous basis of N_t over $\hat{W} \cap X_t$, and the associated patching functions (3) have a very simple form

$$F^A_B = \begin{pmatrix} z^{-1} & 0 \\ 0 & z^{-1} \end{pmatrix}.$$

Since X_t is transverse to D, there is a canonical isomorphic map h from N_t to $D|_{X_t}$ which assigns to an equivalence class in $TY|_{X_t} \bmod TX_t$ a unique representative which lies in $D|_{X_t}$. The matrix-valued function $H^A_B(z,t)$ in (32) is given by

$$h^{-1}(D_A|_{X_t}) = H^B_A(z,t)\, e^t_B.$$

Since

$$\begin{aligned} D_0|_{X_t} \bmod TX_t &= [1 + t_1(\tilde{t}_1 - z\tilde{t}_0)]\, e^t_0 + \tilde{t}_0(\tilde{t}_1 - z\tilde{t}_0)\, e^t_1 \\ D_1|_{X_t} \bmod TX_t &= t_1(t_0 + z t_1)\, e^t_0 + [1 + \tilde{t}_0(t_0 + z t_1)]\, e^t_1, \end{aligned}$$

we find

$$H^A_B = \begin{pmatrix} 1 + t_1(\tilde{t}_1 - z\tilde{t}_0) & t_1(t_0 + z t_1) \\ \tilde{t}_0(\tilde{t}_1 - z\tilde{t}_0) & 1 + \tilde{t}_0(t_0 + z t_1) \end{pmatrix}.$$

Analogously one finds

$$\hat{H}^A_B = \begin{pmatrix} 1 + t_0(\tilde{t}_0 - \hat{z}\,\tilde{t}_1) & -t_0(t_1 + \hat{z}\,t_0) \\ -\tilde{t}_1(\tilde{t}_0 - \hat{z}\,\tilde{t}_1) & 1 + \tilde{t}_1(t_1 + \hat{z}\,t_0) \end{pmatrix}.$$

Note that

$$\det H = \det \hat{H} = 1 + t_0\,\tilde{t}_0 + t_1\,\tilde{t}_1.$$

One may also check that equation (32) is indeed satisfied.

Step 2. Let us solve the equation (24). First we compute the combination

$$\kappa^A_{BC} \equiv \tau^A_{BC} + \tau^A_{CB} - F^A_{BC},$$

where τ^A_{BC} and F^A_{BC} are given by (9) and (5) respectively, and find

$$\begin{aligned}
\kappa^0_{00} &= 2\,\Omega^{-1}\left[-t_0\,\tilde{t}_1^2\,z^{-3} + (t_1 + t_0\,\tilde{t}_0\,\tilde{t}_1)\,z^{-2} - \tilde{t}_0\,(1 + t_1\,\tilde{t}_1)\,z^{-1} + t_1\,\tilde{t}_0^2\right] \\
\kappa^0_{01} &= \Omega^{-1}\left[-2\,t_0^2\,\tilde{t}_1\,z^{-3} + t_1\,(\Omega - 2\,t_0\,\tilde{t}_0)\,z^{-2} - t_0\,(\Omega - 2\,t_1\tilde{t}_1)\,z^{-1} - 2\,t_1^2\,\tilde{t}_0\right] \\
\kappa^0_{11} &= 2\,\Omega^{-1}\left[-t_0^3\,z^{-3} - t_0^2\,t_1\,z^{-2} + t_0\,t_1^2\,z^{-1} + t_1^3\right] \\
\kappa^1_{00} &= 2\,\Omega^{-1}\left[\tilde{t}_1^3\,z^{-3} + \tilde{t}_0\,\tilde{t}_1^2\,z^{-2} - \tilde{t}_1\,\tilde{t}_0^2\,z^{-1} + \tilde{t}_0^3\right] \\
\kappa^1_{01} &= \Omega^{-1}\left[2\,t_0\,\tilde{t}_1^2\,z^{-2} + \tilde{t}_1(\Omega - 2\,t_0\,\tilde{t}_0)\,z^{-2} - \tilde{t}_0(\Omega - 2\,t_1\,\tilde{t}_1)\,z^{-1} - 2\,t_1\,\tilde{t}_0^2\right] \\
\kappa^1_{11} &= 2\,\Omega^{-1}\left[\tilde{t}_1\,t_0^2\,z^{-3} + t_0(1 + t_1\,\tilde{t}_1)\,z^{-2} + t_1(1 + t_0\,\tilde{t}_0)\,z^{-1} + t_1^2\,\tilde{t}_0\right]
\end{aligned}$$

where

$$\Omega = 1 + t_0\,\tilde{t}_0 + t_1\,\tilde{t}_1.$$

The equation (24) has the form

$$\kappa^A_{BC}(z,t) = -z^{-2}\,\hat{\sigma}^A_{BC}(z^{-1},t) + z^{-1}\,\sigma^A_{BC}(z,t),$$

where $\hat{\sigma}^A_{BC}(z^{-1},t)$ and $\sigma^A_{BC}(z,t)$ are entire functions of z^{-1} and z respectively. One easily finds a unique solution

$$\begin{aligned}
\sigma^0_{00} &= 2\,\Omega^{-1}\left[-\tilde{t}_0(1 + t_1\,\tilde{t}_1) + t_1\,\tilde{t}_0^2\,z\right] \\
\sigma^0_{01} &= \Omega^{-1}\left[t_0(\Omega - 2\,t_1\,\tilde{t}_1) - 2\,t_1^2\,\tilde{t}_0\,z\right] \\
\sigma^0_{11} &= 2\,\Omega^{-1}\left[t_0\,t_1^2 + t_1^3\,z\right] \\
\sigma^1_{00} &= 2\,\Omega^{-1}\left[-\tilde{t}_1\,\tilde{t}_0^2 + \tilde{t}_0^3\,z\right] \\
\sigma^1_{01} &= \Omega^{-1}\left[-\tilde{t}_0(\Omega - 2\,t_1\,\tilde{t}_1) - 2\,t_1\,\tilde{t}_0^2\,z\right] \\
\sigma^1_{11} &= 2\,\Omega^{-1}\left[\tilde{t}_1(1 + t_0\,\tilde{t}_0) + t_1^2\,\tilde{t}_0\,z\right].
\end{aligned}$$

Step 3. Now we have all the data necessary to solve the equation

$$\frac{\partial^2\phi^A}{\partial t^\alpha \partial t^\beta} + \sigma^A_{BC}\frac{\partial\phi^B}{\partial t^\alpha}\frac{\partial\phi^C}{\partial t^\beta} - (\frac{\partial H}{\partial t^\alpha}H^{-1})^A_B\frac{\partial\phi^B}{\partial t^\beta} - (\frac{\partial H}{\partial t^\beta}H^{-1})^A_B\frac{\partial\phi^B}{\partial t^\alpha} = \Gamma^\gamma_{\alpha\beta}\frac{\partial\phi^A}{\partial t^\gamma},$$

for coefficients, $\Gamma^\gamma_{\alpha\beta}(t)$, of the connection generated on M_0 by the contact structure D on Y. Here

$$\phi^0(z,t) = t_0 + t_1 z, \qquad \phi^{\bar{1}}(z,t) = -\tilde{t}_1 + \tilde{t}_0 z.$$

As an illustration, let us solve this equation in the case $\alpha = \beta = 0$.

(i) If $A = 0$, one has

$$\sigma^0_{00}\frac{\partial\phi^0}{\partial t_0}\frac{\partial\phi^0}{\partial t_0} - 2[\frac{\partial H}{\partial t_0}H^{-1}]^0_0\frac{\partial\phi^0}{\partial t_0} = \Gamma^0_{00}\frac{\partial\phi^0}{\partial t_0} + \Gamma^1_{00}\frac{\partial\phi^0}{\partial t_1}.$$

Since

$$\frac{\partial H}{\partial t_0}H^{-1} = \Omega^{-1}\begin{pmatrix} -t_1\tilde{t}_0(\tilde{t}_1 - z\tilde{t}_0) & t_1[1 + t_1(\tilde{t}_1 - z\tilde{t}_0)] \\ -\tilde{t}_0^2(\tilde{t}_1 - z\tilde{t}_0) & \tilde{t}_0[1 + t_1(\tilde{t}_1 - z\tilde{t}_0)] \end{pmatrix},$$

we obtain

$$2\Omega^{-1}\left[-\tilde{t}_0(1 + t_1\tilde{t}_1) + t_1\tilde{t}_0^2 z\right] - 2\Omega^{-1}[-t_1\tilde{t}_0(\tilde{t}_1 - z\tilde{t}_0)] = \Gamma^0_{00} + z\Gamma^1_{00}.$$

Hence

$$\Gamma^0_{00} = -2\tilde{t}_0\Omega^{-1}, \qquad \Gamma^1_{00} = 0.$$

(ii) If $A = 1$, one has

$$\sigma^1_{00}\frac{\partial\phi^0}{\partial t_0}\frac{\partial\phi^0}{\partial t_0} - 2[\frac{\partial H}{\partial t_0}H^{-1}]^1_0\frac{\partial\phi^0}{\partial t_0} = \Gamma^{\tilde{0}}_{00}\frac{\partial\phi^1}{\partial\tilde{t}_0} + \Gamma^{\tilde{1}}_{00}\frac{\partial\phi^1}{\partial\tilde{t}_1},$$

or

$$2\Omega^{-1}\left(-\tilde{t}_1\tilde{t}_0^2 + \tilde{t}_0^3 z\right) - 2\Omega^{-1}\left(-\tilde{t}_1\tilde{t}_0^2 + \tilde{t}_0^3 z\right) = \Gamma^{\tilde{0}}_{00}z - \Gamma^{\tilde{1}}_{00}.$$

Hence

$$\Gamma^{\tilde{0}}_{00} = 0, \qquad \Gamma^{\tilde{1}}_{00} = 0.$$

Analogously one finds all the other coefficients of the connection ∇ induced on M_0 in accordance with Corollary 3.9. The result is that ∇ is exactly the Levi-Civita connection of the constant scalar curvature holomorphic metric

$$ds^2 = \Omega^{-1}\left(dt_0\, d\tilde{t}_0 + dt_1\, d\tilde{t}_1\right), \tag{33}$$

on M_0. Besse (1987) and Hitchin (1982) used quite different twistor techniques to induce this metric on M_0.

The involution (Besse 1987)

$$(Z_0, Z_1, Z_2, Z_3) \longrightarrow (\overline{Z_1}, -\overline{Z_0}, \overline{Z_3}, -\overline{Z_2})$$

on Y induces a real structure on the complete moduli space M. Then the metric (33) restricted to the associated real slice S^4 (which is given in our coordinates by the equations $\tilde{t}_0 = \overline{t_0}$ and $\tilde{t}_1 = \overline{t_1}$) is exactly the Cartesian metric on S^4.

In a fully analogous manner one may show that another contact structure (Besse 1987)

$$\theta = Z_0\, dZ_1 - Z_1\, dZ_0 - Z_2\, dZ_3 + Z_3\, dZ_2$$

on Y generates by Corollary 6 the Levi-Civita connection of the hyperbolic metric on the ball $|t_0|^2 + |t_1|^2 < 1$.

Acknowledgments

It is a pleasure to thank Stephen Huggett, Henrik Pedersen, Yat Sun Poon and Paul Tod for valuable discussions and helpful remarks. Thanks are also due to the Department of Mathematics and Computer Science for hospitality and financial support.

References

Arnold, V. I. (1978). Mathematical Methods of Classical Mechanics, Springer, Berlin Heidelberg New York [Russian: Nauka, Moscow, 1974].

Bailey, T. N., and Eastwood, M. G. (1991). Complex paraconformal manifolds – their differential geometry and twistor theory, *Forum Math., 3*: 61-103.

Baston, R. J., and Eastwood, M. G. (1989). The Penrose transform, Oxford University Press, Oxford.

Besse, A. (1987). Einstein manifolds, Springer, Berlin Heidelberg New-York.

Burns, D. (1979). Some background and examples in deformation theory, *Complex Manifolds Techniques in Theoretical Physics*, (D. E. Lerner and P. D. Sommers, eds.), Pitman.

Cartan, E. (1925). Sur les variétés à connexion projective, *Bull. Soc. Math. Fr., 52*: 205-241.

Eastwood, M. G., and LeBrun, C. R. (1986). Thickenings and supersymmetric extensions of complex manifolds, *Amer. J. Math., 108*: 1177-1192.

Eastwood, M. G., and LeBrun, C. R. (1992). Fattening complex manifolds: curvature and Kodaira-Spencer maps, *J. Geom. Phys., 8*: 123-146.

Griffiths, P. A. (1966). The extension problem in complex analysis II: embeddings with positive normal bundle, *Amer. J. Math., 88*: 366-446.

Hitchin, N. (1982). Complex manifolds and Einstein's equations, *Lect. Notes Math., 970*: 73-99.

Kodaira, K. (1962). A theorem of completeness of characteristic systems for analytic families of compact submanifolds of complex manifolds, *Ann. Math., 75*: 146-162.

Kodaira, K. (1986). Complex manifolds and deformations of complex structures, Springer, Berlin Heidelberg New York.

LeBrun, C. R. (1989). Quaternionic-Kähler manifolds and conformal geometry, *Math. Ann., 284*: 353-376.

Manin, Yu. I. (1988). Gauge Field theory and Complex Geometry, Springer, Berlin Heidelberg New York [Russian: Nauka, Moscow, 1984].

Merkulov, S. A. (1992). Deformation theory of compact submanifolds of complex fibred manifolds and affine connections, *Odense preprint.*

Pedersen, H., and Poon, Y. S. (1989), Twistorial construction of quaternionic manifolds, *Proc. VIth Int. Coll. on Diff. Geom. , Cursos y Congresos, Univ. Santiago de Compostela, 61*: 207-218.

Penrose, R. (1976). Non-linear gravitons and curved twistor theory, *Gen. Rel. Grav., 7*: 31-52.

Penrose, R., and Rindler, W. (1986). Spinors and space-time, Vol. 2, Cambridge University Press, Cambridge.

Salamon, S. M. (1982). Quaternionic Kähler manifolds, *Invent. Math., 67*: 143-171.

Ward, R. S. (1977). On self-dual gauge fields, *Phys. Lett., 61A*: 81-82.

Ward, R. S. (1980). Self-dual space-times with cosmological constant, *Commun. Math. Phys., 78*: 1-17.

Ward, R. S., and Wells, R. O., Jr. (1990). Twistor geometry and field theory, Cambridge University Press, Cambridge.

Wells, R. O., Jr. (1980) Differential analysis on complex manifolds, Springer, Berlin Heidelberg New York.

Appendix: Geometry of Relative Deformations. II

Sergey Merkulov

School of Mathematics and Statistics, University of Plymouth

Plymouth, Devon PL4 8AA, United Kingdom

Henrik Pedersen

Department of Mathematics and Computer Science, Odense University,

Campusvej 55, 5230 Odense M, Denmark

1 Introduction

In this paper we study complete moduli spaces $\{X_t \hookrightarrow Y \mid t \in M\}$ of compact complex submanifolds of a complex manifold Y of two types — the spaces of compact complex hypersurfaces and the spaces of compact complex submanifolds $X_t \hookrightarrow Y$ with normal bundle of the form $\mathbb{C}^n \otimes L$ for some holomorphic line bundle L and natural number n. The main conclusion is that in both cases the moduli spaces M often come equipped canonically with rich differential-geometric data. Though this conclusion is very similar to the one made in the previous paper by Merkulov (1994), there is an important difference. If in that paper some additional structures (vector bundles satisfying certain conditions) played a key role in generating canonical affine connections on M, in the present paper we don't ask for any additional data on the ambient manifold Y to draw the conclusion that the moduli space M has distinguished torsion-free affine connections. Therefore the canonical differential-geometric structures on the moduli spaces are intrinsic to the relative deformation problems $X \hookrightarrow Y$ themselves.

The first main result of our paper is Theorem 1 in subsection 2.1 which studies complete analytic families $F \hookrightarrow Y \times M$ of compact complex hypersurfaces in a complex manifold Y such that for some point $t \in M$ the cohomology group $H^1(X_t, \mathcal{O}_{X_t})$ on the corresponding hypersurface $X_t \hookrightarrow Y$ vanishes. The conclusion is that a sufficiently small domain $U \subset M$ of the point t always comes equipped with a distinguished projective connection satisfying some natural integrability conditions. In particular, if the hypersurface X_t is a *rigid* complex manifold (i.e. $H^1(X_t, TX_t) = 0$, see e.g. Kodaira (1986)), then there is a distinguished family of geodesics on the whole complete moduli space M. One of the immediate applications of the theorem on projective connections is in the theory of

3-dimensional Einstein-Weyl manifolds. Hitchin (1982) proved that there is a one-to-one correspondence between local solutions of Einstein-Weyl equations in 3 dimensions and pairs (X, Y), where Y is a complex 2-fold and X is the projective line $\mathbb{CP}^1$ embedded into Y with normal bundle $N \simeq \mathcal{O}(2)$. However the corresponding twistor techniques allowed one to compute only part of the canonical Einstein-Weyl structure induced on the complete moduli space M of relative deformations of X in Y, namely the conformal structure on M. Although the geodesics were formally described and the existence of a connection with special curvature was proved, no explicit formula for the connection was obtained. The theorem on projective connections fills this gap and provides one with a technique which is capable of decoding the full Einstein-Weyl structure from the holomorphic data of the embedding $X \hookrightarrow Y$. In subsection 2.2 we use Theorem 1 to compute explicitly the canonical projective connection and then the canonical Einstein-Weyl structure on the complete moduli space of rational curves embedded into some 2-dimensional complex manifold with normal bundle $N \simeq \mathcal{O}(2)$.

The second main result of the paper is Theorem 3 in section 3 which studies complete analytic families of compact complex submanifolds of complex manifolds with normal bundle of the form $N \simeq \mathbb{C}^n \otimes L$, where L is a holomorphic line bundle and n is a natural number. Let $X \hookrightarrow Y$ be a compact complex rigid submanifold with normal bundle N such that[1] $N \simeq \mathbb{C}^n \otimes L$ and

$$H^1(X, L) = H^1(X, \mathcal{O}_X) = H^1(X, L^*) = H^0(X, L^*) = 0$$

and let $F \hookrightarrow Y \times M$ be the associated complete family of complex submanifolds of Y. Then Theorem 3 states that the moduli space M comes equipped canonically with an almost-Grassmanian structure and a distinguished family of torsion free affine connections.

2 Distinguished projective connections on moduli spaces

2.1 The main theorem on projective connections

In this subsection we prove that very often the complete moduli space M of compact complex hypersurfaces in a complex manifold Y carries a projective structure. In the special case of a rational curve embedded in a 2-dimensional complex manifold with self intersec-

[1]If E_1 and E_2 are holomorphic vector bundles on a complex manifold X, then the notation $E_1 \simeq E_2$ means that these vector bundles are represented in the set $H^1(X, GL(n, \mathcal{O}_X))$ by one and the same cohomology class.

tion number equal to 2 this projective structure has been discovered by Hitchin (1982). We use terminology and notation explained in the previous paper of Merkulov (1994).

Theorem 1 *Let $X \hookrightarrow Y$ be a compact complex submanifold of codimension 1 with normal bundle N such that $H^1(X,N)=0$ and let M be the associated complete moduli space of relative deformations of X in Y. If $H^1(X,\mathcal{O}_X)=0$, then a sufficiently small neighbourhood $M_0 \subset M$ of the point $t_0 \in M$ corresponding to $X = \boldsymbol{\mu} \circ \boldsymbol{\nu}^{-1}(t_0)$, comes equipped canonically with a projective connection such that, for every point $y \in Y' \equiv \cup_{t\in M} X_t$, the associated submanifold $P_y \subseteq \boldsymbol{\nu} \circ \boldsymbol{\mu}^{-1}(y) \cap M_0$ is totally geodesic.*

Proof. An open neighbourhood of the submanifold $X \hookrightarrow Y$ can always be covered by a finite number of coordinate charts $\{W_{\rm i}\}$ with local coordinate functions $(w_{\rm i}, z_{\rm i}^a)$, $a = 1,\dots,n = \dim X$, on each neighbourhood $W_{\rm i}$ such that $X \cap W_{\rm i}$ coincides with the subspace of $W_{\rm i}$ determined by the equation $w_{\rm i}=0$. On the intersection $W_{\rm i}\cap W_{\rm j}$ the coordinates $w_{\rm i}, z_{\rm i}^a$ are holomorphic functions of $w_{\rm j}$ and $z_{\rm j}^b$,

$$w_{\rm i} = f_{\rm ij}(w_{\rm j}, z_{\rm j}^b), \qquad z_{\rm i}^a = g_{\rm ij}^a(w_{\rm j}, z_{\rm j}^b),$$

with $f_{\rm ij}(0, z_{\rm j}^b) = 0$.

Let $U \subset M$ be a coordinate neighbourhood of the point t_0 with coordinate functions $t^\alpha, \alpha = 1,\dots,m = \dim M$. Then the coordinate domains $U\times W_{\rm i}$ with coordinate functions $(w_{\rm i}, z_{\rm i}^a, t^\alpha)$ cover an open neighbourhood of $X\times U$ in the manifold $Y\times U$. For a sufficiently small U, the submanifold $F_U \equiv \boldsymbol{\nu}^{-1}(U) \hookrightarrow Y\times U$ is described in each coordinate chart $W_{\rm i}\times U$ by an equation of the form (Kodaira 1962)

$$w_{\rm i} = \phi_{\rm i}(z_{\rm i}, t),$$

where $\phi_{\rm i}(z_{\rm i},t)$ is a holomorphic function of $z_{\rm i}^a$ and t^α which satisfies the conditions $\phi_{\rm i}(z_{\rm i},t_0)=0$. For each fixed $t\in U$ this equation defines a submanifold $X_t \cap W_{\rm i} \hookrightarrow W_{\rm i}$.

By construction, F_U is covered by a finite number of coordinate neighbourhoods $\{V_{\rm i} \equiv W_{\rm i}\times U\}$ with local coordinate functions $(z_{\rm i}^a, t^\alpha)$ which are related to each other on the intersections $V_{\rm i}\cap V_{\rm j}$ as follows

$$z_{\rm i}^a = g_{\rm ij}^a\left(\phi_{\rm j}(z_{\rm j},t), z_{\rm j}\right).$$

Obviously we have $\qquad \phi_{\rm i}\left(g_{\rm ij}\left(\phi_{\rm j}(z_{\rm j},t),z_{\rm j}\right),t\right) = f_{\rm ij}\left(\phi_{\rm j}(z_{\rm j},t),z_{\rm j}\right).$

The Kodaira map $\boldsymbol{k} : TM|_U \longrightarrow \boldsymbol{\nu}_*^0(N_F|_{F_U})$ can be described in the following way: take any vector field v on U and apply the corresponding 1st-order differential operator $V^\alpha\partial_\alpha$ to each function $\phi_{\rm i}(z_{\rm i},t)$. The result is a collection of holomorphic functions

$$\sigma_{\rm i}(z_{\rm i},t) = V^\alpha \frac{\partial\phi_{\rm i}(z_{\rm i},t)}{\partial t^\alpha}$$

defined respectively on V_i. On the intersection $V_i \cap V_j$ these functions are related to each other according to the formula

$$\sigma_i(z_i, t)|_{z_i = g_{ij}(\phi_j, z_j)} = F_{ij}(z_j, t)\, \sigma_j(z_j, t),$$

where

$$F_{ij} \equiv \left.\frac{\partial f_{ij}}{\partial w_j}\right|_{w_j=\phi_j(z_j,t)} - \left.\frac{\partial \phi_i}{\partial z_i^a}\right|_{z_i=g_{ij}(\phi_j,z_j)} \left.\frac{\partial g_{ij}^a}{\partial w_j}\right|_{w_j=\phi_j(z_j,t)},$$

is the transition matrix of the normal bundle $N_F|_{F_U}$ on the overlap $F_U \cap V_i \cap V_j$. Therefore the 0-cochain

$$\left\{ V^\alpha \frac{\partial \phi_i(z_i, t)}{\partial t^\alpha} \right\}$$

is a Čech 0-cocycle representing a global section $\boldsymbol{k}(v)$ of the normal bundle N_F over F_U.

Let us investigate how second partial derivatives of $\{\phi_i(z_i, t)\}$ and $\{\phi_j(z_j, t)\}$ are related on the intersection $V_i \cap V_j$.

Since

$$\left.\frac{\partial \phi_i(z_i, t)}{\partial t^\alpha}\right|_{z_i=g_{ij}(\phi_j,z_j)} = F_{ij} \frac{\partial \phi_j(z_j, t)}{\partial t^\alpha}$$

we find

$$\left.\frac{\partial^2 \phi_i}{\partial t^\alpha \partial t^\beta}\right|_{z_i=g_{ij}(\phi_j,z_j)} = F_{ij} \frac{\partial^2 \phi_j}{\partial t^\alpha \partial t^\beta} + E_{ij} \frac{\partial \phi_j}{\partial t^\alpha} \frac{\partial \phi_j}{\partial t^\beta} - G_{ij\,\alpha} \frac{\partial \phi_j}{\partial t^\beta} - G_{ij\,\beta} \frac{\partial \phi_j}{\partial t^\alpha}, \tag{1}$$

where

$$\begin{aligned} E_{ij} &= \left.\frac{\partial^2 f_{ij}}{\partial w_j \partial w_j}\right|_{w_j=\phi(z_j,t)} - \left.\frac{\partial \phi_i}{\partial z_i^a}\right|_{z_i=g_{ij}(\phi_j,z_j)} \left.\frac{\partial^2 g_{ij}^a}{\partial w_j \partial w_j}\right|_{w_j=\phi_j(z_j,t)} \\ &\quad - \left.\frac{\partial^2 \phi_i}{\partial z_i^a \partial z_i^b}\right|_{z_i=g_{ij}(\phi_j,z_j)} \left.\left(\frac{\partial g_{ij}^a}{\partial w_j} \frac{\partial g_{ij}^b}{\partial w_j}\right)\right|_{w_j=\phi_j(z_j,t)}, \end{aligned}$$

and

$$G_{ij\,\alpha} = \left.\frac{\partial^2 \phi_i}{\partial z_i^a \partial t^\alpha}\right|_{z_i=g_{ij}(\phi_j,z_j)} \left.\frac{\partial g_{ij}^a}{\partial w_j}\right|_{w_j=\phi_j(z_j,t)}.$$

The collections $\{E_{ij}\}$ and $\{G_{ij\,\alpha}\}$ form 1-cochains with coefficients in N_F^* and $\boldsymbol{\nu}^*(\Omega^1 M)$, respectively. Straightforward calculations reveal the obstructions for these two 1-cochains to be 1-cocycles (Merkulov 1994),

$$\begin{aligned} \delta\{E_{ik}\} &= 2 \frac{\partial F_{ij}(z_j, t)}{\partial z_j^a} \left.\frac{\partial g_{jk}^a}{\partial w_k}\right|_{w_k=\phi_k(z_k,t)} \\ \delta\{G_{ik\,\alpha}\} &= \frac{\partial F_{ij}(z_j, t)}{\partial z_j^a} \left.\frac{\partial g_{jk}^a}{\partial w_k}\right|_{w_k=\phi_k(z_k,t)} \frac{\partial \phi_j(z_j, t)}{\partial t^\alpha}. \end{aligned}$$

From these equations we observe that the 1-cochain $\{\tau_{\mathrm{ik}\,\alpha}\}$, where

$$\tau_{\mathrm{ik}\,\alpha} \equiv \frac{1}{2} E_{\mathrm{ik}} \frac{\partial \phi_{\mathrm{k}}}{\partial t^{\alpha}} - G_{\mathrm{ik}\,\alpha},$$

is actually a 1-cocycle with values in $\boldsymbol{\nu}^*(\Omega^1 M)$. Since $H^1(X, \mathcal{O}_X) = 0$, the semi-continuity principle (Kodaira 1986) implies that $H^1(X_t, \mathcal{O}_{X_t}) = 0$ for all points in some Stein neighbourhood $M_0 \subseteq U$. Hence, by the Leray spectral sequence $H^1(\boldsymbol{\nu}^{-1}(M_0), \boldsymbol{\nu}^*(\Omega^1 M)) = 0$. Therefore, the 1-cocycle $\{\tau_{\mathrm{ik}\,\alpha}\}$ is always a coboundary on $\boldsymbol{\nu}^{-1}(M_0)$,

$$\{\tau_{\mathrm{ij}\,\alpha}\} = \delta\,\{\theta_{\mathrm{i}\,\alpha}\},$$

or more explicitly,

$$\tau_{\mathrm{ij}\,\alpha}(z_{\mathrm{j}}, t) = F_{\mathrm{ij}}(z_j, t)\left(-\,\theta_{\mathrm{i}\,\alpha}(z_{\mathrm{i}}, t)\big|_{z_{\mathrm{i}} = g_{\mathrm{ij}}(\phi_{\mathrm{j}}, z_{\mathrm{j}})} + \theta_{\mathrm{j}\,\alpha}(z_{\mathrm{j}}, t)\right), \tag{2}$$

where $\{\theta_{\mathrm{i}\alpha}(z_{\mathrm{i}}, t)\}$ is a 0-cochain on $\boldsymbol{\nu}^{-1}(M_0)$ with values in $\boldsymbol{\nu}^*(\Omega^1 M)$. However, this 0-cochain is defined non-uniquely — for any global section $\xi = \xi_\alpha\, dt^\alpha$ of $\boldsymbol{\nu}^*(\Omega^1 M)$ over $\boldsymbol{\nu}^{-1}(M_0)$ the 0-cochain

$$\tilde{\theta}_{\mathrm{i}\,\alpha}(z_{\mathrm{i}}, t) = \theta_{\mathrm{i}\,\alpha}(z_{\mathrm{i}}, t) + \xi_\alpha(t)\big|_{\boldsymbol{\nu}^{-1}(M_0)\cap V_{\mathrm{i}}} \tag{3}$$

splits the same 1-cocycle $\{\tau_{\mathrm{ij}\,\alpha}\}$. Note that, due to the compactness of the complex submanifolds $\boldsymbol{\nu}^{-1}(t) \subset F$ for all $t \in M_0$ the components ξ_α of the global section $\xi \in H^0(\boldsymbol{\nu}^{-1}(M_0), \boldsymbol{\nu}^*(\Omega^1 M))$ are constant along the fibers, i.e. $\xi_\alpha \in \boldsymbol{\nu}^{-1}(\mathcal{O}_{M_0})$. As we shall see later, it is this non-uniqueness that is responsible for the fact that the moduli space M comes equipped with a projective connection rather than with an ordinary affine connection.

If we rewrite equation (1) in the form

$$\begin{aligned} \left.\frac{\partial^2 \phi_{\mathrm{i}}(z_{\mathrm{i}}, t)}{\partial t^\alpha \partial t^\beta}\right|_{z_{\mathrm{i}} = g_{\mathrm{ij}}(\phi_{\mathrm{j}}, z_{\mathrm{j}})} &= F_{\mathrm{ij}}(z_{\mathrm{j}}, t)\,\frac{\partial^2 \phi_{\mathrm{j}}(z_{\mathrm{j}}, t)}{\partial t^\alpha \partial t^\beta} \\ &\quad + \tau_{\mathrm{ij}\,\alpha}(z_{\mathrm{j}}, t)\,\frac{\partial \phi_{\mathrm{j}}(z_{\mathrm{j}}, t)}{\partial t^\beta} + \tau_{\mathrm{ij}\,\beta}(z_{\mathrm{j}}, t)\,\frac{\partial \phi_{\mathrm{j}}(z_{\mathrm{j}}, t)}{\partial t^\alpha} \end{aligned}$$

and take equation (2) into account, we obtain the equality

$$\left.\left(\frac{\partial^2 \phi_{\mathrm{i}}}{\partial t^\alpha \partial t^\beta} + \theta_{\mathrm{i}\,\alpha}\frac{\partial \phi_{\mathrm{i}}}{\partial t^\beta} + \theta_{\mathrm{i}\,\beta}\frac{\partial \phi_{\mathrm{i}}}{\partial t^\alpha}\right)\right|_{z_{\mathrm{i}} = g_{\mathrm{ij}}(\phi_{\mathrm{j}}, z_{\mathrm{j}})} = \frac{\partial^2 \phi_{\mathrm{j}}}{\partial t^\alpha \partial t^\beta} + \theta_{\mathrm{j}\,\alpha}\frac{\partial \phi_{\mathrm{j}}}{\partial t^\beta} + \theta_{\mathrm{j}\,\beta}\frac{\partial \phi_{\mathrm{j}}}{\partial t^\alpha}$$

which implies that, for each value of α and β, the holomorphic functions,

$$\Phi_{\mathrm{i}\,\alpha\beta}(z_{\mathrm{i}}, t) \equiv \frac{\partial^2 \phi_{\mathrm{i}}(z_{\mathrm{i}}, t)}{\partial t^\alpha \partial t^\beta} + \theta_{\mathrm{i}\,\alpha}(z_{\mathrm{i}}, t)\,\frac{\partial \phi_{\mathrm{i}}(z_{\mathrm{i}}, t)}{\partial t^\beta} + \theta_{\mathrm{i}\,\beta}(z_{\mathrm{i}}, t)\,\frac{\partial \phi_{\mathrm{i}}(z_{\mathrm{i}}, t)}{\partial t^\alpha},$$

represent a *global section* of the normal bundle N_F over $\boldsymbol{\nu}^{-1}(M_0)$. Since the collections of functions

$$\left\{\Phi_{\mathrm{i}\,\alpha}(z_{\mathrm{i}},t)\equiv\frac{\partial\phi_{\mathrm{i}}(z_i,t)}{\partial t^{\alpha}}\right\}$$

form a Čech representation of a basis for the free $\mathcal{O}_{M_0}$-module $\boldsymbol{\nu}_*^0\left(N_F|_{\boldsymbol{\nu}^{-1}(M_0)}\right)$, the equality

$$\Phi_{\mathrm{i}\alpha\beta}(z_{\mathrm{i}},t)=\Gamma_{\alpha\beta}^{\gamma}(t)\,\Phi_{\mathrm{i}\gamma}(z_{\mathrm{i}},t) \tag{4}$$

must hold for some global holomorphic functions $\Gamma_{\alpha\beta}^{\gamma}$ on $\boldsymbol{\nu}^{-1}(M_0)$. Since all the fibers $\boldsymbol{\nu}^{-1}(t)$, $t\in M_0$, are compact complex manifolds, these functions are actually pull-backs of some holomorphic functions on M_0. A coordinate system $\{t^{\alpha}\}$ on M_0 was used in the construction of $\Gamma_{\alpha\beta}^{\gamma}(t)$. However from (4) it immediately follows that under general coordinate transformations

$$t^{\alpha}\longrightarrow t^{\alpha'}=t^{\alpha'}(t^{\beta})$$

these functions transform according to

$$\Gamma_{\alpha'\beta'}^{\gamma'}=\frac{\partial t^{\gamma'}}{\partial t^{\delta}}\left(\Gamma_{\mu\nu}^{\delta}\,\frac{\partial t^{\mu}}{\partial t^{\alpha'}}\,\frac{\partial t^{\nu}}{\partial t^{\beta'}}+\frac{\partial^2 t^{\delta}}{\partial t^{\alpha'}\partial t^{\beta'}}\right).$$

Thus from any given splitting

$$\{\tau_{\mathrm{ij}\,\alpha}\}=\delta\,\{\theta_{\mathrm{i}\,\alpha}\},$$

of the 1-cocycle $\{\tau_{\mathrm{ij}\,\alpha}\}$ we extract a symmetric affine connection $\Gamma_{\alpha\beta}^{\gamma}(t)$. It is straightforward to check that this connection is independent of the choice of the $(w_{\mathrm{i}},z_{\mathrm{i}}^{a})$-coordinate system used in the construction and thus is well-defined except for the arbitrariness in its construction described by the transformations (3) which, as one can easily check, change the connection as follows

$$\begin{aligned}\theta_{\mathrm{i}\,\alpha}(z_{\mathrm{i}},t)&\longrightarrow\theta_{\mathrm{i}\,\alpha}(z_{\mathrm{i}},t)+\xi_{\alpha}(t)\\ \Gamma_{\alpha\beta}^{\gamma}(t)&\longrightarrow\Gamma_{\alpha\beta}^{\gamma}(t)+\xi_{\alpha}(t)\,\delta_{\beta}^{\gamma}+\xi_{\beta}(t)\,\delta_{\alpha}^{\gamma}.\end{aligned}$$

Therefore we conclude that the neighbourhood M_0 of the point t_0 in the moduli space comes equipped *canonically* with a projective connection.

Let us now prove that for each point $y_0\in Y'=\cup_{t\in M_0}X_t$, the associated submanifold $P_y\subseteq\boldsymbol{\nu}\circ\boldsymbol{\mu}^{-1}(y)\subset M_0$ is totally geodesic relative to the canonical projective connection in M_0 (cf. Merkulov 1994). Suppose that $y_0\in W_{\mathrm{i}}$ for some i. Then $y_0=(w_{\mathrm{i}0},z_{\mathrm{i}0}^{a})$ and the submanifold P_{y_0} is given locally by the equations

$$w_{\mathrm{i}\,0}-\phi_{\mathrm{i}}(z_{\mathrm{i}\,0},t)=0,$$

where $t \in \boldsymbol{\nu} \circ \boldsymbol{\mu}^{-1}(y_0) \setminus \{\text{singular points}\}$. Then a vector field $v(t) = V^\alpha \partial_\alpha|_{P_{y_0}}$ is tangent to P_{y_0} if and only if it satisfies the simultaneous equations

$$V^\alpha \Phi_{i\alpha}(z_{i0}, t) = 0. \tag{5}$$

In order to prove that the submanifold P_{y_0} for arbitrary $y_0 \in Y'$ is totally geodesic relative to the canonical projective connection, we have to show that, for any vector fields $v(t) = V^\alpha \partial_\alpha$ and $w(t) = W^\alpha \partial_\alpha$ on P_{y_0}, the equation

$$\left(W^\beta \partial_\beta V^\alpha + \Gamma^\alpha_{\beta\gamma} V^\gamma W^\beta\right) \bmod TP_{y_0} = 0. \tag{6}$$

holds.

Since $v(t)$ and $w(t)$ are tangent to $P_{y_0} \subset M$, we have the equation

$$W^\beta(t) \frac{\partial}{\partial t^\beta} (V^\alpha \Phi_{i\alpha}(z_{i0}, t)) = 0. \tag{7}$$

By construction, the coefficients, $\Gamma^\gamma_{\alpha\beta}$, of any lift of the induced projective connection to an affine connection satisfy the equation (4). Since

$$V^\alpha \Phi_{i\alpha}(z_{i0}, t) = 0, \qquad W^\beta \Phi_{i\beta}(z_{i0}, t) = 0,$$

equation (4) implies

$$V^\alpha W^\beta \frac{\partial^2 \phi_i(z_{i0}, t)}{\partial t^\alpha \partial t^\beta} = V^\alpha W^\beta \Gamma^\gamma_{\alpha\beta} \frac{\partial \phi_i(z_{i0}, t)}{\partial t^\gamma}.$$

From the latter equation and equation (7) it follows that

$$\left(W^\beta \partial_\beta V^\alpha + \Gamma^\alpha_{\beta\gamma} V^\gamma W^\beta\right) \frac{\partial \phi_i(z_{i0}, t)}{\partial t^\alpha} = 0.$$

By (5) this means that $\left(W^\beta \partial_\beta V^\alpha + \Gamma^\alpha_{\beta\gamma} V^\gamma W^\beta\right) \partial_\alpha \in TP_{y_0}$, and thus equation (6) holds. The proof is completed. ∎

2.2 Rigid hypersurfaces

Let Y be a complex manifold and X a compact complex submanifold with normal bundle N such that $H^1(X, N) = 0$. If X is rigid (e.g. $H^1(X, TX) = 0$), then the corresponding Kodaira complete family $F \hookrightarrow Y \times M$ is locally trivial (Kodaira 1986), i.e. for any point $t \in M$ there is a domain U such that $\boldsymbol{\nu}^{-1}(U) \simeq U \times X$. Therefore, all submanifolds $X_t = \boldsymbol{\mu} \circ \boldsymbol{\nu}^{-1}(t)$, $t \in M$, are biholomorphic to X. For each $y \in Y' \equiv \cup_{t \in M} X_t$ the set $\boldsymbol{\nu} \circ \boldsymbol{\mu}^{-1}(y)$ is a complex analytic subspace of M. Again we let $P_y = \boldsymbol{\nu} \circ \boldsymbol{\mu}^{-1}(y) \setminus \{\text{singular}$

points}.

Theorem 2 *Let $X \hookrightarrow Y$ be a compact complex rigid submanifold of codimension 1 with normal bundle N such that $H^1(X, N) = 0$ and let M be the associated complete moduli space of relative deformations of X in Y. If $H^1(X, \mathcal{O}_X) = 0$, then M comes equipped with a torsion-free projective connection such that, for every point $y \in Y'$, the associated submanifold $P_y = \boldsymbol{\nu} \circ \boldsymbol{\mu}^{-1}(y) \subset M$ is totally geodesic.*

Proof. Let us cover the moduli space M by sufficiently small coordinate neighbourhoods $M = \cup U_i$ and use Theorem 1 to define a canonical projective connection in each U_i. Functorial properties of the construction described in the proof of Theorem 1 immediately imply that these local projective connections glue to form a global projective connection on M which does not depend on the particular covering of M used in the construction. ■

Remark. If we consider $X = \mathbb{CP}^1$ with normal bundle $\mathcal{O}(n)$ then a point t_0 in the moduli space M corresponds to a curve X_{t_0} in the surface Y and a tangent vector v_0 at t_0 corresponds to a section S_{v_0} of $\mathcal{O}(n)$ on X_{t_0}. Let $y_1, \cdots, y_n$ be the n zeros of S_{v_0} and let $P_{y_1}, \cdots, P_{y_n}$ be the associated totally geodesic hypersurfaces in M. Then the intersection $P_{y_1} \cap \cdots \cap P_{y_n}$ is equal to the geodesic (of the projective connection described above) going through t_0 in the direction v_0 and the points on the geodesic correspond to all the curves X_t in Y passing through the points $y_1, \cdots, y_n$. In particular for $n = 2$ we notice that our connection is projectively equivalent to the Weyl connection on M (Hitchin 1982).

2.3 Examples

We seek examples of hypersurfaces X in Y where $H^1(X, N) = H^1(X, \mathcal{O}_X) = H^1(X, TX) = 0$. If $X = \mathbb{CP}^{n-1}$ in $Y = \mathbb{CP}^n$, then we easily see that M is isomorphic to $\mathbb{CP}^n$ with the projective structure given by the projective lines.

Let X be a non-singular quadric in $Y = \mathbb{CP}^3$. The moduli space M of non-singular quadrics is isomorphic to $\mathbb{CP}^9 \backslash Q$, where Q is a non-singular quartic. The geodesics are projective lines lying in $\mathbb{CP}^9 \backslash Q$.

Consider a non-singular curve X of bidegree $(1, n)$ in the quadric $\mathbb{CP}^1 \times \mathbb{CP}^1$. Then X is rational and has normal bundle $\mathcal{O}(2n)$ (Pedersen 1986). The space M of such curves can be described as follows: Let (ζ, η) be affine coordinates on $\mathbb{CP}^1 \times \mathbb{CP}^1$ and consider the graph of a rational function of degree n:

$$\begin{aligned} \eta &= \frac{P(\zeta)}{Q(\zeta)} \\ P(\zeta) &= a_n \zeta^n + a_{n-1} \zeta^{n-1} + \cdots + a_0 \\ Q(\zeta) &= b_n \zeta^n + b_{n-1} \zeta^{n-1} + \cdots + b_0 \end{aligned} \tag{8}$$

The family of such $(1, n)$-curves is parameterized by $\mathbb{CP}^{2n+1}$ and the space M of non-singular curves is $\mathbb{CP}^{2n+1}\backslash R$ where R is the manifold of codimension 1 and degree $2n$ given by the resultant of P and Q. The geodesics of the projective connection are again given by projective lines in $\mathbb{CP}^{2n+1}\backslash R$. We may of course choose to describe the induced structure on the manifold of codimension 1 given by $R = 1$, and for $n = 1$ this corresponds to the standard projective structure on $SL(2, \mathbb{C})$ or on one of its real slices $\mathcal{H}^3$, S^3.

In order to obtain less trivial examples we consider branched coverings. Consider a complex curve C contained in a complex surface S. We want to construct a branched covering of a neighborhood of C branched along C. Choose coordinates (x_i, y_i) on neighborhoods O_i along C such that $O_i \cap C$ is given by $x_i = 0$. Then on overlaps we have

$$x_i = x_j H_{ij}(x_j, y_j)$$

$$y_i = K_{ij}(x_j, y_j)$$

Now, we look for an n-fold cover branched along C: take patches W_i with coordinates (w_i, z_i) and define the covering map

$$(w_i, z_i) \rightarrow (x_i, y_i) = (w_i^n, z_i)$$

This is a branched cover of O_i branched along $O_i \cap C$. We want to identify the neighborhoods W_i along the curve C to obtain a surface Y with a map $\pi : Y \rightarrow S$ which locally has the form above. We get

$$w_i^n = x_i = x_j H_{ij}(z_j, y_j) = w_j^n H_{ij}(w_j^n, z_j)$$

If we make a choice of the nth root and put $\tilde{H}_{ij} = H_{ij}^{\frac{1}{n}}$ we get

$$w_i = w_j \; \tilde{H}_{ij} \; (w_j^n, z_j) = f_{ij}(w_j, z_j).$$

The obstruction for this to work along the curve is the class

$$\tilde{H}_{ij}\tilde{H}_{jk}\tilde{H}_{ki} \in H^2(C, \mathbb{Z}/n).$$

We can identify this obstruction to be the self-intersection number of C modulo n: since $dx_i = H_{ij}(0, y_j)dx_j$ we see that $H_{ij}(0, y_j)$ represents the normal bundle N in $H^1(C, \mathcal{O}^*)$. From the long exact sequence associated with

$$0 \rightarrow \mathbb{Z} \rightarrow \mathcal{O} \overset{\exp}{\rightarrow} \mathcal{O}^* \rightarrow 0$$

we see that the degree of N is equal to $\log\, H_{ij} + \log\, H_{jk} + \log H_{ki}$. Thus, the obstruction to obtain Y is equal to the self-intersection of C, modulo n. Each choice of $H_{ij}^{\frac{1}{n}}$ corresponds to an element in $H^1(C, \mathbb{Z}/n)$. Unless the homology class of C in $H_2(S, \mathbb{Z})$ is divisible by n this local construction along the curve cannot be extended to work globally on S (Atiyah 1969).

Now, let us return to the case where C is a $(1, n)$–curve in $\mathbb{CP}^1 \times \mathbb{CP}^1$. In this case $C \cong \mathbb{CP}^1$, so there is a unique n–fold covering Y branched along C which we cannot extend to all of $\mathbb{CP}^1 \times \mathbb{CP}^1$. The branch locus $X \subseteq Y$ is a copy of C but

$$deg\ N_X = \frac{1}{n}\ deg\ N_C = 2$$

so we may describe an Einstein-Weyl structure on the moduli space of curves in Y (Hitchin 1982) and contrary to earlier attempts we are now able to get the connection $\Gamma^{\gamma}_{\alpha\beta}$ explicitly. Let us concentrate on $(1, 2)$ curves and let C be the curve

$$\eta = \zeta^2$$

The projection π maps the curves in Y onto those $(1, 2)$–curves which meet C in two points to second order. These curves may be given as in (13) with

$$P(\zeta) = \zeta^2 - 2t_0t_1\zeta - t_0^2$$

$$Q(\zeta) = t_2^2\zeta^2 + 2t_1t_2\zeta + 1 + 2t_0t_2 + t_1^2$$

(see Pedersen 1986). In order to describe the lifted curves we introduce the coordinates

$$\begin{array}{ll} x_1 = \eta - \zeta^2 & x_2 = \tilde{\eta} - \tilde{\zeta}^2 \\ y_1 = \zeta & y_2 = \tilde{\zeta} \end{array}$$

where $(\tilde{\zeta}, \tilde{\eta}) = (\frac{1}{\zeta}, \frac{1}{\eta})$. Then C is given by $x_i = 0$. Making the coordinate transformation

$$\begin{array}{rcl} (x_1, y_1) & \longrightarrow & (w, z) = (\sqrt{x_1}, y_1) \\ (x_2, y_2) & \longrightarrow & (\hat{w}, \hat{z}) = (\sqrt{x_1}, y_1) \end{array}$$

we arrive at a covering of Y by two coordinate charts W and $\hat{W}$ which is exactly of the type used in the proof of Theorem 2 and has the transition functions

$$\hat{w} = f(w, z), \quad \hat{z} = g(z),$$

given by

$$\begin{aligned} f(w,z) &= \frac{w}{z\sqrt{w^2+z^2}} \\ g(z) &= z^{-1}. \end{aligned}$$

The complete maximal family of relative deformations of C is described in this chart by the equations (in the notation of the proof of Theorem 1)

$$w = \phi(z,t), \qquad \hat{w} = \hat{\phi}(\hat{z},t),$$

with

$$\phi(z,t) = i\,R(z)\,Q(z)^{-1/2}, \qquad \hat{\phi}(z,t) = i\,R(z)\,P(z)^{-1/2},$$

where

$$R(z) = t_2\,z^2 + t_1\,z + t_0.$$

Note that a useful identity $P = z^2Q - R^2$ holds (Pedersen 1986).

According to the proof of Theorem 1 an affine connection in the canonical projective class induced on the moduli space may be found from the equations

$$\frac{\partial^2\phi(z,t)}{\partial t^\alpha \partial t^\beta} + \theta_\alpha(z,t)\frac{\partial\phi(z,t)}{\partial t^\beta} + \theta_\beta\frac{\partial\phi(z,t)}{\partial t^\alpha} = \Gamma^\gamma_{\alpha\beta}(t)\frac{\partial\phi(z,t)}{\partial t^\gamma}$$

or

$$\frac{\partial^2\hat{\phi}(\hat{z},t)}{\partial t^\alpha \partial t^\beta} + \hat{\theta}_\alpha(\hat{z},t)\frac{\partial\hat{\phi}(\hat{z},t)}{\partial t^\beta} + \hat{\theta}_\beta\frac{\partial\hat{\phi}(\hat{z},t)}{\partial t^\alpha} = \Gamma^\gamma_{\alpha\beta}(t)\frac{\partial\hat{\phi}(\hat{z},t)}{\partial t^\gamma}, \tag{9}$$

where $\Theta = \sum_{\alpha=0}^2 \theta_\alpha(z,t)\,dt_\alpha$ and $\hat{\Theta} = \sum_{\alpha=0}^2 \hat{\theta}_\alpha(\hat{z},t)\,dt_\alpha$ are some (non-unique) 1-forms satisfying the equation (cf. (2))

$$\frac{1}{2}\left.\frac{\partial^2 f}{\partial w^2}\right|_{w=\phi(z,t)} d_t\phi = \left.\frac{\partial f}{\partial w}\right|_{w=\phi(z,t)} \left(-\hat{\Theta}+\Theta\right),$$

which, as easy calculations show, has the form

$$\frac{3}{2}\frac{RS}{QP} = -\hat{\Theta} + \Theta, \tag{10}$$

where

$$\begin{aligned} S &= \sum_{\alpha=0}^{2} S_\alpha dt_\alpha = \sum_{\alpha=0}^{2} \left(\frac{1}{2} R \frac{\partial Q}{\partial t_\alpha} - Q \frac{\partial R}{\partial t_\alpha} \right) dt_\alpha \\ &= -(z\, t_1\, t_2 + t_2\, t_0 + 1 + t_1^2)\, dt_0 + \\ &\quad +(t_0\, t_1 - z(1 + t_0\, t_2))\, dt_1 - P(z,t)\, dt_2. \end{aligned}$$

Since for small t_α the poles of $P(z)$ lie in the domain $W \cap X_t$ while the poles of $Q(z)$ lie in the domain $\hat{W} \cap X_t$, we conclude that Θ and $\hat{\Theta}$ must be of the form $A\, Q^{-1}$ and $B\, P^{-1}$, respectively, where the coefficients of the 1-form $A = \sum_{\alpha=0}^{2} A_\alpha(z,t)\, dt_\alpha$ are holomorphic functions of z, while the coefficients of the 1-form $z^{-2} B = \sum_{\alpha=0}^{2} z^{-2} B_\alpha(z,t)\, dt_\alpha$ are holomorphic functions of z^{-1}. It is also clear that the 1-forms A and B are determined by equation (10) up to a transformation

$$\begin{aligned} A(z,t) &\longrightarrow A(z,t) + \lambda(t)\, Q(z,t) \\ B(z,t) &\longrightarrow B(z,t) + \lambda(t)\, P(z,t), \end{aligned}$$

where λ is an arbitrary 1-form on M. This non-uniqueness is a consequence of the fact that $H^0(X, N \otimes N^*) \neq 0$ (cf. the proof of Theorem 1). We can use this arbitrariness to set

$$\left. \frac{\partial^2 B(z,t)}{\partial z^2} \right|_{z=0} = 0.$$

Then equation (10) has a unique solution

$$\begin{aligned} \Theta &= \frac{3}{2} \left[t_2\, dt_0 + (t_2 z + t_1)\, dt_1 + R(z,t)\, dt_2 \right] Q^{-1}, \\ \hat{\Theta} &= \frac{3}{2} \left[-(t_1 z + t_0)\, dt_0 - z\, t_0\, dt_1 \right] P^{-1}. \end{aligned}$$

Now equation (9) takes the form

$$\left(P \frac{\partial S_\alpha}{\partial t_\beta} - \frac{3}{2} S_\alpha \frac{\partial P}{\partial t_\beta} + B_\alpha\, S_\beta + B_\beta\, S_\alpha \right) P^{-1} = \Gamma^\gamma_{\alpha\beta}(t)\, S_\gamma. \tag{11}$$

Since the right hand side of this equation is a second order polynomial in z,

$$\begin{aligned} \Gamma^\gamma_{\alpha\beta}\, S_\gamma &= -z^2\, \Gamma^2_{\alpha\beta} + z \left[2\, t_0\, t_1\, \Gamma^2_{\alpha\beta} - t_1\, t_2\, \Gamma^0_{\alpha\beta} - (1 + t_0\, t_2)\, \Gamma^1_{\alpha\beta} \right] \\ &\quad + \left[t_0^2\, \Gamma^2_{\alpha\beta} - (1 + t_0\, t_2 + t_1^2)\, \Gamma^0_{\alpha\beta} + t_0\, t_1\, \Gamma^1_{\alpha\beta} \right], \end{aligned}$$

a solution $\Gamma^\gamma_{\alpha\beta}(t)$ exists if and only if the fourth order polynomial in z in the round brackets on the left hand side of equation (11) is divisible by the second order polynomial P. Direct calculations show that this is the case for all values of α and β confirming thus that there does exist a solution of equations (11) in accordance with Theorem 1. As an illustration let us consider the case, say, $\alpha = 0$ and $\beta = 1$ in more detail. Computing the polynomial

$$-\frac{3}{2}\frac{\partial P}{\partial t_1} S_0 + B_1 S_0 + B_0 S_1 = \frac{3}{2}(z^2 t_1 - 2 z t_0 t_1^2 - t_0^2 t_1) = \frac{3}{2} t_1 P,$$

and substituting the result into (11), we obtain the equation

$$\frac{\partial S_0}{\partial t_1} + \frac{3}{2} t_1 = \Gamma^{\gamma}_{01} S_{\gamma},$$

which has a unique solution

$$\Gamma^2_{01} = 0, \qquad \Gamma^0_{01} = t_1 (1 + 3 t_0 t_2)(2 \triangle)^{-1}, \qquad \Gamma^1_{01} = t_2 (2 + t_1^2 + 2 t_0 t_2)(2 \triangle)^{-1},$$

where

$$\triangle = (1 + t_0 t_2)^2 + t_1^2 (1 + 2 t_0 t_2).$$

Note that $\triangle^2 = R$ where R is the resultant of the polynomials in (8).

Analogously one finds

$$\Gamma^2_{\alpha\beta} = \Gamma^{\alpha}_{22} = 0$$

for all α and β, and

$$\begin{aligned}
\Gamma^0_{00} &= t_2 (1 + t_0 t_2) \triangle^{-1}, & \Gamma^1_{00} &= -t_1 t_2^2 \triangle^{-1}, \\
\Gamma^0_{02} &= t_0 (1 + t_0 t_2 + t_1^2)(2 \triangle)^{-1}, & \Gamma^1_{02} &= -t_1 (1 + t_1^2)(2 \triangle)^{-1}, \\
\Gamma^0_{11} &= -t_0 (1 + t_0 t_2) \triangle^{-1}, & \Gamma^1_{11} &= t_0 t_1 t_2 \triangle^{-1} \\
\Gamma^0_{12} &= -t_0^2 t_1 (2 \triangle)^{-1}, & \Gamma^1_{12} &= -t_0 (1 + t_0 t_2 + t_1^2)(2 \triangle)^{-1}
\end{aligned}$$

The conformal structure $[g]$ on M is given by the condition for the curves to meet to second order. Thus we may choose the following metric in the conformal structure (Pedersen 1986)

$$\begin{aligned}
g &= t_1^2 t_2^2 dt_0^2 + (1 + t_0 t_2)^2 dt_1^2 + 4 t_0^2 (1 + t_1^2) dt_2^2 + 2 t_1 t_2 (1 + t_0 t_2) dt_0 dt_1 \\
&- 4(1 + t_1^2)(1 + t_0 t_2) dt_0 dt_2 - 4 t_0^2 t_1 t_2 dt_1 dt_2
\end{aligned} \tag{12}$$

Since our connection ∇ is projectively equivalent to the Weyl connection D it satisfies

$$(\nabla g)_{\alpha\beta\gamma} = a_{\alpha} g_{\beta\gamma} + b_{\beta} g_{\alpha\gamma} + b_{\gamma} g_{\alpha\beta}$$

for some 1-forms $a = \sum_{i=0}^{2} a_{\alpha}\, dt^{\alpha}$ and $b = \sum_{i=0}^{2} b_{\alpha}\, dt^{\alpha}$. We may solve these equations and present the Weyl connection D in terms of the Levi-Civita connection ∇^g and the 1-form $\omega = a - 2b = \sum_\alpha \omega_\alpha dt_\alpha$,

$$D = \nabla^g + \frac{1}{2}\omega^\# g - \omega \odot I$$

see Pedersen and Tod (1992). We get

$$\begin{aligned}
a_0 &= 3t_1^2 t_2 (2\triangle)^{-1}, & a_1 &= -3t_1(1+t_0t_2)(4\triangle)^{-1}, \\
a_2 &= -3t_0(1+t_0t_2+t_1^2)(2\triangle)^{-1}, & & \\
b_0 &= -3t_1^2 t_2 (4\triangle)^{-1}, & b_1 &= -3t_1(1+t_0t_1)(4\triangle)^{-1}, \\
b_2 &= -3t_0(1+t_0t_2+t_1^2)(2\triangle)^{-1}. & &
\end{aligned}$$

Thus using only the methods of the relative deformation theory of compact hypersurfaces we have computed the full Einstein-Weyl structure on the moduli space.

Suppose we blow up a point s on the quadric and take a $(1,n)$-curve passing through the point. Then in the blown up surface the curve will have self-intersection number $2n-1$ and this corresponds to considering all the $(1,n)$-curves passing through s. We may combine this with the branched covering construction. Pedersen and Tod (1992) considered the Einstein-Weyl structure associated to the $(1,3)$-curves: first they considered the 2-fold branched cover which reduced the degree of the normal bundle from 6 to 3 and then they blew up a point on the branch locus to get self-intersection equal to 2. Again we may compute the Weyl connection or compute the connection associated to any combination of blow up and branched cover. This will give non trivial examples with normal bundle $\mathcal{O}(n)$ for any n.

3 Canonical affine connections on moduli spaces

Let X be a compact complex manifold embedded into a complex manifold Y with normal bundle N such that $H^1(X,N)=0$. Then, according to Kodaira's theorem, X belongs to the complete analytic family $F \simeq \{X_t : t \in M\}$ of complex submanifolds X_t of Y with the moduli space M being a complex manifold, and there exists a canonical isomorphism of sheaves $\boldsymbol{k} : TM \longrightarrow \boldsymbol{\nu}_*^0(N_F)$ where N_F is the normal bundle of the embedding $F \hookrightarrow Y \times M$. Denote the point in M corresponding to X by t_0, i.e. $X = \boldsymbol{\mu} \circ \boldsymbol{\nu}^{-1}(t_0)$.

Suppose that there is a holomorphic vector bundle L_F on F such that $N_F \simeq \mathbb{C}^n \otimes L_F$, $n \in \mathbb{N}$. Suppose also that X is rigid. Then the family $F \hookrightarrow Y \times M$ is locally trivial, i.e. each point $t \in M$ has a neighbourhood U such that $\boldsymbol{\nu}^{-1}(U) \simeq U \times X$. If in addition to the rigidity of X the cohomology group $H^1(X, \mathcal{O}_X)$ vanishes, then all the restrictions of the holomorphic line bundle L_F to the submanifolds $X_t = \boldsymbol{\mu} \circ \boldsymbol{\nu}^{-1}(t)$, $t \in M$, are isomorphic to each other. Under these conditions the direct image functor $\boldsymbol{\nu}_*$ applied to the normal bundle N_F factors the tangent bundle TM into the tensor product,

$$TM = \nu^0_*(N_F) = \nu^0_*(N_F \otimes L^*_F) \otimes_{\mathcal{O}_M} \nu^0_*(L_F)) = S \otimes_{\mathcal{O}_M} \tilde{S},$$

of two holomorphic vector bundles $S = \nu^0_*(N_F \otimes L^*_F)$ and $\tilde{S} = \nu^0_*(L_F)$ with rank$S = n$ and rank$\tilde{S} = \dim H^0\left(\nu^{-1}(t_0), L_F|_{\nu^{-1}(t_0)}\right)$. Therefore the moduli space M comes equipped canonically with an almost-Grassmanian structure (Akivis 1980) which has been worked out fairly thoroughly by Manin (1988) and Bailey and Eastwood (1991).

Since $TM = S \otimes \tilde{S}$, there is a natural monomorphism of locally free $\mathcal{O}_M$-modules

$$\boldsymbol{i} : \mathrm{End}\, S \longrightarrow \mathrm{End}\, TM$$

given by

$$\begin{array}{ccc} \mathrm{End}\, S & \longrightarrow & \mathrm{End}\, S \otimes \mathrm{End}\, \tilde{S} = \mathrm{End}\, TM \\ a & \longrightarrow & a \otimes id. \end{array}$$

Therefore M has a distinguished Ξ-structure (Merkulov 1994) with $\Xi = \boldsymbol{i}(\mathrm{End}\, S)$.

Theorem 3 *Let $X \hookrightarrow Y$ be a compact complex rigid submanifold with normal bundle $N \simeq \mathbb{C}^n \otimes L$ for some holomorphic line bundle L on X and natural number $n \geq 2$. If*

$$H^1(X, L) = H^1(X, \mathcal{O}_X) = H^1(X, L^*) = H^0(X, L^*) = 0,$$

the Kodaira complete moduli space M of relative deformations of X comes equipped canonically with

(i) an almost-Grassmanian structure

$$TM = S \otimes \tilde{S}$$

with rank $S = n$ *and* rank $\tilde{S} = \dim H^0(X, L)$, *and*

(ii) a torsion-free $\boldsymbol{i}(\mathrm{End}\, S)$-connection such that, for every point $y \in Y' = \cup_{t\in M} X_t$, the associated submanifold $P_y \subseteq \boldsymbol{\nu} \circ \boldsymbol{\mu}^{-1}(y) \subset M$ is totally geodesic.

Proof. We shall prove this theorem by explicit calculations in the coordinate charts $\{w^A_{\mathrm{i}}, z^a_{\mathrm{i}}, t^\alpha\}$ on $Y \times M$ introduced in proof of Theorem 1 in (Merkulov 1994). We assume for simplicity that the moduli space M is biholomorphic to an open ball in $\mathbb{C}^m$, $m = \dim M$, and F is globally trivial, i.e. $F \simeq M \times X$. Functoriality of the construction of the $\boldsymbol{i}(\mathrm{End}\, S)$-connection on M given below makes it obvious how to deal with the general case. We have (cf. Merkulov 1993)

$$\left.\frac{\partial^2\phi_{\rm i}^A}{\partial t^\alpha\partial t^\beta}\right|_{z_{\rm i}=g_{\rm ij}(\phi_{\rm j},z_{\rm j})} = F_{{\rm ij}\,B}^{\;A}\frac{\partial^2\phi_{\rm j}^B}{\partial t^\alpha\partial t^\beta} + E_{{\rm ij}\,BC}^{\;A}\frac{\partial\phi_{\rm j}^B}{\partial t^\alpha}\frac{\partial\phi_{\rm j}^C}{\partial t^\beta} - G_{{\rm ij}\,\alpha B}^{\;A}\frac{\partial\phi_{\rm j}^B}{\partial t^\beta} - G_{{\rm ij}\,\beta B}^{\;A}\frac{\partial\phi_{\rm j}^B}{\partial t^\alpha}, \quad (13)$$

where

$$\begin{aligned} E_{{\rm ij}\,BC}^{\;A} &= \left.\frac{\partial^2 f_{\rm ij}^A}{\partial w_{\rm j}^B\partial w_{\rm j}^C}\right|_{w_{\rm j}=\phi(z_{\rm j},t)} - \left.\frac{\partial\phi_{\rm i}^A}{\partial z_{\rm i}^a}\right|_{z_{\rm i}=g_{\rm ij}(\phi_{\rm j},z_{\rm j})}\left.\frac{\partial^2 g_{\rm ij}^a}{\partial w_{\rm j}^B\partial w_{\rm j}^C}\right|_{w_{\rm j}=\phi_{\rm j}(z_{\rm j},t)} \\ &\quad - \left.\frac{\partial^2\phi_{\rm i}^A}{\partial z_{\rm i}^a\partial z_{\rm i}^b}\right|_{z_{\rm i}=g_{\rm ij}(\phi_{\rm j},z_{\rm j})}\left.\left(\frac{\partial g_{\rm ij}^a}{\partial w_{\rm j}^B}\frac{\partial g_{\rm ij}^b}{\partial w_{\rm j}^C}\right)\right|_{w_{\rm j}=\phi_{\rm j}(z_{\rm j},t)}, \end{aligned}$$

$$G_{{\rm ij}\,\alpha B}^{\;A} = \left.\frac{\partial^2\phi_{\rm i}^A}{\partial z_{\rm i}^a\partial t^\alpha}\right|_{z_{\rm i}=g_{\rm ij}(\phi_{\rm j},z_{\rm j})}\left.\frac{\partial g_{\rm ij}^a}{\partial w_{\rm j}^B}\right|_{w_{\rm j}=\phi_{\rm j}(z_{\rm j},t)},$$

and

$$F_{{\rm ij}\,B}^{\;A} = \left.\frac{\partial f_{\rm ij}^A}{\partial w_{\rm j}^B}\right|_{w_{\rm j}=\phi(z_{\rm j},t)} - \left.\frac{\partial\phi_{\rm i}^A}{\partial z_{\rm i}^a}\right|_{z_{\rm i}=g_{\rm ij}(\phi_{\rm j},z_{\rm j})}\left.\frac{\partial g_{\rm ij}^a}{\partial w_{\rm j}^B}\right|_{w_{\rm j}=\phi_{\rm j}(z_{\rm j},t)}.$$

The latter are transition functions for the normal bundle N_F of $F \hookrightarrow Y \times M$. The collections $\{E_{{\rm ij}\,BC}^{\;A}\}$ and $\{G_{{\rm ij}\,\alpha B}^{\;A}\}$ form 1-cochains with coefficients in $N_F \otimes \odot^2 N_F^*$ and $N_F \otimes N_F^* \otimes \nu^*(\Omega^1 M_0)$ respectively, relative to the covering $\{V_{\rm i}\}$. The obstructions for these two 1-cochains to be 1-cocycles have the form (Merkulov 1994),

$$\delta\left\{E_{{\rm ik}\,BC}^{\;A}\right\} = \tau_{{\rm ijk}\,BC}^{\;A} + \tau_{{\rm ijk}\,CB}^{\;A}, \qquad \delta\left\{G_{{\rm ik}\,\alpha B}^{\;A}\right\} = \tau_{{\rm ijk}\,BC}^{\;A}\frac{\partial\phi_{\rm j}^C}{\partial t^\alpha}, \quad (14)$$

where the functions

$$\tau_{{\rm ijk}\,BC}^{\;A} = \left.\frac{\partial F_{{\rm ij}\,C}^{\;A}(z_{\rm j},t)}{\partial z_{\rm j}^a}\right|_{z_{\rm j}=g_{\rm jk}(\phi_{\rm k},z_{\rm k})}\frac{\partial g_{\rm jk}^a}{\partial w_{\rm k}^B}$$

are defined on triple intersections $V_{\rm i}\cap V_{\rm j}\cap V_{\rm k}$.

Note that the 1-cochain

$$\left\{\frac{1}{2}E_{{\rm ik}\,BC}^{\;A}\frac{\partial\phi_{\rm k}^C}{\partial t^\alpha} - G_{{\rm ik}\,\alpha B}^{\;A}\right\}$$

is no longer a 0-cocycle with coefficients in $N_F\otimes N_F^*\otimes\nu^*(\Omega^1 M)$ if the codimension of the embedding $X \hookrightarrow Y$ is greater than or equal to 2. Therefore the basic argument of the proof of Theorem 1 does not work in the case of a higher codimensional embedding.

By assumption, there is a holomorphic line bundle L on the submanifold X such that $N \simeq \mathbb{C}^n \otimes L$. Since X is rigid and

$$H^1(X, N\otimes N^*) = \mathbb{C}^n\otimes\mathbb{C}^n\otimes H^1(X,\mathcal{O}_X) = 0,$$

each compact submanifold $X_t \hookrightarrow Y$, $t\in M$, is biholomorphic to X and has normal bundle $N_t \simeq \mathbb{C}^n \otimes L_t$, where the holomorphic line bundle L_t is isomorphic to L. Since $F \simeq M \times X$, there is a natural projection $\pi : F \longrightarrow X$. Then, defining $L_F = \pi^*(L)$, we have $N_F \simeq \mathbb{C}^n \otimes L_F$ and $L_F|_{\nu^{-1}(t)} \simeq L_t$.

Suppose that L_F is trivialized over each neighbourhood V_i and denote by $q_{ij}(z_j, t)$ the corresponding Čech 1-cocycle representing L_F as an element of $H^1(F, \mathcal{O}_F^*)$. There exists a 0-cochain, $\{P_i\}$, of non-singular holomorphic $p \times p$ matrices defined on V_i such that

$$F_{ij}{}^A_B = q_{ij}\,(P_i^{-1})^A_C\, P_j{}^C_B. \tag{15}$$

Such a 0-cochain $\{P_i\}$ is determined non-uniquely: for any *global* non-singular holomorphic section S of $N_F \otimes N_F^*$ and 0-cochain $\{g_i\}$ of non-vanishing holomorphic functions the cochains $\left\{\tilde{P}_i{}^A_B = g_i\, P_i{}^A_C\, S^C_B\big|_{V_i}\right\}$ and $\{\tilde{q}_{ij} = g_i\, q_{ij}\, g_j^{-1}\}$ satisfy an analogous equation

$$F_{ij}{}^A_B = \tilde{q}_{ij}\,(\tilde{P}_i^{-1})^A_C\, \tilde{P}_j{}^C_B.$$

Although in our construction of the distinguished family of torsion-free affine connections on the moduli space M we shall use some fixed set of functions q_{ij} and P_i which split the transition matrix F_{ij} as described above, we want to emphasize that, due to the fact that N_F is projectively flat and the fibers of the surjection $\nu : F \longrightarrow M$ are compact complex manifolds, the final answer (that is the distinguished $\boldsymbol{i}\,(\mathrm{End}\, S)$-connection on M whose existence we want to prove) will evidently be independent of the choice of the trivialization of L_F over F and the chosen splitting (15) of F_{ij} used in the calculations.

We use the non-singular matrices P_i to define new trivializations of N_F over F in such a way that the new transition matrices acquire the form $q_{ij} \otimes id$, where id is the unit $p \times p$ matrix. All quantities associated with this new trivialization will be equipped with a hat over the kernel symbol. Thus defining

$$\hat{\Phi}_i{}^A_{\alpha\beta} = P_i{}^A_B\, \frac{\partial^2 \phi_i^B(z_i, t)}{\partial t^\alpha \partial t^\beta}, \quad \hat{\Phi}^A_{i\alpha} = P_i{}^A_B\, \frac{\partial \phi_i^B(z_i, t)}{\partial t^\alpha},$$

we rewrite equation (13) in the form

$$\hat{\Phi}_i{}^A_{\alpha\beta}\Big|_{z_i = g_{ij}(\phi_j, z_j)} = q_{ij}\, \hat{\Phi}_j{}^A_{\alpha\beta} + \hat{E}_{ij}{}^A_{BC}\, \hat{\Phi}^B_{j\alpha}\, \hat{\Phi}^C_{j\beta} - \hat{G}_{ij}{}^A_{\alpha B}\, \hat{\Phi}^B_{j\beta} - \hat{G}_{ij}{}^A_{\beta B}\, \hat{\Phi}^B_{j\alpha}, \tag{16}$$

where

$$\hat{E}_{ij}{}^A_{BC} = P_i{}^A_D\, E_{ij}{}^D_{EF}\, (P_j^{-1})^E_B (P_j^{-1})^F_C, \quad \hat{G}_{ij}{}^A_{\alpha B} = P_i{}^A_D\, G_{ij}{}^D_{\alpha E}\, (P_j^{-1})^E_B.$$

From equations (14) it follows that

$$\delta\left\{\hat{E}_{ik}{}^A_{BC}\right\} = \hat{\tau}_{ijk}{}^A_{BC} + \hat{\tau}_{ijk}{}^A_{CB}, \qquad \delta\left\{\hat{G}_{ik}{}^A_{\alpha B}\right\} = \hat{\tau}_{ijk}{}^A_{BC}\, \hat{\Phi}^C_{j\alpha} \tag{17}$$

where

$$\hat{\tau}_{\mathrm{ijk}}{}^{A}{}_{BC} = -q_{\mathrm{ik}} \left[\frac{\partial P_{\mathrm{i}}}{\partial z_{\mathrm{i}}^{b}} P_{\mathrm{i}}^{-1}\right]^{A}_{C} \left(\frac{\partial g_{\mathrm{ij}}^{b}}{\partial z_{\mathrm{j}}^{a}} + \frac{\partial g_{\mathrm{ij}}^{b}}{\partial w_{\mathrm{j}}^{D}} \frac{\partial \phi_{\mathrm{j}}^{D}}{\partial z_{\mathrm{j}}^{a}}\right) \frac{\partial g_{\mathrm{jk}}^{a}}{\partial w_{\mathrm{k}}^{E}} \left(P_{\mathrm{k}}^{-1}\right)^{E}_{B}$$
$$+ q_{\mathrm{ik}} \left[\frac{\partial P_{\mathrm{j}}}{\partial z_{\mathrm{j}}^{a}} P_{\mathrm{j}}^{-1}\right]^{A}_{C} \frac{\partial g_{\mathrm{jk}}^{a}}{\partial w_{\mathrm{k}}^{E}} \left(P_{\mathrm{k}}^{-1}\right)^{E}_{B} + \delta_{C}^{A} q_{\mathrm{jk}} \frac{\partial q_{\mathrm{ij}}}{\partial z_{\mathrm{j}}^{a}} \frac{\partial g_{\mathrm{jk}}^{a}}{\partial w_{\mathrm{k}}^{E}} \left(P_{\mathrm{k}}^{-1}\right)^{E}_{B}.$$

Consider a 1-cochain, $\left\{\hat{\sigma}_{ik}{}^{A}{}_{BC}\right\}$, with coefficients in $N_F \otimes \odot^2 N_F^*$ relative to the covering $\{V_{\mathrm{i}}\}$, where

$$\hat{\sigma}_{\mathrm{ik}}{}^{A}{}_{BC} = q_{\mathrm{ik}} \left[\frac{\partial P_{\mathrm{i}}}{\partial z_{\mathrm{i}}^{a}} P_{\mathrm{i}}^{-1}\right]^{A}_{C} \frac{\partial g_{\mathrm{ik}}^{a}}{\partial w_{\mathrm{k}}^{E}} \left(P_{\mathrm{k}}^{-1}\right)^{E}_{B}.$$

Straightforward though very tedious calculations give the obstruction for this 1-cochain be a 1-cocycle,

$$\delta\left\{\hat{\sigma}_{\mathrm{ik}}{}^{A}{}_{BC}\right\} = q_{\mathrm{ik}} \left[\frac{\partial P_{\mathrm{i}}}{\partial z_{\mathrm{i}}^{b}} P_{\mathrm{i}}^{-1}\right]^{A}_{C} \left(\frac{\partial g_{\mathrm{ij}}^{b}}{\partial z_{\mathrm{j}}^{a}} + \frac{\partial g_{\mathrm{ij}}^{b}}{\partial w_{\mathrm{j}}^{D}} \frac{\partial \phi_{\mathrm{j}}^{D}}{\partial z_{\mathrm{j}}^{a}}\right) \frac{\partial g_{\mathrm{jk}}^{a}}{\partial w_{\mathrm{k}}^{E}} \left(P_{\mathrm{k}}^{-1}\right)^{E}_{B}$$
$$- q_{\mathrm{ik}} \left[\frac{\partial P_{\mathrm{j}}}{\partial z_{\mathrm{j}}^{a}} P_{\mathrm{j}}^{-1}\right]^{A}_{C} \frac{\partial g_{\mathrm{jk}}^{a}}{\partial w_{\mathrm{k}}^{E}} \left(P_{\mathrm{k}}^{-1}\right)^{E}_{B}$$
$$= -\hat{\tau}_{\mathrm{ijk}}{}^{A}{}_{BC} + \delta_{C}^{A} q_{\mathrm{jk}} \frac{\partial q_{\mathrm{ij}}}{\partial z_{\mathrm{j}}^{a}} \frac{\partial g_{\mathrm{jk}}^{a}}{\partial w_{\mathrm{k}}^{E}} \left(P_{\mathrm{k}}^{-1}\right)^{E}_{B}. \tag{18}$$

If we now define a 1-cochain $\left\{\hat{\Omega}_{\mathrm{ik}}{}^{A}{}_{BC}\right\}$,

$$\hat{\Omega}_{\mathrm{ik}}{}^{A}{}_{BC} = \hat{E}_{\mathrm{ik}}{}^{A}{}_{BC} + \hat{\sigma}_{\mathrm{ik}}{}^{A}{}_{BC} + \hat{\sigma}_{\mathrm{ik}}{}^{A}{}_{CB},$$

with coefficients in $N_F \otimes \odot^2 N_F^*$, then from equations (17) and (18) we infer that

$$\delta\left\{\hat{\Omega}_{\mathrm{ik}}{}^{A}{}_{BC}\right\} = \delta_{B}^{A} q_{\mathrm{jk}} \frac{\partial q_{\mathrm{ij}}}{\partial z_{\mathrm{j}}^{a}} \frac{\partial g_{\mathrm{jk}}^{a}}{\partial w_{\mathrm{k}}^{E}} \left(P_{\mathrm{k}}^{-1}\right)^{E}_{C} + \delta_{C}^{A} q_{\mathrm{jk}} \frac{\partial q_{\mathrm{ij}}}{\partial z_{\mathrm{j}}^{a}} \frac{\partial g_{\mathrm{jk}}^{a}}{\partial w_{\mathrm{k}}^{E}} \left(P_{\mathrm{k}}^{-1}\right)^{E}_{B}.$$

Decomposing the tensor $\hat{\Omega}_{\mathrm{ik}}{}^{A}{}_{BC}$ into irreducible parts,

$$\hat{\Omega}_{\mathrm{ik}}{}^{A}{}_{BC} = \hat{\Psi}_{\mathrm{ik}}{}^{A}{}_{BC} + \delta_{B}^{A} \hat{\omega}_{\mathrm{ik}\,C} + \delta_{C}^{A} \hat{\omega}_{\mathrm{ik}\,B},$$

$\hat{\Psi}_{\mathrm{ik}}{}^{A}{}_{BC}$ being a trace-free part of $\hat{\Omega}_{\mathrm{ik}}{}^{A}{}_{BC}$, we conclude that the 1-cochain $\left\{\hat{\Psi}_{\mathrm{ik}}{}^{A}{}_{BC}\right\}$ is actually a 1-*cocycle*

$$\delta\left\{\hat{\Psi}_{\mathrm{ik}}{}^{A}{}_{BC}\right\} = 0,$$

while

$$\delta\left\{\hat{\omega}_{\mathrm{ik}\,B}\right\} = q_{\mathrm{jk}} \frac{\partial q_{\mathrm{ij}}}{\partial z_{\mathrm{j}}^{a}} \frac{\partial g_{\mathrm{jk}}^{a}}{\partial w_{\mathrm{k}}^{E}} \left(P_{\mathrm{k}}^{-1}\right)^{E}_{B}. \tag{19}$$

Now we can rewrite equation (16) in the form

$$\hat{\Phi}_{\mathrm{i}\,\alpha\beta}^{\;A}\Big|_{z_\mathrm{i}=g_{\mathrm{ij}}(\phi_\mathrm{j},z_\mathrm{j})} = q_{\mathrm{ij}}\,\hat{\Phi}_{\mathrm{j}\,\alpha\beta}^{\;A} + \hat{\Psi}_{\mathrm{ij}\,BC}^{\;A}\,\hat{\Phi}_{\mathrm{j}\alpha}^{B}\,\hat{\Phi}_{\mathrm{j}\beta}^{C} - \hat{H}_{\mathrm{ij}\,\alpha B}^{\;A}\,\hat{\Phi}_{\mathrm{j}\beta}^{B} - \hat{H}_{\mathrm{ij}\,\beta B}^{\;A}\,\hat{\Phi}_{\mathrm{j}\alpha}^{B}, \tag{20}$$

where

$$\hat{H}_{\mathrm{ij}\,\alpha B}^{\;A} = -\,\hat{\omega}_{\mathrm{ij}\,B}\,\hat{\Phi}_{\mathrm{j}\alpha}^{A} + \hat{G}_{\mathrm{ij}\,\alpha B}^{\;A} + \hat{\sigma}_{\mathrm{ij}\,BC}^{\;A}\,\hat{\Phi}_{\mathrm{j}\alpha}^{C}.$$

From (17) and (19) it follows that

$$\delta\left\{\hat{H}_{\mathrm{ij}\,\alpha B}^{\;A}\right\} = 0.$$

i.e. the 1-cochain $\left\{\hat{H}_{\mathrm{ij}\,\alpha B}^{\;A}\right\}$ with coefficients in $N_F \otimes N_F^* \otimes \boldsymbol{\nu}^*(\Omega^1 M)$ is actually a 1-*cocycle* which represents thus some element of the cohomology group $H^1\left(F, N_F \otimes N_F^* \otimes \boldsymbol{\nu}^*(\Omega^1 M)\right)$. Since $H^1(X, \mathcal{O}_X) = 0$ and $N \simeq \mathbb{C}^n \otimes L$, the group $H^1(X, N \otimes N^*)$ vanishes. Since $F \simeq X \times M$ and M is a Stein manifold, the Künneth theorem (Burns 1979) implies that $H^1(F, N_F \otimes N_F^* \otimes \boldsymbol{\nu}^*(\Omega^1 M)) = 0$. Therefore there exists a 0-cochain $h_{\mathrm{i}\,\alpha B}^{A}$ with coefficients in $N_F \otimes N_F^* \otimes \boldsymbol{\nu}^*(\Omega^1 M)$ which splits this 1-cocycle, $\left\{\hat{H}_{\mathrm{ij}\,\alpha B}^{\;A}\right\} = \delta\left\{h_{\mathrm{i}\,\alpha B}^{\;A}\right\}$, or, more explicitly,

$$\hat{H}_{\mathrm{ij}\,\alpha B}^{\;A} = q_{\mathrm{ij}}\,h_{\mathrm{i}\,\alpha B}^{\;A} - h_{\mathrm{j}\,\alpha B}^{\;A}\,q_{\mathrm{ij}}. \tag{21}$$

This splitting is non-unique: given any global section $\{\xi_{\mathrm{i}\,\alpha B}^{\;A}\}$ of $N_F \otimes N_F^* \otimes \boldsymbol{\nu}^*(\Omega^1 M)$, the 0-cochain

$$\tilde{h}_{\mathrm{i}\,\alpha B}^{\;A} = h_{\mathrm{i}\,\alpha B}^{\;A} + \xi_{\mathrm{i}\,\alpha B}^{\;A} \tag{22}$$

splits the same 1-cocycle $\left\{\hat{H}_{\mathrm{ij}\,\alpha B}^{\;A}\right\}$.

Now let us return back to the constructed 1-cocycle $\left\{\hat{\Psi}_{\mathrm{ik}\,BC}^{\;A}\right\}$ with coefficients in $N_F \otimes \odot^2 N_F^*$. Since $H^1(X, L^*) = H^0(X, L^*) = 0$, the cohomology groups $H^1(X, N \otimes \odot^2 N^*) = H^0(X, N \otimes \odot^2 N^*) = 0$. Therefore the triviality of F and the Künneth theorem imply $H^1(F, N_F \otimes \odot^2 N_F^*) = H^0(F, N_F \otimes \odot^2 N_F^*) = 0$. Thus, the 1-cocycle $\left\{\hat{\Psi}_{\mathrm{ik}\,BC}^{\;A}\right\}$ is actually a coboundary, $\left\{\hat{\Psi}_{\mathrm{ik}\,BC}^{\;A}\right\} = \delta\left\{\hat{\psi}_{\mathrm{i}\,BC}^{\;A}\right\}$, for some 0-cochain $\left\{\hat{\psi}_{\mathrm{i}\,BC}^{\;A}\right\}$ with coefficients in $N_F \otimes \odot^2 N_F^*$. More explicitly,

$$\hat{\Psi}_{\mathrm{ij}\,BC}^{\;A} = -\hat{\psi}_{\mathrm{i}\,BC}^{\;A}\,(q_{\mathrm{ij}})^2 + q_{\mathrm{ij}}\,\hat{\psi}_{j\,BC}^{\;A}. \tag{23}$$

This time the functions $\hat{\psi}_{\mathrm{i}\,BC}^{\;A}$ are defined *uniquely* due to the fact that the cohomology group $H^0(F_0, N_F \otimes N_F^* \odot N_F^*)$ is zero.

Substituting (21) and (23) into (20) we obtain the equality

$$\left(\hat{\Phi}_{\mathrm{i}\,\alpha\beta}^{\;A} + \hat{\psi}_{\mathrm{i}\,BC}^{\;A}\,\hat{\Phi}_{\mathrm{i}\,\alpha}^{\;B}\,\hat{\Phi}_{\mathrm{i}\,\beta}^{\;C} + \hat{h}_{\mathrm{i}\,\alpha B}^{\;A}\,\hat{\Phi}_{\mathrm{i}\,\beta}^{\;B} + \hat{h}_{\mathrm{i}\,\beta B}^{\;A}\,\hat{\Phi}_{\mathrm{i}\,\alpha}^{\;B}\right)\Big|_{z_\mathrm{i}=g_{\mathrm{ij}}(\phi_\mathrm{j},z_\mathrm{j})}$$

$$= q_{ij} \left(\hat{\Phi}_{j\,\alpha\beta}^{\;A} + \hat{\psi}_{j\,BC}^{\;A} \hat{\Phi}_{j\,\alpha}^{\;B} \hat{\Phi}_{j\,\beta}^{\;C} + \hat{h}_{j\,\alpha B}^{\;A} \hat{\Phi}_{j\,\beta}^{\;B} + \hat{h}_{j\,\beta B}^{\;A} \hat{\Phi}_{j\,\alpha}^{\;B} \right)$$

which implies that, for each value of α and β, the vector-valued functions,

$$\hat{\Lambda}_{i\,\alpha\beta}^{\;A} \equiv \hat{\Phi}_{i\,\alpha\beta}^{\;A} + \hat{\psi}_{i\,BC}^{\;A} \hat{\Phi}_{i\,\alpha}^{\;B} \hat{\Phi}_{i\,\beta}^{\;C} + \hat{h}_{i\,\alpha B}^{\;A} \hat{\Phi}_{i\,\beta}^{\;B} + \hat{h}_{i\,\beta B}^{\;A} \hat{\Phi}_{i\,\alpha}^{\;B},$$

represent a global section of the normal bundle N_F over F. Since the sections

$$\hat{\Phi}_{i\,\alpha}^{A}(z_i, t) \equiv P_{i\,B}^{A} \frac{\partial \phi_i^B(z_i, t)}{\partial t^\alpha}$$

form a basis of the free $\mathcal{O}_M$-module of global sections of N_F over F, the equality

$$\hat{\Lambda}_{i\,\alpha\beta}^{\;A}(z_i, t) = \Gamma_{\alpha\beta}^{\gamma}(t)\, \hat{\Phi}_{i\,\gamma}^{\;A}(z_i, t). \tag{24}$$

must hold for some global holomorphic functions $\Gamma_{\alpha\beta}^{\gamma}$ on F. Since all fibers $\boldsymbol{\nu}^{-1}(t)$, $t \in M$ are compact complex manifolds, these functions are actually pull-backs of some holomorphic functions on M. A coordinate system $\{t^\alpha\}$ on M_0 was used in the construction of $\Gamma_{\alpha\beta}^{\gamma}(t)$. However from (24) it immediately follows that under general coordinate transformations $t^\alpha \longrightarrow t^{\alpha'} = t^{\alpha'}(t^\beta)$ these functions transform as coefficients of an affine connection on M. Thus from any given splitting

$$\left\{ \hat{H}_{ij\,\alpha B}^{\;A} \right\} = \delta \left\{ \hat{h}_{i\,\alpha B}^{\;A} \right\},$$

of the 1-cocycle $\left\{ \hat{H}_{ij\,\alpha B}^{\;A} \right\}$ we can extract a symmetric affine connection $\Gamma_{\alpha\beta}^{\gamma}(t)$. It is straightforward to check that this connection is independent of

1. the choice of the (w_i^A, z_i^a)-coordinate system on Y;
2. the transformations

$$\begin{aligned} q_{ij}(z_j, t) &\longrightarrow g_i(z_i, t)\, q_{ij}(z_j, t)\, g_j^{-1}(z_j, t), \\ P_{i\,B}^{A}(z_i, t) &\longrightarrow g_i(z_i, t)\, P_{i\,B}^{A}(z_i, t), \end{aligned}$$

   where $\{g_i(z_i, t)\}$ is any 0-cochain of non-vanishing holomorphic functions;
3. the transformations

$$P_{i\,B}^{A}(z_i, t) \longrightarrow P_{i\,C}^{A}(z_i, t)\, S_B^C(t)$$

   where $S_B^C(t)$ is an arbitrary section of the trivial vector bundle $N_F \otimes N_F^* \simeq \mathbb{C}^n \otimes \mathbb{C}^n \otimes \mathcal{O}_F$ over F.

The invariance under transformations 2 means that the definition of the connection γ is independent of the particular trivialization of the bundle $L_F \longrightarrow F$ used in the construction. In order to prove this statement it is enough to check that Čhech 1-cocycles

$\{\hat{W}_{ij\,BC}^{\;\;A}\}$ and $\{\hat{H}_{ij\,\alpha B}^{\;\;A}\}$ are invariant under these transformations. We leave this simple calculation to the reader as an exercise.

The invariance under transformation 3 means that the definition of γ does not depend on the particular isomorphism $N_F \xrightarrow{\sim} \mathbb{C}^n \otimes L_F$ which was tacitly assumed in the construction. Therefore it is only the equivalence of the vector bundles N_F and $\mathbb{C}^n \otimes L_F$ that is relevant to the definition of γ. The key point in checking this invariance is the fact that a global section of a trivial vector bundle over F is constant along the fibers of the surjection $\boldsymbol{\nu} : F \longrightarrow M$.

Therefore the connection γ we have constructed is well-defined except for the arbitrariness involved in the splittings of the cocycle $\left\{\hat{H}_{ij\,\alpha B}^{\;\;A}\right\}$ — another splitting (22) gives another affine connection $\tilde{\gamma}$. It is clear that $\tilde{\gamma} - \gamma \in \mathbf{sym}(\boldsymbol{i}(\mathrm{End}(S)) \otimes \Omega^1 M)$. Thus we conclude that the the moduli space M comes equipped canonically with the family of torsion-free connections which have the property that the difference between any two connections of the family is a section of the bundle $\boldsymbol{i}\,(\mathrm{End}\,S)$. Therefore we have proved that M has a distinguished $\boldsymbol{i}\,(\mathrm{End}\,S)$-connection. The proof of the fact that, for any $y \in Y'$, the submanifold $P_y \subseteq \boldsymbol{\nu} \circ \boldsymbol{\mu}^{-1}(y) \subset M$ is totally geodesic relative to the canonical $\boldsymbol{i}\,(\mathrm{End}\,S)$-connection relies on exactly the same arguments as those used in the proof of Theorems 3.1 and 3.4 in (Merkulov 1994) and we refer the reader to that paper for details. The proof is completed. ■

Example. The basic example, which was in fact one of the main motivations behind our present research project, is the twistor description of quaternionic Kähler manifolds (Bailey and Eastwood 1991, Hitchin 1982, LeBrun 1989, Pedersen and Poon 1989, Salamon 1982, Ward 1980). Let X be a rational curve in a $(2m+1)$-dimensional contact manifold which is transverse to the contact structure and has normal bundle isomorphic to $\mathbb{C}^{2m} \otimes \mathcal{O}(1)$. Then Theorem 3 implies that an associated $4m$-dimensional moduli space comes equipped with a complexified almost quaternionic structure and a torsion-free $\boldsymbol{i}\,(\mathrm{End}\,S)$-connection $\hat{\gamma}$ such that every $2m$-dimensional submanifold $P_y = \boldsymbol{\nu} \circ \boldsymbol{\mu}^{-1}(y)$, $y \in Y'$, is totally geodesic relative to $\hat{\gamma}$. If one takes into account the contact structure on Y (which has not been used in the construction of $\hat{\gamma}$), then one can construct a much more informative geometric object on the moduli space — the Levi-Civita connection consistent with some complex Riemannian metric on M. In the case $m = 1$ this metric satisfies the self-dual Einstein equations with non-zero scalar curvature, while in the case $m \geq 2$ it represents a complexified quaternionic Kähler structure induced on the moduli space M.

Acknowledgments

It is a pleasure to thank Stephen Huggett, Claude LeBrun, Yat Sun Poon and Paul

Tod for valuable discussions and remarks. One of the authors (SM) is also grateful to the Department of Mathematics of Odense University for hospitality and financial support.

References

Akivis, M. A. (1980). Webs and almost-Grassmanian structures, *Dokl. Akad. Nauk USSR, 252*: 267-270.

Atiyah, M. F. (1969). The signature of fibre bundles, *Global Analysis*, Papers in honor of K. Kodaira (D. C. Spencer and S. Yanaga, eds.), Princeton Univ. Press, Princeton, 73–84.

Bailey, T. N., and Eastwood, M. G. (1991). Complex paraconformal manifolds – their differential geometry and twistor theory, *Forum Math., 3*: 61-103.

Burns, D. (1979). Some background and examples in deformation theory, *Complex Manifold Techniques in Theoretical Physics*, (D. E. Lerner and P. D. Sommers, eds.), Pitman Publishes.

Hitchin, N. (1982). Complex manifolds and Einstein's equations, *Lect. Notes Math., 970*: 73-99.

Kodaira, K. (1962). A theorem of completeness of characteristic systems for analytic families of compact submanifolds of complex manifolds, *Ann. Math., 75*: 146-162.

Kodaira, K. (1986). Complex manifolds and deformations of complex structures, Springer, Berlin Heidelberg New York.

LeBrun, C. R. (1989). Quaternionic-Kähler manifolds and conformal geometry, *Math. Ann., 284*: 353-376.

Manin, Yu. I. (1988). Gauge Field theory and Complex Geometry, Springer, Berlin Heidelberg New York [Russian: Nauka, Moscow, 1984].

Merkulov, S. A. (1994). Geometry of relative deformations. I. Article in this volume.

Pedersen, H. (1986). Einstein-Weyl Spaces and $(1, n)$-Curves in the Quadric Surface, *Ann. Global Anal. Geom., 4*: 89-120.

Pedersen, H., and Poon, Y. S. (1989), Twistorial construction of quaternionic manifolds, *Proc. VIth Int. Coll. on Diff. Geom., Cursos y Congresos, Univ. Santiago de Compostela, 61*: 207-218.

Pedersen, H., and Tod, K. P. (1992). Three-Dimensional Einstein-Weyl Geometry, *Adv. Math., 97*: 74-109.

Salamon, S. M. (1982). Quaternionic Kähler manifolds, *Invent. Math.*, *67*: 143-171.

Ward, R. S. (1980). Self-dual space-times with cosmological constant, *Commun. Math. Phys.*, *78*: 1-17.

Index

D

E

F

G

L

M

N

O

P

Q

R

S

T

V

W

X

Y

Z

Printed and bound by CPI Group (UK) Ltd, Croydon, CR0 4YY
22/10/2024
01777634-0017